A flexible approach to the modern microbiology lab

S0-AAM-192

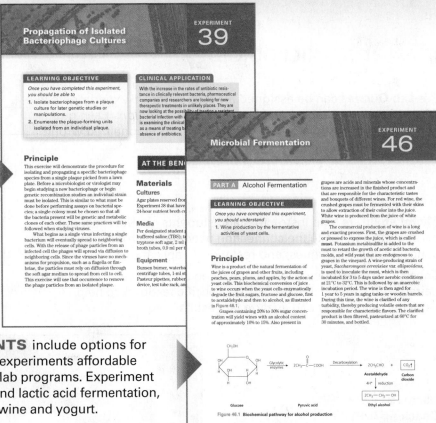

NEW! "Propagation of Isolated Bacteriophage Cultures" experiment has been added to the **Eleventh Edition**. This experiment (39) guides students to isolate bacteriophages for genetic manipulation, an important technique in current clinical research as a possible way to treat antibiotic-resistant bacterial infections.

REVISED EXPERIMENTS include options for alternate media, making the experiments affordable and accessible to all sizes of lab programs. Experiment 46 now includes both wine and lactic acid fermentation, looking at the production of wine and yogurt.

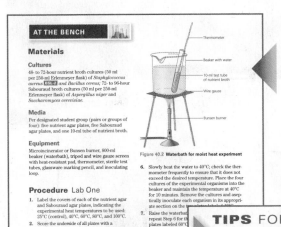

NEW! BioSafety Levels (BSLs) alert students to appropriate safety techniques. The organisms within this manual are mostly BSL-1 organisms, with any BSL-2 organisms now marked within the text. The Eleventh Edition also reflects the most up to date safety protocols from governing bodies such as the EPA, ASM, and AOAC, better preparing students for professional lab work.

TIPS FOR SUCCESS

- Gram stain your unknown culture first and then determine which tests would be useful in identifying your bacteria. For example, the oxidase test and the citrate test would be of no use in identifying a Gram positive cocci bacteria.
- Since many of the tests utilize agars that are similar in appearance, be sure to label all tubes and plates to ensure that results are collected for the correct test.

NEW! Tips for Success appear throughout the experiments and draw attention to common mistakes and stumbling blocks in the lab. Each tip explains why specific techniques are necessary to yield accurate results and helps guide students on how to perform crucial procedural steps correctly.

MasteringMicrobiology prepares students for the modern microbiology lab

MasteringMicrobiology®

Pre-Lab Quizzes can be assigned for each of the 76 experiments in *Microbiology: A Laboratory Manual Eleventh Edition.* Each quiz consists of 10 multiple-choice questions with personalized wrong answer feedback.

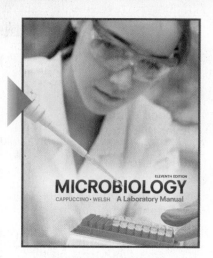

MicroLab Tutors help instructors and students get the most out of lab time and make the connection between microbiology concepts, lab techniques, and real-world applications.

These tutorials combine live-action video and molecular animation paired with assessment and answer-specific feedback to help students to interpret and analyze lab results.

MicroLab Tutor Coaching Activities include the following topics:

- ▶ Use and Application of the Acid-Fast Stain
- ▶ Multitest Systems— API 20E
- ▶ Aseptic Transfer of Bacteria
- ▶ ELISA
- ▶ Gram Stain
- ▶ Use and Application of Microscopy
- ▶ Polymerase Chain Reaction (PCR)
- ▶ Safety in the Microbiology Laboratory
- ▶ Quantifying Bacteria with Serial Dilutions and Pour Plates
- ▶ Smear Preparation and Fixation
- ▶ Streak Plate Technique
- ▶ Survey of Protozoa
- ▶ Identification of Unknown Bacteria

MasteringMicrobiology®

Lab Technique Videos give students an opportunity to see techniques performed correctly and quiz themselves on lab procedures both before and after lab time. Lab Technique videos can be assigned as pre-lab quizzes in MasteringMicrobiology and include coaching and feedback.

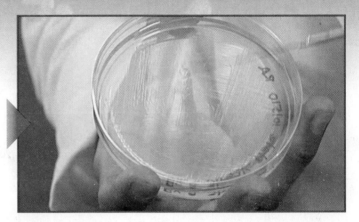

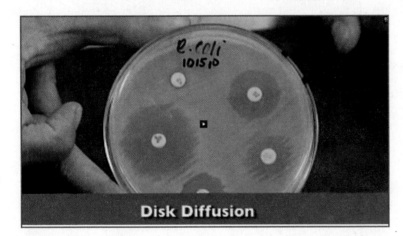

Disk Diffusion

Lab Technique Videos include:
- ▶ **NEW!** The Scientific Method
- ▶ **NEW!** How to Write a Lab Report
- ▶ Acid-fast Staining
- ▶ Amylase Production
- ▶ Carbohydrate Catabolism
- ▶ Compound Microscope
- ▶ Differential and Selective Media
- ▶ Disk-diffusion Assay
- ▶ ELISA
- ▶ Gram Stain
- ▶ Hydrogen Sulfide Production
- ▶ Litmus Milk Reactions
- ▶ Negative Staining
- ▶ Respiration
- ▶ Serial Dilutions
- ▶ Simple Staining
- ▶ Smear Preparation
- ▶ Structural Stains
- ▶ Safety in the Microbiology Laboratory

MicroLab Practical: Biochemical Tests - Voges-Proskauer test (1 of 2)

Interpreting biochemical test results

Part A

Two different bacterial samples, A and B, were analyzed with the Voges-Poskauer (VP) test. The results are pictured here

A   B

View the image in greater detail.

Select ALL appropriate statements regarding the pictured oxidase test results.

- ☐ Specimen A fermented glucose and formed acetoin (acetylmethylcarbinol).
- ☐ Specimen A fermented glucose and formed neutral end-products.
- ☐ Specimen A had a positive result for the VP test.
- ☐ Specimen B fermented glucose and formed acetoin (acetylmethylcarbinol).
- ☐ Specimen B is fermented glucose and formed neutral end-products.
- ☐ Specimen B had a positive result for the VP test.

Submit   My Answers **Give Up**

MicroLab Practical Activities assess students' observation skills and give them extra practice to analyze important lab tests, procedures, and results.

Instructors: Tailor this lab manual to perfectly fit your course!

NEW! **Easy-to-adapt Lab Reports** include blank spaces for individual course customization. Instructors can select their preferred organisms.

NEW! **Revised Experiments** include options for alternate media, reduced volumes, and fewer bacteria, making the experiments affordable and accessible to any-sized lab program.

REVISED! **Instructor's Guide for** *Microbiology: A Laboratory Manual* by James G. Cappuccino, Chad T. Welsh © 2017 | 0-13-429869-1 • 978-0-13-429869-6

Updated to reflect changes in the lab manual, this guide is a valuable teaching aid for instructors and provides:

- ▶ NEW! Recommended readings for each experiment
- ▶ Detailed lists of required materials
- ▶ Tables for calculating the amount of media and equipment needed for your class
- ▶ Procedural points to emphasize
- ▶ Suggestions for optional procedural additions or modifications
- ▶ Helpful tips for preparing or implementing each experiment
- ▶ Answers to the Review Questions in the lab manual
- ▶ Information on laboratory safety protocol for instructional and technical staff

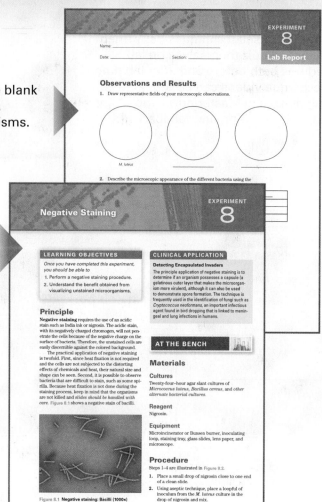

The Pearson Custom Library for the Biological Sciences can assist you in creating a lab manual that meets your specific course organization and content needs. View and select just the labs you need, in the sequence you want. Your students pay only for the materials you choose, and you can also include your own original content, like your syllabus, course schedule, or unique labs. Then preview your custom lab manual online and request an evaluation copy to examine before placing your order.

To create the perfect lab manual, visit www.pearsoncustomlibrary.com.

MICROBIOLOGY

A LABORATORY MANUAL

ELEVENTH EDITION

James G. Cappuccino

SUNY Rockland Community College

Chad Welsh

Lindenwood University

PEARSON

Acquisitions Editor: Kelsey Churchman
Project Manager: Arielle Grant
Program Manager: Chriscelle Palaganas
Development Editor: Laura Cheu
Editorial Assistant: Ashley Williams
Program Management Team Lead: Mike Early
Project Management Team Lead: Nancy Tabor
Production Management, Interior Design, and
    Composition: Integra Software Services Pvt Ltd.
Design Manager: Marilyn Perry
Cover Designer: Elise Lansdon
Rights & Permissions Project Manager: Donna Kalal
Photo Researcher: Kristin Piljay
Manufacturing Buyer: Stacey Weinberger
Executive Marketing Manager: Lauren Harp
Cover Photo Credit: Science Photo/Shutterstock

Copyright © 2017, 2014, 2011 Pearson Education, Inc. All rights reserved. Printed in the United States of America. This publication is protected by copyright, and permission should be obtained from the publisher prior to any prohibited reproduction, storage in a retrieval system, or transmission in any form or by any means, electronic, mechanical, photocopying, recording, or otherwise. For information regarding permissions, request forms and the appropriate contacts within the Pearson Education Global Rights & Permissions department, please visit www.pearsoned.com/permissions/.

Acknowledgements of third party content appear on page 531, which constitutes an extension of this copyright page.

Unless otherwise indicated herein, any third-party trademarks that may appear in this work are the property of their respective owners and any references to third-party trademarks, logos or other trade dress are for demonstrative or descriptive purposes only. Such references are not intended to imply any sponsorship, endorsement, authorization, or promotion of Pearson's products by the owners of such marks, or any relationship between the owner and Pearson Education, Inc. or its affiliates, authors, licensees or distributors..

Library of Congress Cataloging-in-Publication Data

Cappuccino, James G.
  Microbiology: a laboratory manual / James G. Cappuccino and Chad Welsh.—Edition 11.
    pages cm
  Includes index.
  ISBN 978-0-13-409863-0—ISBN 0-13-409863-3
  1. Microbiology–Laboratory manuals.   I. Welsh, Chad.   II. Title.
  QR63.C34 2015
  579—dc23

                              2015030070

Student edition
ISBN 10:   0-134-09863-3
ISBN 13: 978-0-134-09863-0

Instructor's Review Copy
ISBN 10:   0-134-29868-3
ISBN 13: 978-0-134-29868-9

www.pearsonhighered.com

1 2 3 4 5 6 7 8 9 10 11 12—V303—20 19 18 17 16

Contents

Microbiology is a dynamic science. It is constantly evolving as more information is added to the continuum of knowledge, and as microbiological techniques are rapidly modified and refined. The eleventh edition of *Microbiology: A Laboratory Manual* continues to provide a blend of traditional methodologies with more contemporary procedures to meet the pedagogical needs of all students studying microbiology. As in previous editions, this laboratory manual provides a wide variety of critically selected and tested experiments suitable for undergraduate students in allied health programs, as well as elementary and advanced general microbiology courses.

Our Approach

This laboratory manual is designed to guide students in the development of manipulative skills and techniques essential for understanding the biochemical structure and function of a single cell. Its main goal is to encourage students to apply these laboratory skills in the vocational field of applied microbiology and allied health or to further pursue the study of life at the molecular level.

In this manual, comprehensive introductory material is given at the beginning of each major area of study, and specific explanations and detailed directions precede each experiment. This approach augments, enhances, and reinforces course lectures, enabling students to comprehend more readily the concepts and purposes of each experiment. This also provides a review aid if the laboratory and lecture sections are not taught concurrently. The manual should also reduce the time required for explanations at the beginning of each laboratory session and thus make more time available for performing the experiments. Finally, care has been taken to design experimental procedures so that the supplies, equipment, and instrumentation commonly found in undergraduate institutions will suffice for their successful execution.

Organization

This manual consists of 72 experiments arranged into 16 parts. The experiments progress from those that are basic and introductory, requiring minimal manipulations, to those that are more complex, requiring more sophisticated skills. The format of each experiment is intended to facilitate presentation of the material by the instructor and to maximize the learning experience. To this end, each experiment is designed with the following components.

Learning Objectives

This introductory section defines the specific principles and/or techniques to be mastered.

Principle

This is an in-depth discussion of the microbiological concept or technique and the specific experimental procedure.

Clinical Application

Clinical or medical applications that appear within each experiment help students connect what they are learning in lecture with what they are doing in the lab. For students who intend to have careers as nurses or in other allied health fields, Clinical Applications explain the relevance of each lab technique to their career plans.

At the Bench

This section signals the beginning of the experiment, and includes the materials, notes of caution, and procedural instructions—all of the things students will need to know at the bench, during the course of the experiment.

Materials

This comprehensive list helps students and instructors prepare for each laboratory session. Materials appear under one of the following headings:

Cultures These are the selected test organisms that have been chosen to demonstrate effectively the experimental principle or technique under study. The choice is also based on their ease of cultivation and maintenance in stock culture. A complete listing of the experimental cultures and prepared slides is presented in Appendix 6.

Media These are the specific media and their quantities per designated student group. Appendix 3 lists the composition and method of preparation of all the media used in this manual.

Reagents These include biological stains as well as test reagents. The chemical composition and preparation of the reagents are presented in Appendices 4 and 5.

Equipment Listed under this heading are the supplies and instrumentation that are needed during the laboratory session. The suggested equipment was selected to minimize expense while reflecting current laboratory technique.

Procedure

This section provides explicit instructions, augmented by diagrams, that aid in the execution and interpretation of the experiment.

⚠ A caution icon has been placed in experiments that may use potentially pathogenic materials. The instructor may wish to perform some of these experiments as demonstrations.

Lab Report

These tear-out sheets, located at the end of each experiment, facilitate interpretation of data and subsequent review by the instructor. The Observations and Results portion of the report provides tables for recording observations and results, and helps the students draw conclusions from and interpret their data. The Review Questions aid the instructor in determining the student's ability to understand the experimental concepts and techniques. Questions that call for more critical thinking are indicated by the brain icon.

New to the Eleventh Edition

For this eleventh edition, the primary aim was to build upon and enrich the student experience. The changes described below are intended to impart the relevance of microbiological lab techniques to published standard protocols, and to enhance student understanding in the validity of each of the microbiological procedures as they apply laboratories in both the educational and industrial setting.

New Tips for Success

The eleventh edition includes new Tips for Success that will help students avoid common mistakes while they learn laboratory techniques. These tips alert students to potential issues or mistakes that other students have encountered while doing the same experiments. Warning students about potential issues before they begin experiments will reduce the number of procedural mistakes and maximize the number of successful experiments.

New Experiment 39: Propagation of Isolated Bacteriophage Cultures

This experiment builds on previous experiments by utilizing procedures for the cultivation and enumeration of coliophages isolated from individual plaques produced in experiment 38. These techniques when combined should allow students to isolate, cultivate, and further propagate bacteriophages from commercial or environmental sources.

New Experiment 46: Microbial Fermentation

The previous version of this experiment examined alcohol fermentation by yeast cells. The current experiment has been expanded to include lactic acid fermentation. Students will produce yogurt in an experimental setting, examining changes in culture pH and liquid consistency over time as a means to study bacterial lactic acid production during anaerobic metabolism.

Information Concerning Governing Bodies

Where appropriate, information concerning governing bodies, such as the Environmental Protection Agency (EPA), the Clinical and

Laboratory Standards Institute (CLSI), and AOAC International, has been included in the introductory material for some experiments. By drawing attention to governing bodies beyond the American Society for Microbiology (ASM) that have published laboratory standards, students will be introduced to the various industry standards that regulate microbiology laboratories.

Updates and Revisions

Throughout the manual, updates and revisions have been made to background information, terminology, equipment, and procedural techniques, including the following:

- Added a new procedure for a streak plate method in Experiment 2.
- Updated protocols in many experiments, including Experiment 20, to utilize micro-volume procedures.
- Added an alternate protocol in Experiment 12 that uses the McFarland Standards to culture preparations for each lab.
- Protocols for the utilization of plate readers in measuring turbidity and bacterial growth have been added to Experiment 12 and other experiments.
- Results tables have been updated for many experiments (e.g., Experiments 20 and 21) to allow instructors to modify or customize the lab to include organisms of interest not listed in the protocol.
- Added biosafety level (BSL) markers throughout the text. Organisms are clearly labeled with biosafety level 2 if they require additional precautions (BSL-2).

Instructor Resources

The Instructor Guide (ISBN 978-0-134-29869-6) is a valuable teaching aid for instructors. It has been updated to reflect changes in the main text, and provides:

- Laboratory safety protocol for the instructional staff
- Laboratory safety protocol for the technical staff
- New Additional Reading research articles for each experiment
- Detailed lists of required materials, procedural points to emphasize, suggestions for optional

procedural additions or modifications, helpful tips for preparing or implementing each experiment, and answers to the Review Questions in the student manual
- Appendices with the formulas for the preparation of all media, test reagents, and microbiological stains, as well as the micro-organisms required for the performance of each procedure

Acknowledgments

I wish to express my sincere gratitude to the following instructors for their manuscript reviews of the tenth edition. Their comments and direction contributed greatly to the eleventh edition.

Sue Katz Amburn, *Rogers State University*
Qasim K. Beg, *Boston University*
Karen E. Braley, *Daytona State College - Flagler/Palm Coast Campus*
Tanya L. Crider, *East Central Community College*
John L. Dahl, *University of Minnesota Duluth*
Melanie Eldridge, *University of New Haven*
Karla Lightfield, *University of Kentucky*
Sergei A. Markov, *Austin Peay State University*
Michelle H. McGowan, *Temple University*
Karen Meysick, *University of Oklahoma*
Michael F. Minnick, *University of Montana*
Alicia Musser, *Lansing Community College*
Ines Rauschenbach, *Rutgers University* and *Union County College*
Michael W. Ruhl, *Vernon College*
Gene M. Scalarone, *Idaho State University*
Steven J. Scott, *Merritt College*

I would like to express my sincere gratitude to Dr. James Cappuccino for the opportunity to work with him on this laboratory manual that has been his work for the past 20+ years. My hope is that with this eleventh edition we will begin a long and rewarding collaboration.

I also wish to extend my appreciation to the staff at Pearson who helped me through the creation of this manual. Specifically I would like to thank Kelsey Churchman, Senior Acquisitions Editor, Chriscelle Palaganas, Program Manager, and Arielle Grant, Project Manager.

Chad Welsh

Laboratory Safety

General Rules and Regulations

A rewarding laboratory experience demands strict adherence to prescribed rules for personal and environmental safety. The former reflects concern for your personal safety in terms of avoiding laboratory accidents. The latter requires that you maintain a scrupulously clean laboratory setting to prevent contamination of experimental procedures by microorganisms from exogenous sources.

Because most microbiological laboratory procedures require the use of living organisms, an integral part of all laboratory sessions is the use of aseptic techniques. Although the virulence of microorganisms used in the academic laboratory environment has been greatly diminished because of their long-term maintenance on artificial media, all microorganisms should be treated as potential pathogens (organisms capable of producing disease). Thus, microbiology students must develop aseptic techniques (free of contaminating organisms) in the preparation of pure cultures that are essential in the industrial and clinical marketplaces.

The following basic steps should be observed at all times to reduce the ever-present microbial flora of the laboratory environment.

1. Upon entering the laboratory, place coats, books, and other paraphernalia in specified locations—never on bench tops.

2. Keep doors and windows closed during the laboratory session to prevent contamination from air currents.

3. At the beginning and termination of each laboratory session, wipe bench tops with a disinfectant solution provided by the instructor.

4. Do not place contaminated instruments, such as inoculating loops, needles, and pipettes, on bench tops. Loops and needles should be sterilized by incineration, and pipettes should be disposed of in designated receptacles.

5. On completion of the laboratory session, place all cultures and materials in the disposal area as designated by the instructor.

6. Rapid and efficient manipulation of fungal cultures is required to prevent the dissemination of their reproductive spores in the laboratory environment.

To prevent accidental injury and infection of yourself and others, observe the following regulations:

1. Wash your hands with liquid detergent, rinse with 95% ethyl alcohol, and dry them with paper towels upon entering and prior to leaving the laboratory.

2. Always use the appropriate safety equipment as determined by your instructor:

   a. A laboratory coat or apron may be necessary while working in the laboratory. Lab coats protect clothing from contamination or accidental discoloration by staining solutions.

   b. You may be required to wear gloves while performing the lab exercises. Gloves shield your hands from contamination by microorganisms. They also prevent the hands from coming in direct contact with stains and other reagents.

   c. Masks and safety goggles may be required to prevent materials from coming in contact with your eyes.

3. Wear a paper cap or tie back long hair to minimize its exposure to open flames.

4. Wear closed shoes at all times in the laboratory setting.

5. Never apply cosmetics or insert contact lenses in the laboratory.

6. Do not smoke, eat, or drink in the laboratory. These activities are absolutely prohibited.

7. Carry cultures in a test-tube rack when moving around the laboratory. Likewise, keep cultures in a test-tube rack on the bench tops when not in use. This serves a dual purpose: to prevent accidents and to avoid contamination of yourself and the environment.

8. Never remove media, equipment, or especially, microbial cultures from the laboratory. Doing so is absolutely prohibited.

9. Immediately cover spilled cultures or broken culture tubes with paper towels and then saturate them with disinfectant solution. After 15 minutes of reaction time, remove the towels and dispose of them in a manner indicated by the instructor.

10. Report accidental cuts or burns to the instructor immediately.

11. Never pipette by mouth any broth cultures or chemical reagents. Doing so is strictly prohibited. Pipetting is to be carried out with the aid of a mechanical pipetting device only.

12. Do not lick labels. Use only self-stick labels for the identification of experimental cultures.

13. Speak quietly and avoid unnecessary movement around the laboratory to prevent distractions that may cause accidents.

The following specific precautions must be observed when handling body fluids of unknown origin due to the possible transmission of human immunodeficiency virus (HIV) and hepatitis B virus in these test specimens.

1. Wear disposable gloves during the manipulation of test materials such as blood, serum, and other body fluids.

2. Immediately wash hands if contact with any of these fluids occurs and also on removal of the gloves.

3. Wear masks, safety goggles, and laboratory coats if an aerosol might be formed or splattering of these fluids is likely to occur.

4. Decontaminate spilled body fluids with a 1:10 dilution of household bleach, covered with paper toweling, and allowed to react for 10 minutes before removal.

5. Place test specimens and supplies in contact with these fluids into a container of disinfectant prior to autoclaving.

I have read the above laboratory safety rules and regulations and agree to abide by them.

Name: _____ Date: _____

Laboratory Protocol

Student Preparation for Laboratory Sessions

The efficient performance of laboratory exercises mandates that you attend each session fully prepared to execute the required procedures. Read the assigned experimental protocols to effectively plan and organize the related activities. This will allow you to maximize use of laboratory time.

Preparation of Experimental Materials

Microscope Slides: Meticulously clean slides are essential for microscopic work. Commercially precleaned slides should be used for each microscopic slide preparation. However, wipe these slides with dry lens paper to remove dust and finger marks prior to their use. With a glassware marking pencil, label one end of each slide with the abbreviated name of the organism to be viewed.

Labeling of Culture Vessels: Generally, microbiological experiments require the use of a number of different test organisms and a variety of culture media. To ensure the successful completion of experiments, organize all experimental cultures and sterile media at the start of each experiment. Label culture vessels with non–water-soluble glassware markers and/or self-stick labels prior to their inoculation. The labeling on each of the experimental vessels should include the name of the test organism, the name of the medium, the dilution of sample (if any), your name or initials, and the date. Place labeling directly below the cap of the culture tube. When labeling Petri dish cultures, only the name of the organism(s) should be written on the bottom of the plate, close to its periphery, to prevent obscuring observation of the results. The additional information for the identification of the culture should be written on the cover of the Petri dish.

Inoculation Procedures

Aseptic techniques for the transfer or isolation of microorganisms, using the necessary transfer instruments, are described fully in the experiments in Part 1 of the manual. Technical skill will be acquired through repetitive practice.

Inoculating Loops and Needles: It is imperative that you incinerate the entire wire to ensure absolute sterilization. The shaft should also be briefly passed through the flame to remove any dust or possible contaminants. To avoid killing the cells and splattering the culture, cool the inoculating wire by tapping the inner surface of the culture tube or the Petri dish cover prior to obtaining the inoculum, or touch the edge of the medium in the plate.

When performing an aseptic transfer of microorganisms, a minute amount of inoculum is required. If an agar culture is used, touch only a single area of growth with the inoculating wire to obtain the inoculum. Never drag the loop or needle over the entire surface, and take care not to dig into the solid medium. If a broth medium is used, first tap the bottom of the tube against the palm of your hand to suspend the microorganisms. Caution: Do not tap the culture vigorously as this may cause spills or excessive foaming of the culture, which may denature the proteins in the medium.

Pipettes: Use only sterile, disposable pipettes or glass pipettes sterilized in a canister. The practice of pipetting by mouth has been discontinued to eliminate the possibility of autoinfection by accidentally imbibing the culture or infectious body fluids. Instead, use a mechanical pipetting device to obtain and deliver the material to be inoculated.

Incubation Procedure

Microorganisms exhibit a wide temperature range for growth. However, for most used in this manual, optimum growth occurs at 37°C over a period of 18 to 24 hours. Unless otherwise indicated in specific exercises, incubate all cultures under the conditions cited above. Place culture tubes in a rack for incubation. Petri dishes may be stacked; however, they must always be incubated in an inverted position (top down) to prevent water condensation from dropping onto the surface of the culture medium. This excess moisture could allow the spread of the microorganisms on the surface of the culture medium, producing confluent rather than discrete microbial growth.

Procedure for Recording Observations and Results

The accurate accumulation of experimental data is essential for the critical interpretation of the observations upon which the final results will be based. To achieve this end, it is imperative that you complete all the preparatory readings that are necessary for your understanding of the basic principles underlying each experiment. Meticulously record all the observed data in the Lab Report of each experiment.

In the experiments that require drawings to illustrate microbial morphology, it will be advantageous to depict shapes, arrangements, and cellular structures enlarged to 5 to 10 times their actual microscopic size, as indicated by the following illustrations. For this purpose a number 2 pencil is preferable. Stippling may be used to depict different aspects of cell structure (e.g., endospores or differences in staining density).

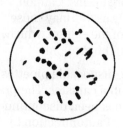 

Microscopic drawing   Enlarged drawing

Review Questions

The review questions are designed to evaluate the student's understanding of the principles and the interpretations of observations in each experiment. Completion of these questions will also serve to reinforce many of the concepts that are discussed in the lectures. At times, this will require the use of ancillary sources such as textbooks, microbiological reviews, or abstracts. The designated critical-thinking questions are designed to stimulate further refinement of cognitive skills.

Procedure for Termination of Laboratory Sessions

1. Return all equipment, supplies, and chemical reagents to their original locations.

2. Neatly place all capped test tube cultures and closed Petri dishes in a designated collection area in the laboratory for subsequent autoclaving.

3. Place contaminated materials, such as swabs, disposable pipettes, and paper towels, in a biohazard receptacle prior to autoclaving.

4. Carefully place hazardous biochemicals, such as potential carcinogens, into a sealed container and store in a fume hood prior to their disposal according to the institutional policy.

5. Wipe down table tops with recommended disinfectant.

6. Wash hands before leaving the laboratory.

PART 1

Basic Laboratory Techniques for Isolation, Cultivation, and Cultural Characterization of Microorganisms

Once you have completed the experiments in this section, you should be familiar with

1. The types of laboratory equipment and culture media needed to develop and maintain pure cultures.

2. The types of microbial flora that live on the skin and the effect of hand washing on them.

3. The concept of aseptic technique and the procedures necessary for successful subculturing of microorganisms.

4. Streak-plate and spread-plate inoculation of microorganisms in a mixed microbial population for subsequent pure culture isolation.

5. Cultural and morphological characteristics of microorganisms grown in pure culture.

Introduction

Microorganisms are ubiquitous. They are found in soil, air, water, food, sewage, and on body surfaces. In short, every area of our environment is replete with them. The microbiologist separates these mixed populations into individual species for study. A culture containing a single unadulterated species of cells is called a **pure culture**. To isolate and study microorganisms in pure culture, the microbiologist requires basic laboratory apparatus and the application of specific techniques, as illustrated in **Figure P1.1**.

Media

The survival and continued growth of microorganisms depend on an adequate supply of nutrients and a favorable growth environment. For survival, most microbes must use soluble low-molecular-weight substances that are frequently derived from the enzymatic degradation of complex nutrients. A solution containing these nutrients is a **culture medium**. Basically, all culture media are liquid, semisolid, or solid. A liquid medium lacks a solidifying agent and is called a **broth medium**. A broth medium is useful for the cultivation of high numbers of bacterial cells in a small volume of medium, which is particularly helpful when an assay requires a high number of healthy bacterial cells. A broth medium supplemented with a solidifying agent called **agar** results in a solid or semisolid medium. Agar, an extract of seaweed, is a complex carbohydrate composed mainly of galactose, and is without nutritional value. Agar serves as an excellent solidifying agent because it liquefies at 100°C and solidifies at 40°C. Because of these properties, organisms, especially pathogens, can be cultivated at temperatures of 37.5°C or slightly higher without fear of the medium liquefying. A completely solid medium requires an agar concentration of 1.5% to 1.8%. A concentration of less than 1% agar results in a **semisolid medium**. A semisolid medium is useful for testing a cell's ability to grow within the agar at lower oxygen levels and

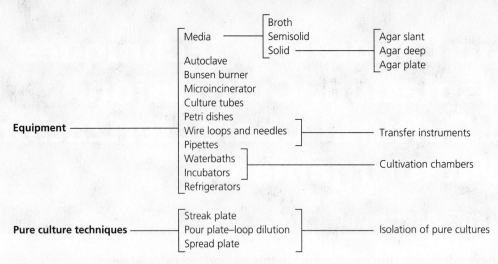

Figure P1.1 **Laboratory apparatus and culture techniques**

for testing the species' motility. A solid medium has the advantage that it presents a hardened surface on which microorganisms can be grown using specialized techniques for the isolation of discrete colonies. Each **colony** is a cluster of cells that originates from the multiplication of a single cell and represents the growth of a single species of microorganism. Such a defined and well-isolated colony is a pure culture. Also, while in the lique-fied state, solid media can be placed in test tubes, which are then allowed to cool and harden in a slanted position, producing **agar slants**. These are useful for maintaining pure cultures. The slanted surface of the agar maximizes the available surface area for microorganism growth while minimizing the amount of medium required. Similar tubes that, following preparation, are allowed to harden in the upright position are designated as **agar deep tubes**. Agar deep tubes are used primarily for the study of the gaseous requirements of microorganisms since gas exchange between the agar at the butt of the test tube and the exter-nal environment is impeded by the height of the agar. Liquid agar medium can also be poured into Petri dishes, producing **agar plates**, which provide large surface areas for the isolation and study of microorganisms. The various forms of solid media are illustrated in Figure P1.2.

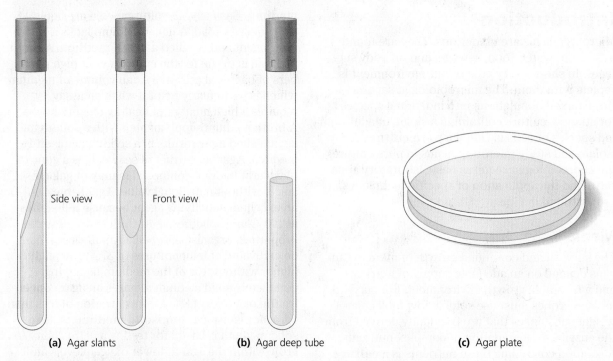

(a) Agar slants   **(b)** Agar deep tube   **(c)** Agar plate

Figure P1.2 **Forms of solid (agar) media**

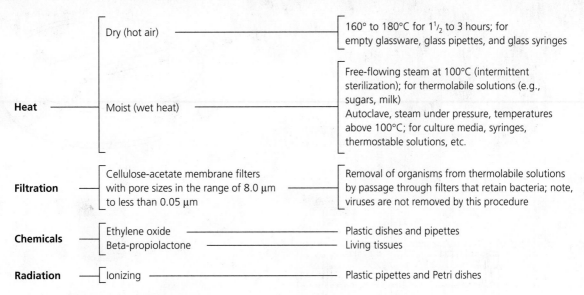

Figure P1.3 **Sterilization techniques**

In addition to nutritional needs, the environmental factors must also be regulated, including proper pH, temperature, gaseous requirements, and osmotic pressure. A more detailed explanation is presented in Part 4, which deals with cultivation of microorganisms; for now, you should simply bear in mind that numerous types of media are available.

Aseptic Technique

Sterility is the hallmark of successful work in the microbiology laboratory, and **sterilization** is the process of rendering a medium or material free of all forms of life. To achieve sterility, it is mandatory that you use sterile equipment and employ **aseptic techniques** when handling bacterial cultures. Using correct aseptic techniques minimizes the likelihood that bacterial cultures will be contaminated, and reduces the opportunity for students to be exposed to potential pathogens. Although a more detailed discussion is presented in Part 9, which describes the control of microorganisms, Figure P1.3 is a brief outline of the routine techniques used in the microbiology laboratory.

Culture Tubes and Petri Dishes

Glass **test tubes** and glass or plastic **Petri dishes** are used to cultivate microorganisms. A suitable nutrient medium in the form of broth or agar may be added to the tubes, while only a solid medium is used in Petri dishes. A sterile environment is maintained in culture tubes by various types of closures. Historically, the first type, a cotton plug, was developed by Schröeder and von Dusch in the nineteenth century. Today most laboratories use

sleeve-like caps (Morton closures) made of metal, such as stainless steel, or heat-resistant plastics. The advantage of these closures over the cotton plug is that they are labor-saving and, most of all, slip on and off the test tubes easily.

Petri dishes provide a larger surface area for growth and cultivation. They consist of a bottom dish portion that contains the medium and a larger top portion that serves as a loose cover. Petri dishes are manufactured in various sizes to meet different experimental requirements. For routine purposes, dishes approximately 15 cm in diameter are used. The sterile agar medium is dispensed to previously sterilized dishes from molten agar deep tubes containing 15 ml to 20 ml of medium, or from a molten sterile medium prepared in bulk and contained in 250-, 500-, and 1000-ml flasks, depending on the volume of medium required. When cooled to 40°C, the medium will solidify. Remember that *after inoculation, Petri dishes are incubated in an inverted position* (top down) to prevent condensation formed on the cover during solidification from dropping down onto the surface of the hardened agar. For this reason, Petri dishes should be labeled on the bottom of the dish. This makes it easier to read the label and minimizes confusion if two Petri dish covers are interchanged. Figure P1.4 illustrates some of the culture vessels used in the laboratory. Built-in ridges on tube closures and Petri dishes provide small gaps necessary for the exchange of air.

Transfer Instruments

Microorganisms must be transferred from one vessel to another or from stock cultures to various media for maintenance and study. This transfer

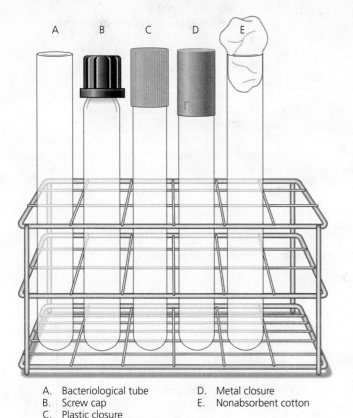

A. Bacteriological tube   D. Metal closure
B. Screw cap   E. Nonabsorbent cotton
C. Plastic closure

(a) Test tube rack with tubes showing various closures

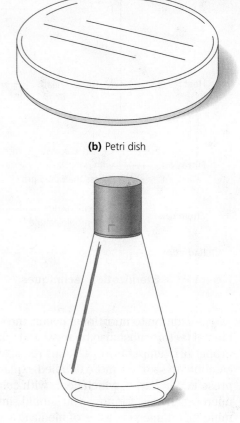

(b) Petri dish

(c) DeLong shaker flask with closure

Figure P1.4 Culture vessels

is called **subculturing** and must be carried out under aseptic conditions to prevent possible contamination.

Wire loops and needles are made from inert metals such as Nichrome or platinum and are inserted into metal shafts that serve as handles. They are extremely durable instruments and are easily sterilized by incineration in the blue (hottest) portion of the Bunsen burner flame. A wire loop is useful for transferring a small volume of bacteria onto the surface of an agar plate or slant. A needle is used primarily to inoculate a culture into a broth medium or into an agar deep tube.

A **pipette** is another instrument used for aseptic transfers. Pipettes are similar in function to straws; that is, they draw up liquids. They are glass or plastic and drawn out to a tip at one end, with a mouthpiece forming the other end. They are calibrated to deliver different volumes depending on requirements. Pipettes may be sterilized in bulk inside canisters, or they may be wrapped individually in brown paper and sterilized in an autoclave or dry-heat oven. A micropipette

(commonly called a "pipetter") with a disposable, single-use plastic tip is useful for transferring small volumes of liquid (less than ≤ 1 ml).

Figure P1.5 illustrates these transfer instruments. Your instructor will demonstrate the proper procedure for using pipettes.

> ⚠ **Pipetting by mouth is not permissible!**
> **Pipetting must be performed with mechanical**
> **pipette aspirators.**

Cultivation Chambers

The specific temperature requirements for growth are discussed in detail in Part 4. However, a prime requirement for the cultivation of microorganisms is that they be grown at their optimum temperature. An **incubator** is used to maintain optimum temperature during the necessary growth period. It resembles an oven and is thermostatically

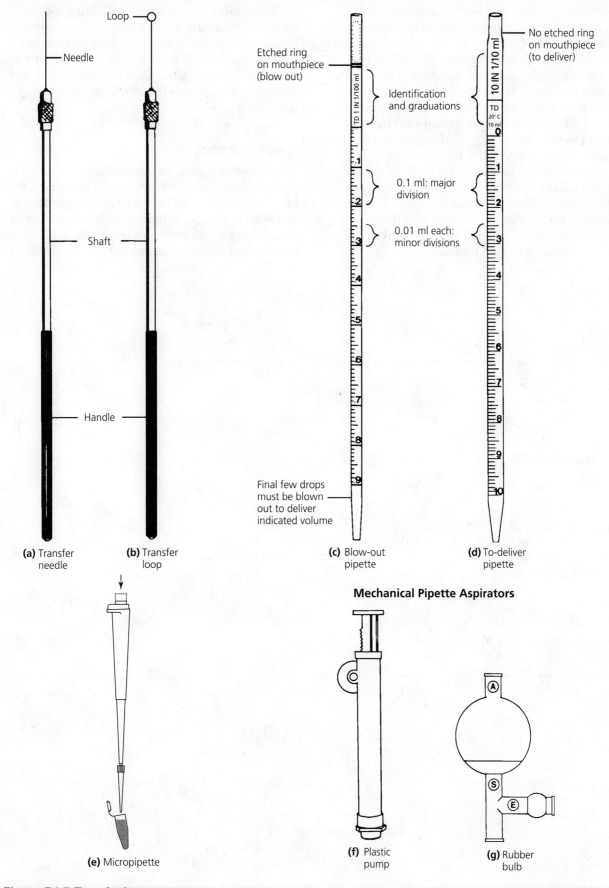

(a) Transfer needle

(b) Transfer loop

(c) Blow-out pipette

(d) To-deliver pipette

(e) Micropipette

(f) Plastic pump

(g) Rubber bulb

Loop

Needle

Shaft

Handle

Etched ring on mouthpiece (blow out)

Identification and graduations

0.1 ml: major division

0.01 ml each: minor divisions

Final few drops must be blown out to deliver indicated volume

No etched ring on mouthpiece (to deliver)

Mechanical Pipette Aspirators

Figure P1.5 Transfer instruments

controlled so that temperature can be varied depending on the requirements of specific microorganisms. Most incubators use dry heat. Moisture is supplied by placing a beaker of water in the incubator during the growth period. A moist environment retards dehydration of the medium and thereby avoids misleading experimental results.

A thermostatically controlled **shaking waterbath** is another piece of apparatus used to cultivate microorganisms. Its advantage is that it provides a rapid and uniform transfer of heat to the culture vessel, and its agitation provides increased aeration, resulting in acceleration of growth. The primary disadvantage of this instrument is that it can be used only for cultivation of organisms in a broth medium.

Many laboratories also use shaking incubators that utilize dry air incubation to promote aeration of the broth medium. This method has a distinct advantage over a shaking waterbath since there is no chance of cross contamination from microorganisms that might grow in the waterbath.

Refrigerator

A refrigerator is used for a wide variety of purposes such as maintenance and storage of stock cultures between subculturing periods, and storage of sterile media to prevent dehydration. It is also used as a repository for thermolabile solutions, antibiotics, serums, and biochemical reagents.

Culture Transfer Techniques

LEARNING OBJECTIVES

Once you have completed this experiment, you should be able to

1. Carry out the technique for aseptic removal and transfer of microorganisms for subculturing.

2. Correctly sterilize inoculating instruments in a microincinerator or the flame of a Bunsen burner.

3. Correctly remove and replace the test tube closure.

Principle

Microorganisms are transferred from one medium to another by **subculturing**. This technique is of basic importance and is used routinely in preparing and maintaining stock cultures, as well as in microbiological test procedures.

Microorganisms are always present in the air and on laboratory surfaces, benches, and equipment. These ambient microorganisms can serve as a source of external contamination and interfere with experimental results unless proper aseptic techniques are used during subculturing. Described below are essential steps that you must follow for aseptic transfer of microorganisms. The complete procedure is illustrated in Figure 1.1.

1. Label the tube to be inoculated with the name of the organism and your initials.

2. Hold the stock culture tube and the tube to be inoculated in the palm of your hand, secure with your thumb, and separate the two tubes to form a V in your hand.

3. Sterilize an inoculating needle or loop by holding it in the microincinerator or the hottest portion of the Bunsen burner flame, until the wire becomes red hot. Once sterilized, the loop is held in the hand and allowed to cool for 10 to 20 seconds; it is never put down.

4. Uncap each tube by grasping the first cap with your little finger and the second cap with your next finger and lifting the closure upward. *Note: Once removed, these caps must be kept in the hand that holds the sterile inoculating loop or needle; the inner aspects of the caps point away from the palm of the hand.* The caps must never be placed on the laboratory bench because that would compromise the aseptic procedure.

5. After removing the caps, flame the necks and mouths of the tubes by briefly passing them through the opening of the microincinerator or through the Bunsen burner flame two to three times rapidly. The sterile transfer instrument is further cooled by touching it to the sterile inside wall of the culture tube before removing a small sample of the inoculum.

6. Depending on the culture medium, a loop or needle is used for removal of the inoculum. Loops are commonly used to obtain a sample from a broth culture. Either instrument can be used to obtain the inoculum from an agar slant culture by carefully touching the surface of the solid medium in an area exhibiting growth so as not to gouge the agar. A straight needle is always used when transferring microorganisms to an agar deep tube from both solid and liquid cultures.

   a. For a slant-to-broth transfer, obtain inoculum from the slant and lightly shake the loop or needle in the broth culture to dislodge the microorganisms.

   b. For a broth-to-slant transfer, obtain a loopfull of broth and place at the base of an agar slant medium. Lightly draw the loop over the hardened surface in a straight or zig-zag line, from the base of the agar slant to the top.

   c. For a slant-to-agar deep tube transfer, obtain the inoculum from the agar slant. Insert a straight needle to the bottom of the tube in a straight line and rapidly withdraw along the line of insertion. This is called a stab inoculation.

7. Following inoculation, remove the instrument and reheat or reflame the necks of the tubes.

PROCEDURE

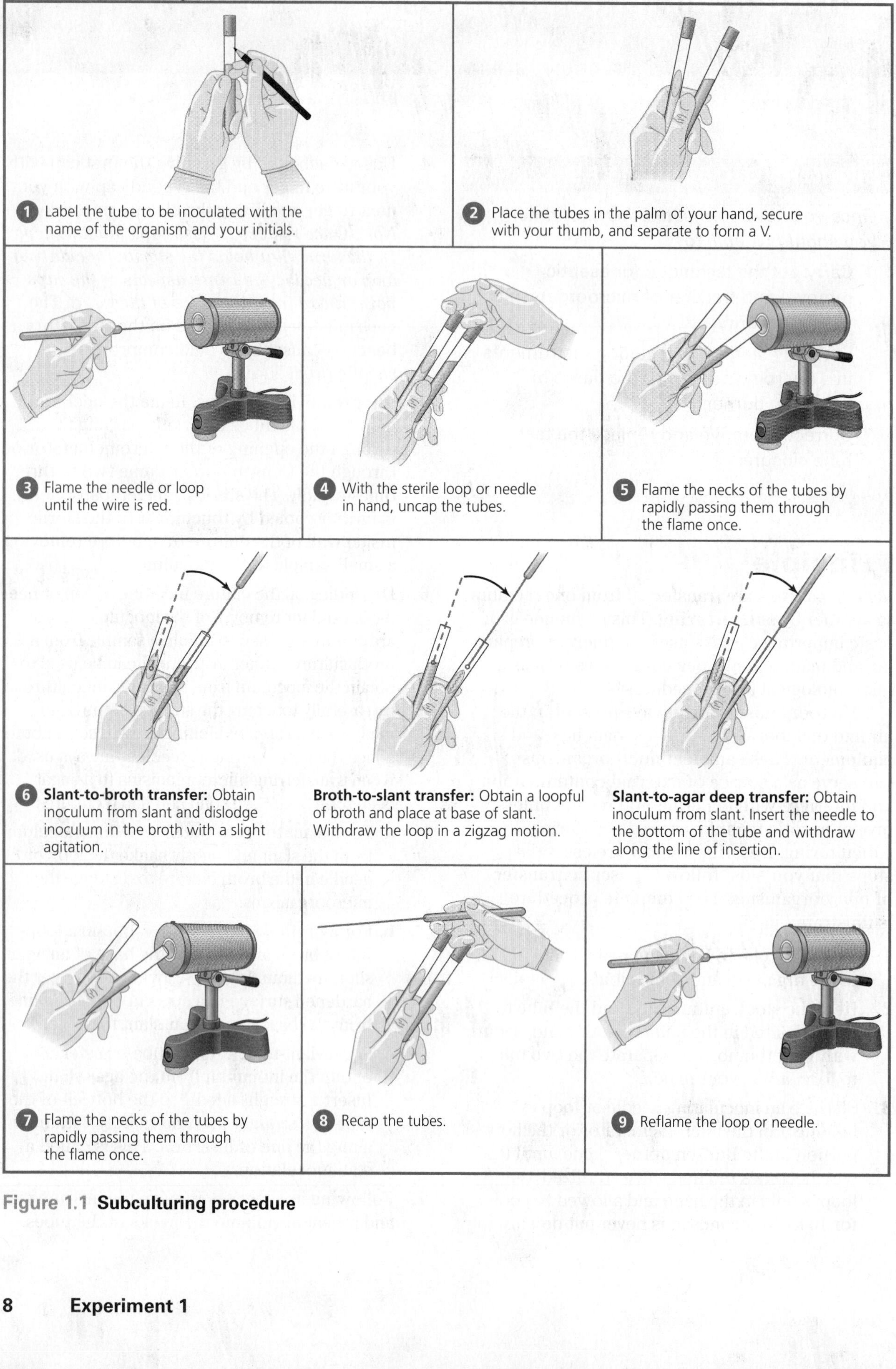

1 Label the tube to be inoculated with the name of the organism and your initials.

2 Place the tubes in the palm of your hand, secure with your thumb, and separate to form a V.

3 Flame the needle or loop until the wire is red.

4 With the sterile loop or needle in hand, uncap the tubes.

5 Flame the necks of the tubes by rapidly passing them through the flame once.

6 **Slant-to-broth transfer:** Obtain inoculum from slant and dislodge inoculum in the broth with a slight agitation.

Broth-to-slant transfer: Obtain a loopful of broth and place at base of slant. Withdraw the loop in a zigzag motion.

Slant-to-agar deep transfer: Obtain inoculum from slant. Insert the needle to the bottom of the tube and withdraw along the line of insertion.

7 Flame the necks of the tubes by rapidly passing them through the flame once.

8 Recap the tubes.

9 Reflame the loop or needle.

Figure 1.1 Subculturing procedure

8. Replace the caps on the same tubes from which they were removed.

9. Resterilize the loop or needle to destroy any remaining organisms.

In this experiment, you will master the manipulations required for aseptic transfer of microorganisms in broth-to-slant, slant-to-broth, and slant-to-agar deep tubes. You will be using a positive and a negative control to test your ability to maintain aseptic techniques while transferring cultures. The technique for transfer to and from agar plates is discussed in Experiment 2.

CLINICAL APPLICATION

Aseptic Inoculation and Transfer

It is mandatory that microbiology laboratory workers learn and perfect the skill of inoculating bacterial specimens on agar plates, in liquid broth, or in semisolid medium, and be able to subculture the organism from one medium to another. A sterile inoculating needle or loop is the basic instrument of transfer. Keep in mind that, transferring bacterial cultures requires aseptic or sterile techniques at all times, especially if you are working with pathogens. Do not contaminate what you are working with and do not contaminate yourself.

AT THE BENCH

Materials

Cultures

Twenty-four–hour nutrient broth and nutrient agar slant cultures of *Serratia marcescens* and a sterile tube of nutrient broth. The nutrient broth tubes will be labeled "A" and "B," and the contents will be known only by the instructor.

Media

Per student: three nutrient broth tubes, three nutrient agar slants, and three nutrient agar deep tubes.

Equipment

Microincinerator or Bunsen burner, inoculating loop and needle, and glassware marking pencil.

Procedure Lab One

1. Label all tubes of sterile media as described in the Laboratory Protocol section on page xv.

2. Following the procedure outlined and illustrated previously (Figure 1.1), perform the following transfers:

    a. *Broth culture "A" to a nutrient agar slant, nutrient agar deep tube, and nutrient broth.*

    b. *Broth culture "B" to a nutrient agar slant, nutrient agar deep tube, and nutrient broth.*

    c. *S. marcescens agar slant culture to a nutrient agar slant, nutrient agar deep tube and nutrient broth.*

3. Incubate all cultures at 25°C for 24 to 48 hours.

Procedure Lab Two

1. Examine all cultures for the appearance of growth, which is indicated by turbidity in the broth culture and the appearance of an orange-red growth on the surface of the slant and along the line of inoculation in the agar deep tube.

2. Record your observations in the chart provided in the Lab Report.

3. Confirm your results with the instructor to determine the negative control tube.

▶ TIPS FOR SUCCESS

1. **It is imperative that you maintain sterility and utilize aseptic techniques at all times.** If you allow a contaminating organism into your bacterial culture, you will see a positive growth in media that was inoculated with the negative control.

Techniques for Isolation of Pure Cultures

In nature, microbial populations do not segregate themselves by species, but exist with a mixture of many other cell types. In the laboratory, these populations can be separated into **pure cultures**. These cultures contain only one type of organism and are suitable for the study of their cultural, morphological, and biochemical properties.

In this experiment, you will first use one of the techniques designed to produce discrete colonies. Colonies are individual, macroscopically visible masses of microbial growth on a solid medium surface, each representing the multiplication of a single organism. Once you have obtained these discrete colonies, you will make an aseptic transfer onto nutrient agar slants for the isolation of pure cultures.

PART A Isolation of Discrete Colonies from a Mixed Culture

LEARNING OBJECTIVE

Once you have completed this experiment, you should be able to

1. Perform the streak-plate and/or the spread-plate inoculation procedure to separate the cells of a mixed culture so that discrete colonies can be isolated.

Principle

The techniques commonly used for isolation of discrete colonies initially require that the number of organisms in the inoculum be reduced. The resulting diminution of the population size ensures that, following inoculation, individual cells will be sufficiently far apart on the surface of the agar medium to separate the different species. The following are techniques that can be used to accomplish this necessary dilution:

1. The **streak-plate** method is a rapid qualitative isolation method. It is essentially a dilution technique that involves spreading a loopful of culture over the surface of an agar plate. Although many types of procedures are performed, the four-way, or quadrant, streak is described. Refer to Figure 2.1, which schematically illustrates this technique.

   a. Place a loopful of culture on the agar surface in Area 1. Flame the loop, cool it by touching an unused part of the agar surface close to the periphery of the plate, and then drag it rapidly several times across the surface of Area 1.

   b. Reflame and cool the loop, and turn the Petri dish 90°. Then touch the loop to a corner of the culture in Area 1 and drag it several times across the agar in Area 2. The loop should never enter Area 1 again.

   c. Reflame and cool the loop and again; turn the dish 90°. Streak Area 3 in the same manner as Area 2.

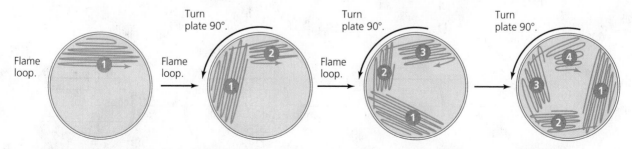

Figure 2.1 **Four-way streak-plate technique**

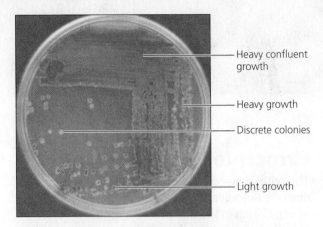

Figure 2.2 **Four-way streak-plate inoculation with *Serratia marcescens***

— Heavy confluent growth

— Heavy growth

— Discrete colonies

— Light growth

d. Without reflaming the loop, again turn the dish 90° and then drag the culture from a corner of Area 3 across Area 4, using a wider streak. Don't let the loop touch any of the previously streaked areas. The flaming of the loop at the points indicated is to dilute the culture so that fewer organisms are streaked in each area, resulting in the final desired separation. A photograph of a streak-plate inoculation is shown in Figure 2.2.

2. An alternative streak-plate method is for students new to the laboratory who have yet to master the necessary lab skills that would allow them to use the rapid method listed above. This alternative method involves spreading a loopful of culture over the surface of an agar plate that has the quadrants laid out visibly for quick reference. Refer to Figure 2.3, which illustrates this technique.

a. Using a marker, draw two bisecting lines on the bottom of the Petri dish to divide the plate into 4 equal parts. Label each quadrant 1 through 4, starting with the top right quadrant and labeling counter-clockwise. Sterilizing the loop at the points indicated is to dilute the culture due to fewer organisms available to be streaked into each area, resulting in the final desired separation.

b. Turn the Petri dish over and place a loopful of culture on the agar surface in quadrant 1. Using the edge of the loop and holding the loop at a shallow angle so as not to gouge the agar, quickly spread the bacteria throughout the quadrant.

c. Reflame and cool the loop, and turn the Petri dish 90°. Then touch the loop into an area that has been streaked in quadrant 1 and drag it across the agar into quadrant 2, repeat this twice without flaming the loop.

(a)

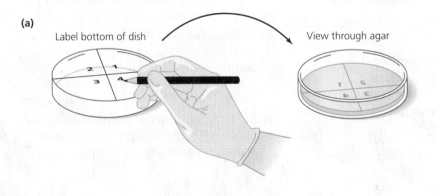

Label bottom of dish

View through agar

(b)

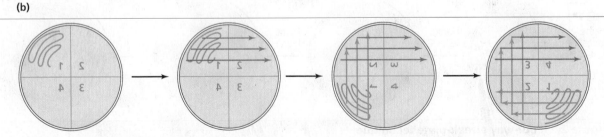

Figure 2.3 **Alternate streak-plate method**

d. Reflame and cool the loop and again turn the dish 90°. Streak the bacteria into quadrant 3 in the same manner used for quadrant 2.

e. Reflame and cool the loop and again turn the dish 90°. Streak the bacteria into quadrant 4 in the same manner used for quadrant 3.

3. The **spread-plate** technique requires that a previously diluted mixture of microorganisms be used. During inoculation, the cells are spread over the surface of a solid agar medium with a sterile, L-shaped bent glass rod while the Petri dish is spun on a "lazy Susan" turntable. The step-by-step procedure for this technique is as follows:

a. Place the bent glass rod into a beaker and add a sufficient amount of 95% ethyl alcohol to cover the lower, bent portion.

b. Place an appropriately labeled nutrient agar plate on the turntable. With a sterile pipette, place one drop of sterile water on the center of the plate, followed by a sterile loopful of *Micrococcus luteus*. Mix gently with the loop and replace the cover.

c. Remove the glass rod from the beaker, and pass it through the Bunsen burner flame with the bent portion of the rod pointing downward to prevent the burning alcohol from running down your arm. Allow the alcohol to burn off the rod completely. Cool the rod for 10 to 15 seconds.

d. Remove the Petri dish cover and spin the turntable.

e. While the turntable is spinning, lightly touch the sterile bent rod to the surface of the agar and move it back and forth. This will spread the culture over the agar surface.

f. When the turntable comes to a stop, replace the cover. Immerse the rod in alcohol and reflame.

g. In the absence of a turntable, turn the Petri dish manually and spread the culture with the sterile bent glass rod.

4. The **pour-plate** technique requires a serial dilution of the mixed culture by means of a loop or pipette. The diluted inoculum is then added to a molten agar medium in a Petri dish, mixed, and allowed to solidify. The serial dilution and pour-plate procedures are outlined in Experiment 18.

CLINICAL APPLICATION

Isolation of Cultures as a Diagnostic Technique

The isolation of pure cultures is the most important diagnostic tool used in a clinical or research laboratory to uncover the cause of an infection or disease. Before any biochemical or molecular techniques may be used to identify or characterize the causative organism, an individual bacterial colony must be isolated for testing. The isolation of *Staphylococcus aureus* from cultures taken from abscesses or *Streptococcus pyogenes* from a throat culture are two examples of clinical applications of this technique.

AT THE BENCH

Materials

Cultures

24- to 48-hour nutrient broth cultures of a mixture of one part *Serratia marcescens* and three parts *Micrococcus luteus* and a mixture of one part *Escherichia coli* and ten parts *Micrococcus luteus*.

Sources of mixed cultures from the environment could include cultures from a tabletop, bathroom sink, water fountain, or inside of an incubator. Each student should obtain a mixed culture from one of the environmental sources listed above.

Media

Three Trypticase™ soy agar plates per designated student group for each inoculation technique to be performed.

Equipment

Microincinerator or Bunsen burner, inoculating loop, turntable, glassware marking pencil, culture tubes containing 1 ml of sterile water, test tube rack, and sterile cotton swabs.

Procedure Lab One

1. Following the procedures previously described, prepare a spread-plate and/or streak-plate inoculation of each test culture on an appropriately labeled plate.

2. Prepare an environmental mixed culture.

   a. Dampen a sterile cotton swab with sterile water. Wring out the excess water by pressing the wet swab against the walls of the tube.

   b. With the moistened cotton swab, obtain your mixed-culture specimen from one of the selected environmental sources listed in the section on cultures.

   c. Place the contaminated swab back into the tube of sterile water. Mix gently and let stand for 5 minutes.

   d. Perform spread-plate and/or streak-plate inoculation on an appropriately labeled plate.

3. **Incubate all plates in an inverted position** for 48 to 72 hours at 25°C.

Procedure Lab Two

1. Examine all agar plate cultures to identify the distribution of colonies. In the charts provided in Part A of the Lab Report, complete the following:

   a. Draw the distribution of colonies appearing on each of the agar plate cultures.

   b. On each of the agar plate cultures, select two discrete colonies that differ in appearance. Using Figure 3.1 on page 26 as a reference, describe each colony as to its

      Form: Circular, irregular, or spreading.

      Elevation: Flat, slightly raised, or markedly raised.

      Pigmentation.

      Size: Pinpoint, small, medium, or large.

2. Retain the mixed-culture plates to perform Part B of this experiment.

▶ TIPS FOR SUCCESS

1. **An isolation plate has isolated distinct, individual colonies.** If your technique results in isolated colonies in a quadrant that was not the last one to be streaked, that is okay. The point of using this method is to get those individual colonies somewhere on the plate.
2. **Pay attention to how well you sterilize your loop and maintain your aseptic technique.** If the loop is not properly sterilized between streaks, or your aseptic technique is not maintained, the resulting plate will not exhibit a decrease in bacteria leading to individual colonies. With that in mind, if a plate you have streaked or poured does not exhibit a decrease in bacterial colonies area-to-area, you may want to re-examine your technique for maintaining sterilization.

PART B   Isolation of Pure Cultures from a Spread-Plate or Streak-Plate Preparation

LEARNING OBJECTIVE

Once you have completed this experiment, you should be able to

1. Prepare a stock culture of an organism using isolates from mixed cultures prepared on an agar streak plate and/or spread plate.

Principle

Once discrete, well-separated colonies develop on the surface of a nutrient agar plate culture, each may be picked up with a sterile needle and transferred to separate nutrient agar slants. Each of these new slant cultures represents the growth of a single bacterial species and is designated as a **pure** or **stock culture**.

CLINICAL APPLICATION

Transferring a Colony of Bacteria Daughter Cells

For identification of a bacterial pathogen, a discrete bacterial colony must be transferred from a streak or spread plate to the new testing media. This new culture will consist of daughter cells that are genetic and metabolic clones of the original bacterial cells that were transferred to the plate. This will allow for identification of the unknown bacterial species through its biochemical and molecular characteristics.

AT THE BENCH

Materials

Cultures

Mixed-culture, nutrient agar streak-plate and/or spread-plate preparations of *S. marcescens* and *M. luteus*, *M. luteus* and *E. coli*, and the environmental specimen plate from Part A.

Media

Four Trypticase™ soy agar slants per designated student group.

Equipment

Microincinerator or Bunsen burner, inoculating needle, and glassware marking pencil.

Procedure Lab One

1. Aseptically transfer, from visibly discrete colonies, the yellow *M. luteus*, the white *E. coli*, the red *S. marcescens*, and a discrete colony from the environmental agar plate specimen to the appropriately labeled agar slants as shown in Figure 2.4.

2. Incubate all slants at 37°C for 18 to 24 hours.

Procedure Lab Two

1. In the chart provided in Part B of the Lab Report, complete the following:

   a. Draw and indicate the type of growth of each pure-culture isolate, using Figure 3.1 on page 26 as a reference.

   b. Observe the color of the growth and record its pigmentation.

   c. Indicate the name of the isolated organisms.

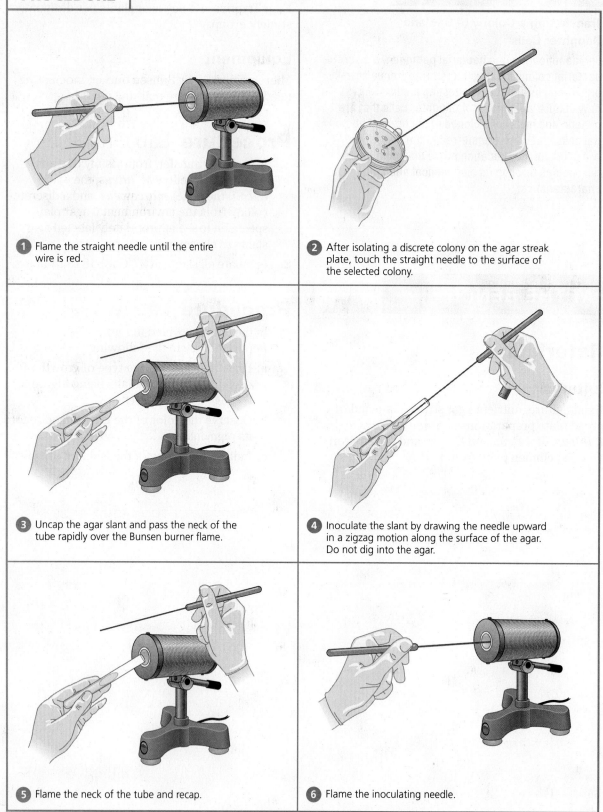

1 Flame the straight needle until the entire wire is red.

2 After isolating a discrete colony on the agar streak plate, touch the straight needle to the surface of the selected colony.

3 Uncap the agar slant and pass the neck of the tube rapidly over the Bunsen burner flame.

4 Inoculate the slant by drawing the needle upward in a zigzag motion along the surface of the agar. Do not dig into the agar.

5 Flame the neck of the tube and recap.

6 Flame the inoculating needle.

Figure 2.4 **Procedure for the preparation of a pure culture**

Cultural Characteristics of Microorganisms

LEARNING OBJECTIVE

Once you have completed this experiment, you should be able to

1. Determine the cultural characteristics of microorganisms as an aid in identifying and classifying organisms into taxonomic groups.

Principle

When grown on a variety of media, microorganisms will exhibit differences in the macroscopic appearance of their growth. These differences, called **cultural characteristics**, are used as a basis for separating microorganisms into taxonomic groups. The cultural characteristics for all known microorganisms are contained in *Bergey's Manual of Systematic Bacteriology*. They are determined by culturing the organisms on nutrient agar slants and plates, in nutrient broth, and in nutrient gelatin. The patterns of growth to be considered in each of these media are described below, and some are illustrated in Figure 3.1.

Nutrient Agar Slants

These have a single straight line of inoculation on the surface and are evaluated in the following manner:

1. **Abundance of growth:** The amount of growth is designated as none, slight, moderate, or large.

2. **Pigmentation:** Chromogenic microorganisms may produce intracellular pigments that are responsible for the coloration of the organisms as seen in surface colonies. Other organisms produce extracellular soluble pigments that are excreted into the medium and also produce a color. Most organisms, however, are nonchromogenic and will appear white to gray.

3. **Optical characteristics:** Optical characteristics may be evaluated on the basis of the amount of light transmitted through the growth. These characteristics are described as **opaque** (no light transmission), **translucent** (partial transmission), or **transparent** (full transmission).

4. **Form:** The appearance of the single-line streak of growth on the agar surface is designated as:
   a. **Filiform:** Continuous, threadlike growth with smooth edges.
   b. **Echinulate:** Continuous, threadlike growth with irregular edges.
   c. **Beaded:** Nonconfluent to semiconfluent colonies.
   d. **Effuse:** Thin, spreading growth.
   e. **Arborescent:** Treelike growth.
   f. **Rhizoid:** Rootlike growth.

5. **Consistency:**
   a. **Dry:** Free from moisture.
   b. **Buttery:** Moist and shiny.
   c. **Mucoid:** Slimy and glistening.

Nutrient Agar Plates

These demonstrate well-isolated colonies and are evaluated in the following manner:

1. **Size:** Pinpoint, small, moderate, or large.
2. **Pigmentation:** Color of colony.
3. **Form:** The shape of the colony is described as follows:
   a. **Circular:** Unbroken, peripheral edge.
   b. **Irregular:** Indented, peripheral edge.
   c. **Rhizoid:** Rootlike, spreading growth.
4. **Margin:** The appearance of the outer edge of the colony is described as follows:
   a. **Entire:** Sharply defined, even.
   b. **Lobate:** Marked indentations.
   c. **Undulate:** Wavy indentations.

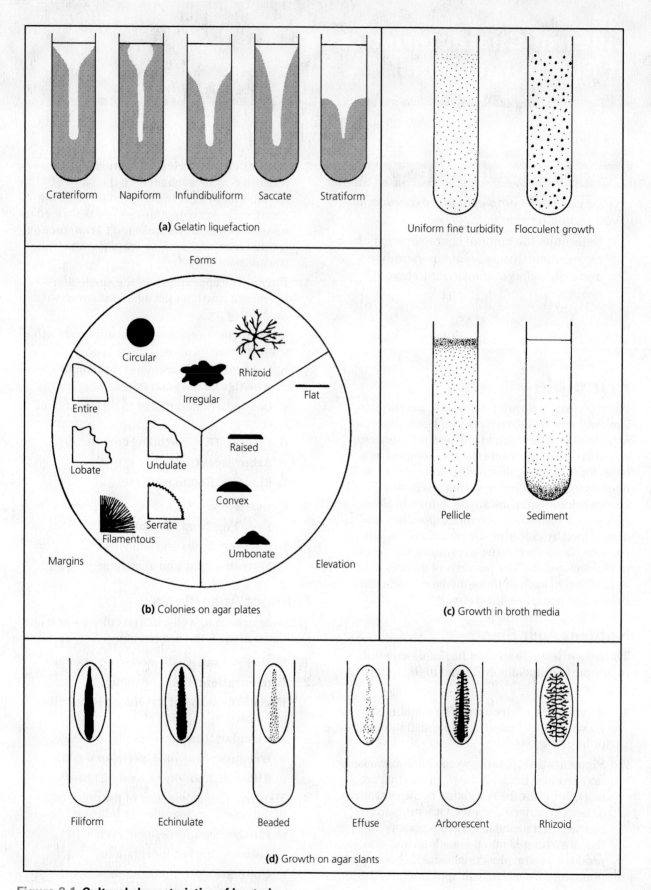

Figure 3.1 Cultural characteristics of bacteria

d. Serrate: Toothlike appearance.

e. Filamentous: Threadlike, spreading edge.

5. **Elevation:** The degree to which colony growth is raised on the agar surface is described as follows:

   a. Flat: Elevation not discernible.

   b. Raised: Slightly elevated.

   c. Convex: Dome-shaped elevation.

   d. Umbonate: Raised, with elevated convex central region.

Nutrient Broth Cultures

These are evaluated as to the distribution and appearance of the growth as follows:

1. **Uniform fine turbidity:** Finely dispersed growth throughout.

2. **Flocculent:** Flaky aggregates dispersed throughout.

3. **Pellicle:** Thick, padlike growth on surface.

4. **Sediment:** Concentration of growth at the bottom of broth culture may be granular, flaky, or flocculent.

Nutrient Gelatin

This solid medium may be liquefied by the enzymatic action of gelatinase. Liquefaction occurs in a variety of patterns:

1. **Crateriform:** Liquefied surface area is saucer-shaped.

2. **Napiform:** Bulbous-shaped liquefaction at surface.

3. **Infundibuliform:** Funnel-shaped.

4. **Saccate:** Elongated, tubular.

5. **Stratiform:** Complete liquefaction of the upper half of the medium.

CLINICAL APPLICATION

Examining Colony Growth Characteristics to Aid Identification

Bacterial species each have a characteristic pattern of colony growth in a liquid culture or on a solid medium. While not truly a diagnostic tool, recognition of these patterns of characteristics will aid in a clinical lab setting by helping to minimize the list of potential bacterial species to test for.

AT THE BENCH

Materials

Cultures

Twenty-four–hour nutrient broth cultures of *Pseudomonas aeruginosa* BSL-2 , *Bacillus cereus, Micrococcus luteus,* and *Escherichia coli.* Seventy-two– to 96-hour Trypticase™ soy broth culture of *Mycobacterium smegmatis.*

Media

Per designated student group: five each of nutrient agar slants, nutrient agar plates, nutrient broth tubes, and nutrient gelatin tubes.

Equipment

Microincinerator or Bunsen burner, inoculating loop and needle, and glassware marking pencil.

Procedure Lab One

1. Using aseptic technique, inoculate each of the appropriately labeled media listed below in the following manner:

   a. Nutrient agar slants: With a sterile needle, make a single-line streak of each of the cultures provided, starting at the butt and drawing the needle up the center of the slanted agar surface.

   b. Nutrient agar plates: With a sterile loop, prepare a streak-plate inoculation of each of the cultures for the isolation of discrete colonies.

   c. Nutrient broth cultures: Using a sterile loop, inoculate each organism into a tube of nutrient broth. Shake the loop a few times to dislodge the inoculum.

   d. Nutrient gelatin: Using a sterile needle, prepare a stab inoculation of each of the cultures provided.

2. Incubate all cultures at 37°C for 24 to 48 hours.

Procedure Lab Two

1. Before beginning observation of all the cultures, place the gelatin cultures in a refrigerator for 30 minutes or in a beaker of crushed ice for a few minutes. The gelatin culture will be the last to be observed.

2. Refer to Figure 3.1 on page 26 and the descriptions presented in the introductory section of Experiment 3 while making the following observations:

a. Nutrient agar slants: Observe each of the nutrient agar slant cultures for the amount, pigmentation, form, and consistency of the growth. Record your observations in the chart provided in the Lab Report.

b. Nutrient agar plates: Observe a single, well-isolated colony on each of the nutrient agar plate cultures and identify its size, elevation, margin, form, and pigmentation.

Record your observations in the chart provided in the Lab Report.

c. Nutrient broth cultures: Observe each of the nutrient broth cultures for the appearance of growth (flocculation, turbidity, sediment, or pellicle). Record your observations in the chart provided in the Lab Report.

d. Nutrient gelatin: Remove gelatin cultures from the refrigerator or beaker of crushed ice, and observe whether liquefaction of the medium has developed and whether the organism has produced gelatinase. Record your observations in the chart provided in the Lab Report.

PART 2

Microscopy

Once you have completed the experiments in this section, you should be

1. Familiar with the history and diversity of microscopic instruments.
2. Able to understand the components, use, and care of the brightfield microscope.
3. Able to correctly use the microscope for observation and measurement of microorganisms.

Introduction

Microbiology, the branch of science that has so vastly extended and expanded our knowledge of the living world, owes its existence to Antoni van Leeuwenhoek. In 1673, with the aid of a crude microscope consisting of a biconcave lens enclosed in two metal plates, Leeuwenhoek introduced the world to the existence of microbial forms of life. Over the years, microscopes have evolved from the simple, single-lens instrument of Leeuwenhoek, with a magnification of 300×, to the present-day electron microscopes capable of magnifications greater than 250,000×.

Microscopes are designated as either light microscopes or electron microscopes. The former use visible light or ultraviolet rays to illuminate specimens. They include brightfield, darkfield, phase-contrast, and fluorescent instruments. Fluorescent microscopes use ultraviolet radiations whose wavelengths are shorter than those of visible light and are not directly perceptible to the human eye. Electron microscopes use electron beams (instead of light rays) and magnets (instead of lenses) to observe submicroscopic particles.

Essential Features of Various Microscopes

Brightfield Microscope This instrument contains two-lens systems for magnifying specimens: the ocular lens in the eyepiece and the objective lens located in the nosepiece. The specimen is illuminated by a beam of tungsten light focused on it by a substage lens called a condenser; the result is a specimen that appears dark against a bright background. A major limitation of this system is the absence of contrast between the specimen and the surrounding medium, which makes it difficult to observe living cells. Therefore, most brightfield observations are performed on nonviable, stained preparations.

Darkfield Microscope This is similar to the ordinary light microscope; however, the condenser system is modified so that the specimen is not illuminated directly. The condenser directs the light obliquely so that the light is deflected or scattered from the specimen, which then appears bright against a dark background. Living specimens may be observed more readily with darkfield than with brightfield microscopy.

Phase-Contrast Microscope Observation of microorganisms in an unstained state is possible with this microscope. The optics include special objectives and a condenser that make visible cellular components that differ only slightly in their refractive indexes. As light is transmitted through a specimen with a refractive index different from that of the surrounding medium, a portion of the light is refracted (bent) due to slight variations in density and thickness of the cellular components. The special optics convert the difference between transmitted light and refracted rays, resulting in a significant variation in the intensity of light and thereby producing a discernible image of the structure under study. The image appears dark against a light background.

Fluorescent Microscope This microscope is used most frequently to visualize specimens that are chemically tagged with a fluorescent dye. The source of illumination is an ultraviolet (UV) light obtained from a high-pressure mercury lamp or hydrogen quartz lamp. The ocular lens is fitted with a filter that permits the longer ultraviolet wavelengths to pass, while the shorter wavelengths are blocked or eliminated. Ultraviolet radiations are absorbed by the fluorescent label, and the energy is re-emitted in the form of a different wavelength in the visible light range. The fluorescent dyes absorb at wavelengths between 230 and 350 nanometers (nm) and emit orange, yellow, or greenish light. This microscope is used primarily for the detection of antigen-antibody reactions. Antibodies are conjugated with a fluorescent dye that becomes excited in the presence of ultraviolet light, and

the fluorescent portion of the dye becomes visible against a black background.

Electron Microscope This instrument provides a revolutionary method of microscopy, with magnifications up to 1 million ×. This permits visualization of submicroscopic cellular particles as well as viral agents. In the electron microscope, the specimen is illuminated by a beam of electrons rather than light, and the focusing is carried out by electromagnets instead of a set of optics. These components are sealed in a tube in which a complete vacuum is established. Transmission electron microscopes require specimens that are prepared as thin filaments, fixed and dehydrated for the electron beam to pass freely through them. As the electrons pass through the specimen, images are formed by directing the electrons onto photographic film, thus making internal cellular structures visible. Scanning electron microscopes are used for visualizing surface characteristics rather than intracellular structures. A narrow beam of electrons scans back and forth, producing a three-dimensional image as the electrons are reflected off the specimen's surface.

While scientists have a variety of optical instruments with which to perform routine laboratory procedures and sophisticated research, the compound brightfield microscope is the "workhorse" and is commonly found in all biological laboratories. Although you should be familiar with the basic principles of microscopy, you probably have not been exposed to this diverse array of complex and expensive equipment. Therefore, only the compound brightfield microscope will be discussed in depth and used to examine specimens.

Microscopic Examination of Stained Cell Preparations

LEARNING OBJECTIVES

Once you have completed this experiment, you should be familiar with them

1. Theoretical principles of brightfield microscopy.

2. Component parts of the compound microscope.

3. Use and care of the compound microscope.

4. Practical use of the compound microscope for visualization of cellular morphology from stained slide preparations.

Principle

Microbiology is a science that studies living organisms that are too small to be seen with the naked eye. Needless to say, such a study must involve the use of a good compound microscope. Although there are many types and variations, they all fundamentally consist of a two-lens system, a variable but controllable light source, and mechanical adjustable parts for determining focal length between the lenses and specimen (Figure 4.1).

Components of the Microscope

Stage A fixed platform with an opening in the center allows the passage of light from an illuminating source below to the lens system above the stage. This platform provides a surface for the placement of a slide with its specimen over the central opening. In addition to the fixed stage, most microscopes have a **mechanical stage** that can be moved vertically or horizontally by means of adjustment controls. Less sophisticated microscopes have clips on the fixed stage, and the slide must be positioned manually over the central opening.

Illumination The light source is positioned in the base of the instrument. Some microscopes are equipped with a built-in light source to provide direct illumination. Others are provided with a reversible mirror that has one side flat and the other concave. An external light source, such as a lamp, is placed in front of the mirror to direct the light upward into the lens system. The flat side of the mirror is used for artificial light, and the concave side for sunlight.

Abbé Condenser This component is found directly under the stage and contains two sets of lenses that collect and concentrate light as it passes upward from the light source into the lens systems. The condenser is equipped with an **iris diaphragm**, a shutter controlled by a lever that is used to regulate the amount of light entering the lens system.

Body Tube Above the stage and attached to the arm of the microscope is the body tube. This structure houses the lens system that magnifies the specimen. The upper end of the tube contains the **ocular** or **eyepiece** lens. The lower portion consists of a movable **nosepiece** containing the **objective lenses**. Rotation of the nosepiece positions objectives above the stage opening. The body tube may be raised or lowered with the aid of **coarse-adjustment** and **fine-adjustment knobs** that are located above or below the stage, depending on the type and make of the instrument.

Theoretical Principles of Microscopy

To use the microscope efficiently and with minimal frustration, you should understand the basic principles of microscopy: magnification, resolution, numerical aperture, illumination, and focusing.

Magnification Enlargement, or magnification, of a specimen is the function of a two-lens system; the **ocular lens** is found in the eyepiece, and the **objective lens** is situated in a revolving nosepiece. These lenses are separated by the **body tube**. The objective lens is nearer the specimen and magnifies it, producing the **real image** that is projected up into the focal plane and then magnified by the ocular lens to produce the final image.

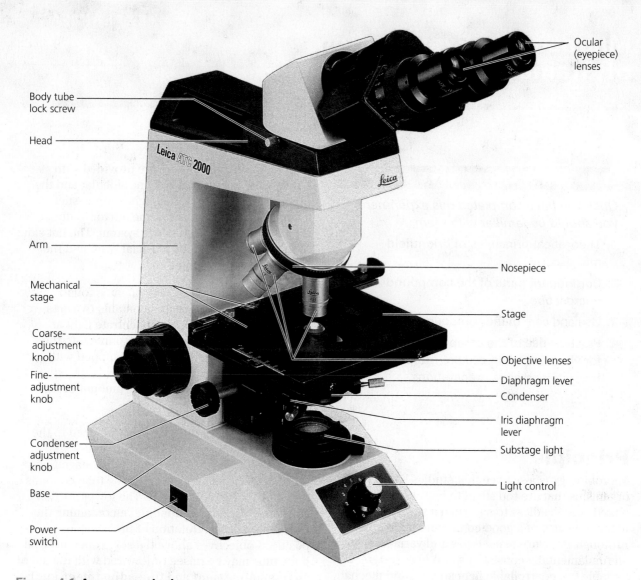

Figure 4.1 **A compound microscope**

The most commonly used microscopes are equipped with a revolving nosepiece containing four objective lenses, each possessing a different degree of magnification. When these are combined with the magnification of the ocular lens, the total or overall linear magnification of the specimen is obtained. This is shown in **Table 4.1**.

Resolving Power or Resolution Although magnification is important, you must be aware that unlimited enlargement is not possible by merely increasing the magnifying power of the lenses or by using additional lenses, because lenses are limited by a property called **resolving power**. By definition, resolving power is how far apart two adjacent objects must be before a given lens shows them as discrete entities. When a lens

cannot discriminate, that is, when the two objects appear as one, it has lost resolution. Increased magnification will not rectify the loss and will, in fact, blur the object. The resolving power of a lens is dependent on the wavelength of light used and the **numerical aperture**, which is a characteristic of each lens and imprinted on each objective. The numerical aperture is defined as a function of the diameter of the objective lens in relation to its focal length. It is doubled by use of the substage condenser, which illuminates the object with rays of light that pass through the specimen obliquely as well as directly. Thus, resolving power is expressed mathematically as follows:

$$\text{resolving power} = \frac{\text{wavelength of light}}{2 \times \text{numerical aperture}}$$

TABLE 4.1 Overall Linear Magnification

MAGNIFICATION		TOTAL MAGNIFICATION
OBJECTIVE LENSES	OCULAR LENS	OBJECTIVE MULTIPLIED BY OCULAR
Scanning 4×	10×	40×
Low-power 10×	10×	100×
High-power 40×	10×	400×
Oil-immersion 100×	10×	1000×

Based on this formula, the shorter the wavelength, the greater the resolving power of the lens. Thus, for the same numerical aperture, short wavelengths of the electromagnetic spectrum are better suited for higher resolution than are longer wavelengths.

However, as with magnification, resolving power also has limits. Decreasing the wavelength will not automatically increase the resolving power of a lens, because the visible portion of the electromagnetic spectrum is very narrow and borders on the very short wavelengths found in the ultraviolet portion of the spectrum.

The relationship between wavelength and numerical aperture is valid only for increased resolving power when light rays are parallel. Therefore, the resolving power is also dependent on another factor, the **refractive index**. This is the bending power of light passing through air from the glass slide to the objective lens. The refractive index of air is lower than that of glass; as light rays pass from the glass slide into the air, they are bent or refracted so that they do not pass into the objective lens. This would cause a loss of light, which would reduce the numerical aperture and diminish the resolving power of the objective lens. Loss of refracted light can be compensated for by interposing mineral oil, which has the same refractive index as glass, between the slide and the objective lens. In this way, decreased light refraction occurs and more light rays enter directly into the objective lens, producing a vivid image with high resolution (Figure 4.2).

Illumination Effective illumination is required for efficient magnification and resolving power. Since the intensity of daylight is an uncontrolled variable, artificial light from a tungsten lamp is the most commonly used light source in microscopy. The light is passed through the condenser located beneath the stage. The condenser contains two lenses that are necessary to produce a maximum numerical aperture. The height of the condenser

can be adjusted with the **condenser knob**. Always keep the condenser close to the stage, especially when using the oil-immersion objective.

Between the light source and the condenser is the iris diaphragm, which can be opened and closed by means of a lever, thereby regulating the amount of light entering the condenser. Excessive illumination may actually obscure the specimen because of lack of contrast. The amount of light entering the microscope differs with each objective lens used. A rule of thumb is that *as the magnification of the lens increases, the distance between the objective lens and slide, called working distance, decreases, whereas the numerical aperture of the objective lens increases* (Figure 4.3).

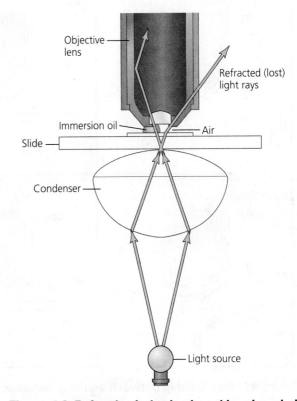

Figure 4.2 Refractive index in air and in mineral oil

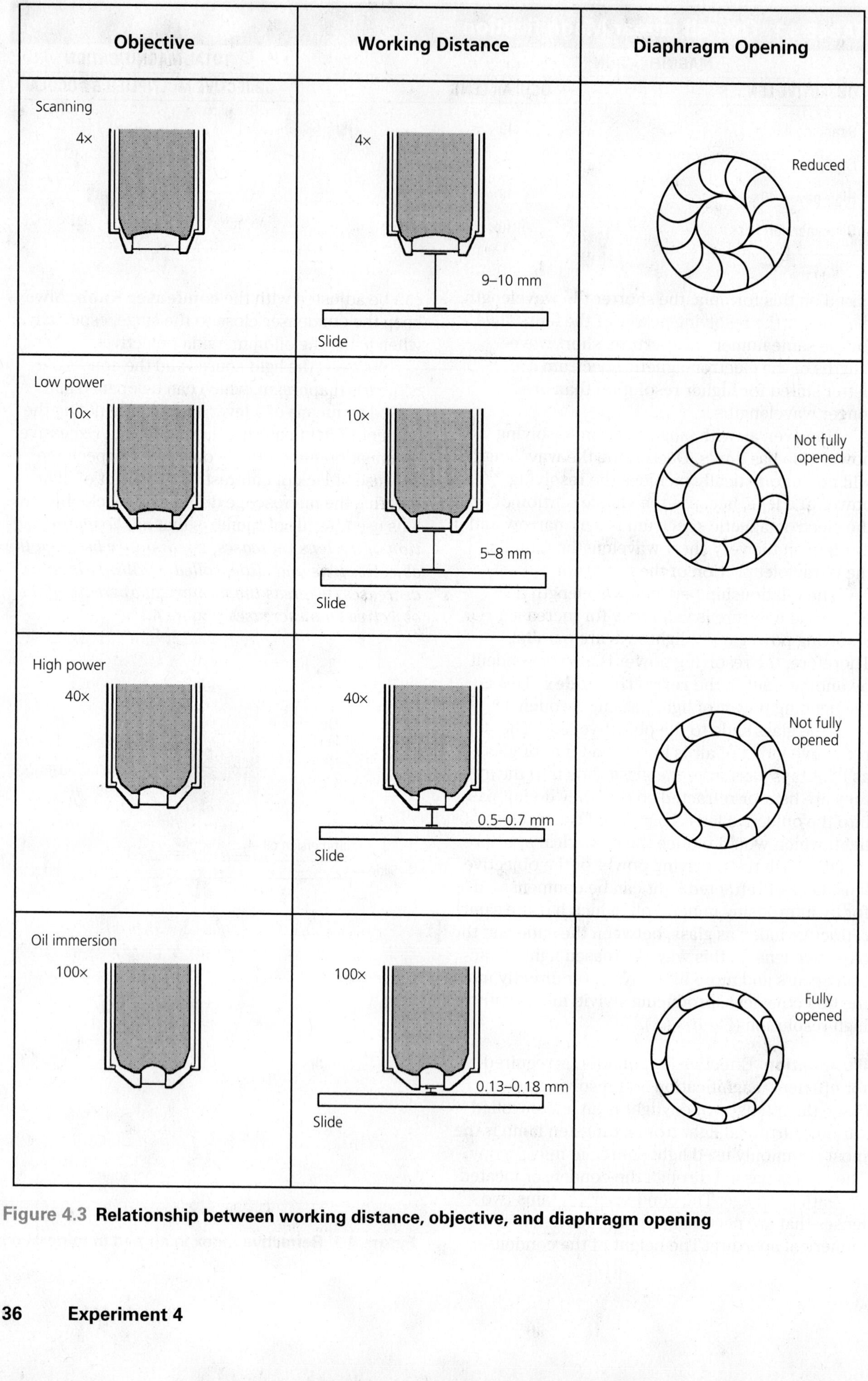

Objective	Working Distance	Diaphragm Opening
Scanning 4×	4× 9–10 mm Slide	Reduced
Low power 10×	10× 5–8 mm Slide	Not fully opened
High power 40×	40× 0.5–0.7 mm Slide	Not fully opened
Oil immersion 100×	100× 0.13–0.18 mm Slide	Fully opened

Figure 4.3 **Relationship between working distance, objective, and diaphragm opening**

Use and Care of the Microscope

You will be responsible for the proper care and use of microscopes. Since microscopes are expensive, you must observe the following regulations and procedures.

The instruments are housed in special cabinets and must be moved by users to their laboratory benches. The correct and only acceptable way to do this is to grip the microscope arm firmly with the right hand and the base with the left hand, and lift the instrument from the cabinet shelf. Carry it close to the body and gently place it on the laboratory bench. This will prevent collision with furniture or coworkers and will protect the instrument against damage.

Once the microscope is placed on the laboratory bench, observe the following rules:

1. Remove all unnecessary materials (including books, papers, purses, and hats) from the laboratory bench.

2. Uncoil the microscope's electric cord and plug it into an electrical outlet.

3. Clean all lens systems; the smallest bit of dust, oil, lint, or eyelash will decrease the efficiency of the microscope. The ocular, scanning, low-power, and high-power lenses may be cleaned by wiping several times with acceptable lens tissue. Never use paper toweling or cloth on a lens surface. If the oil-immersion lens is gummy or tacky, a piece of lens paper moistened with xylol is used to wipe it clean. The xylol is immediately removed with a tissue moistened with 95% alcohol, and the lens is wiped dry with lens paper. *Note: This xylol cleansing procedure should be performed only by the instructor and only if necessary; consistent use of xylol may loosen the lens.*

The following routine procedures must be followed to ensure correct and efficient use of the microscope.

1. Place the microscope slide with the specimen within the stage clips on the fixed stage. Move the slide to center the specimen over the opening in the stage directly over the light source.

2. Raise the microscope stage up as far as it will go. Rotate the scanning lens or low-power lens into position. Lower the body tube with the coarse-adjustment knob to its lowest position. *Note: Never lower the body tube while looking through the ocular lens.*

3. While looking through the ocular lens, use the fine-adjustment knob, rotating it back and forth slightly, to bring the specimen into sharp focus.

4. Adjust the substage condenser to achieve optimal focus.

5. Routinely adjust the light source by means of the light-source transformer setting, and/or the iris diaphragm, for optimum illumination for each new slide and for each change in magnification.

6. Most microscopes are **parfocal**, which means that when one lens is in focus, other lenses will also have the same focal length and can be rotated into position without further major adjustment. In practice, however, usually a half-turn of the fine-adjustment knob in either direction is necessary for sharp focus.

7. Once you have brought the specimen into sharp focus with a low-powered lens, preparation may be made for visualizing the specimen under oil immersion. Place a drop of oil on the slide directly over the area to be viewed. Rotate the nosepiece until the oil-immersion objective locks into position. *Note: Care should be taken not to allow the high-power objective to touch the drop of oil.* The slide is observed from the side as the objective is rotated slowly into position. This will ensure that the objective will be properly immersed in the oil. The fine-adjustment knob is readjusted to bring the image into sharp focus.

8. During microscopic examination of microbial organisms, it is always necessary to observe several areas of the preparation. This is accomplished by scanning the slide without the application of additional immersion oil. *Note: This will require continuous, very fine adjustments by the slow, back-and-forth rotation of the fine-adjustment knob only.*

On completion of the laboratory exercise, return the microscope to its cabinet in its original condition. The following steps are recommended:

1. Clean all lenses with dry, clean lens paper. *Note: Use xylol to remove oil from the stage only.*

2. Place the low-power objective in position and lower the body tube completely.

3. Center the mechanical stage.

4. Coil the electric cord around the body tube and the stage.

5. Carry the microscope to its position in its cabinet in the manner previously described.

CLINICAL APPLICATION

Using Microscopic Examination in the Diagnosis of Tuberculosis

The visualization of stained bacterial cells using a compound light microscope can be the first step in diagnosing microbial infections. For example, a rapid diagnosis for tuberculosis can be made by identifying the unique characteristics of *Mycobacterium tuberculosis* in a stained sample of patient sputum.

AT THE BENCH

Materials

Slides

Commercially prepared slides of *Staphylococcus aureus*, *Bacillus subtilis*, *Aquaspirillum itersonii*, and other alternate slides.

Equipment

Compound microscope, lens paper, and immersion oil.

Procedure

1. Review the parts of the microscope, making sure you know the names and understand the function of each of these components.

2. Review instructions for the use of the microscope, giving special attention to the use of the oil-immersion objective.

3. Examine the prepared slides, noting the shapes and the relative sizes of the cells under the high-power (also called high-dry, because it is the highest power that does not use oil) and oil-immersion objectives.

4. Record your observations in the Lab Report.

Microscopic Examination of Living Microorganisms Using a Hanging-Drop Preparation or a Wet Mount

LEARNING OBJECTIVES

Once you have completed this experiment, you should know how to

1. Microscopically examine living microorganisms.

2. Make a hanging-drop preparation or wet mount to view living microorganisms.

Principle

Bacteria, because of their small size and a refractive index that closely approximates that of water, do not lend themselves readily to microscopic examination in a living, unstained state. Examination of living microorganisms is useful, however, to do the following:

1. Observe cell activities such as motility and binary fission.

2. Observe the natural sizes and shapes of the cells, considering that **heat fixation** (the rapid passage of the smear over the Bunsen burner flame) and exposure to chemicals during staining cause some degree of distortion.

In this experiment, you will use individual cultures of *Pseudomonas aeruginosa, Bacillus cereus, Staphylococcus aureus,* and *Proteus vulgaris* for a hanging-drop preparation or a wet mount. Hay infusion or pond water may be substituted or used in addition to the above organisms. Figure 5.1 illustrates several organisms commonly found in pond water and hay infusions.

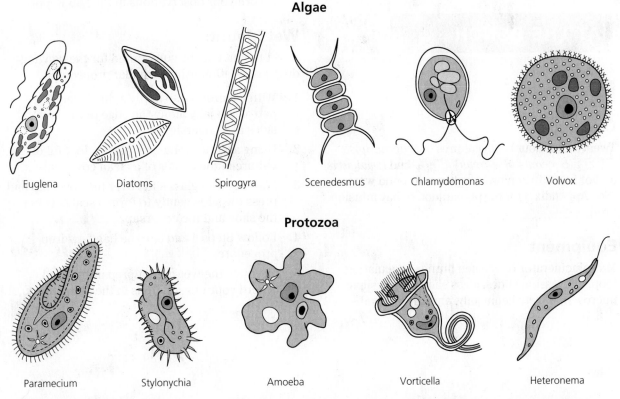

Algae

Euglena Diatoms Spirogyra Scenedesmus Chlamydomonas Volvox

Protozoa

Paramecium Stylonychia Amoeba Vorticella Heteronema

Figure 5.1 Algae and protozoa commonly found in natural infusions and pond water (drawings are not to scale)

You will observe the preparation(s) microscopically for differences in the sizes and shapes of the cells, as well as for motility, a self-directed movement. It is essential to differentiate between actual motility and **Brownian movement**, a vibratory movement of the cells due to their bombardment by water molecules in the suspension. Hanging-drop preparations and wet mounts make the movement of microorganisms easier to see because they slow down the movement of water molecules.

CLINICAL APPLICATION

Observation of Living Bacteria and the Diagnosis of Syphilis

Some microorganisms are difficult or unable to be stained. One of these bacteria is *Treponema pallidum,* the causative agent for syphilis. Special stains must be used to stain this bacterium; however, it can be viewed unstained and alive using a darkfield microscope. Under those conditions, its characteristic shape and motility can be observed, leading to a diagnosis of syphilis.

AT THE BENCH

Materials

Cultures

Twenty-four–hour broth cultures of *P. aeruginosa* BSL-2 , *B. cereus*, *S. aureus* BSL-2 , and *P. vulgaris*; and/or hay infusion broth cultures or pond water. (See Appendix 3 for the preparation of hay infusion broth.)

Equipment

Microincinerator or Bunsen burner, inoculating loop, depression slides, glass slides, coverslips, microscope, petroleum jelly, and cotton swabs.

Procedure

Hanging-Drop Preparation

Perform the following steps for each culture provided in this experiment. Steps 1–4 are illustrated in Figure 5.2.

1. With a cotton swab, apply a ring of petroleum jelly around the concavity of the depression slide.

2. Using aseptic technique, place a loopful of the culture in the center of a clean coverslip.

3. Place the depression slide, with the concave surface facing down, over the coverslip so that the depression covers the drop of culture. Press the slide gently to form a seal between the slide and the coverslip.

4. Quickly turn the slide right side up so that the drop continues to adhere to the inner surface of the coverslip.

5. For microscopic examination, first focus on the drop culture under the low-power objective (10X) and reduce the light source by adjusting the Abbé condenser. Repeat using the high-power objective (40X).

6. Examine the hanging-drop preparation and record your observations in the Lab Report.

Wet Mount

A wet mount may be substituted for the hanging-drop preparation using a similar procedure:

1. With a cotton swab apply a thin layer of petroleum jelly along the edge of the four sides of a coverslip.

2. Using aseptic technique, place a loopful of the culture in the center of a clean coverslip.

3. Place a clean glass slide over the coverslip and press the slide gently to form a seal between the slide and the coverslip.

4. Follow Steps 4 and 5 in the hanging-drop procedure.

5. Examine the wet-mount preparation and record your observations in the Lab Report.

PROCEDURE

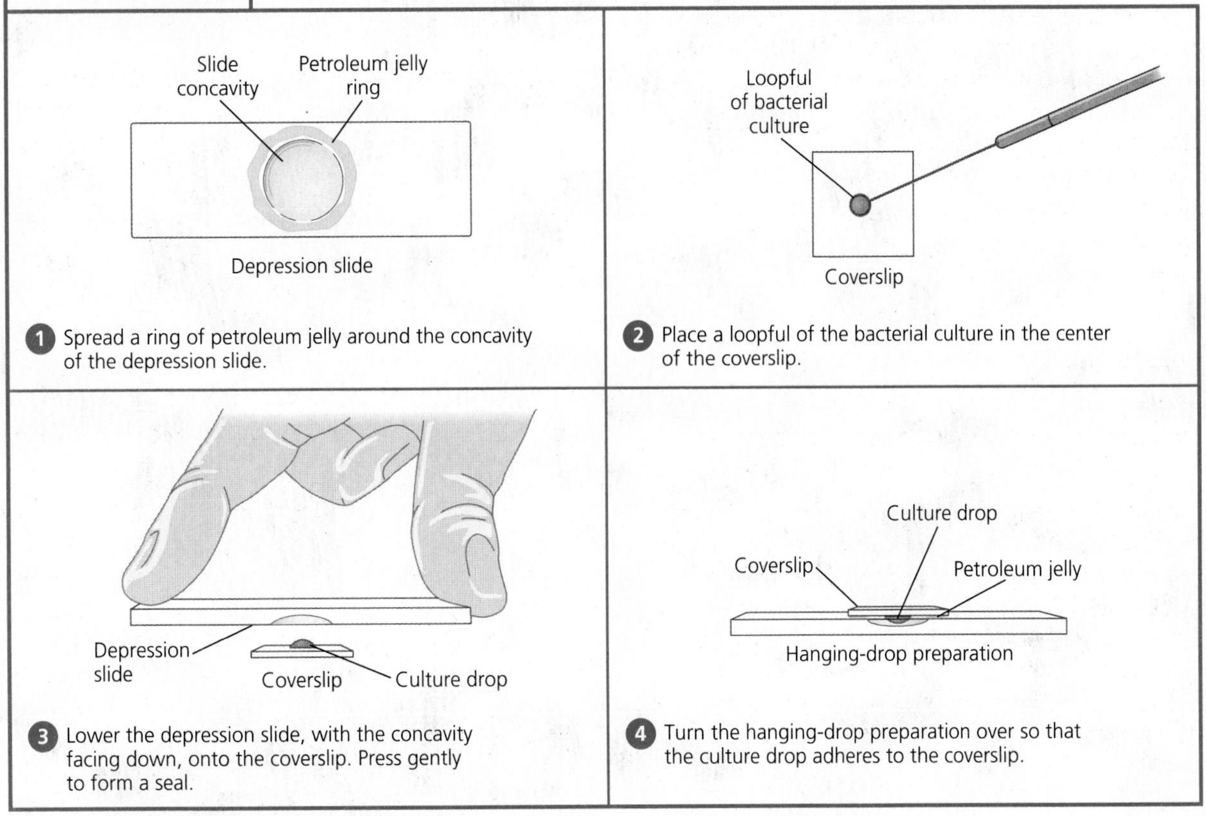

Slide concavity   Petroleum jelly ring

Depression slide

1 Spread a ring of petroleum jelly around the concavity of the depression slide.

Loopful of bacterial culture

Coverslip

2 Place a loopful of the bacterial culture in the center of the coverslip.

Depression slide   Coverslip   Culture drop

3 Lower the depression slide, with the concavity facing down, onto the coverslip. Press gently to form a seal.

Culture drop

Coverslip   Petroleum jelly

Hanging-drop preparation

4 Turn the hanging-drop preparation over so that the culture drop adheres to the coverslip.

Figure 5.2 **Hanging-drop preparation**

Name: _____

Date: _____ Section: _____

Observations and Results

1. Examine the hanging-drop or wet-mount preparation to determine shape and motility of the different bacteria present. Record your results in the chart below.

Organisms	Shape	True Motility or Brownian Movement?
S. aureus		
P. aeruginosa		
B. cereus		
P. vulgaris		

2. Draw a representative field of each of the above organisms.

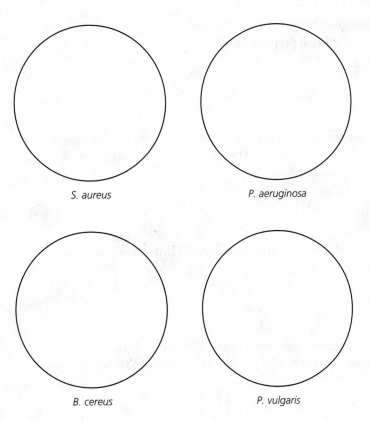

S. aureus   P. aeruginosa

B. cereus   P. vulgaris

3. Draw representative fields of pond water and hay infusion if you used them. Try to identify some of the organisms that you see by referring to Figure 5.1. Note the shape and type of movement in the chart below.

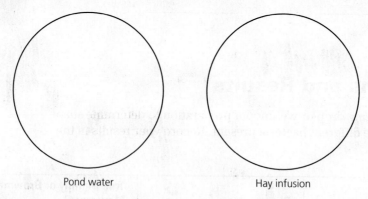

Pond water   Hay infusion

	Pond Water			Hay Infusion		
Shape						
True motility or Brownian movement?						
Organism						

Review Questions

1. Why are living, unstained bacterial preparations more difficult to observe microscopically than stained preparations?

2. What is the major advantage of using living cell preparations (hanging-drop or wet-mount) rather than stained preparations?

3. How do you distinguish between true motility and Brownian movement?

4. During the microscopic observation of a drop of stagnant pond water, what criteria would you use to distinguish viable organisms from nonviable suspended debris?

PART 3

Bacterial Staining

Once you have completed the experiments in this section, you should be familiar with

1. The chemical and theoretical basis of biological staining.
2. Manipulative techniques of smear preparation.
3. Procedures for simple staining and negative staining.
4. The method for performing differential staining procedures, such as the Gram, acid-fast, capsule, and spore stains.

Introduction

Visualization of microorganisms in the living state is quite difficult, not only because they are minute, but also because they are transparent and practically colorless when suspended in an aqueous medium. To study their properties and to divide microorganisms into specific groups for diagnostic purposes, biological stains and staining procedures in conjunction with light microscopy have become major tools in microbiology.

Chemically, a stain (dye) may be defined as an organic compound containing a benzene ring plus a chromophore and an auxochrome group (Figure P3.1).

The stain picric acid may be used to illustrate this definition (Figure P3.2).

The ability of a stain to bind to macromolecular cellular components such as proteins or nucleic acids depends on the electrical charge found on the chromogen portion, as well as on the cellular component to be stained.

Acidic stains are anionic, which means that, on ionization of the stain, the chromogen portion exhibits a negative charge and therefore has a strong affinity for the positive constituents of the cell. Proteins, positively charged cellular components, will readily bind to and accept the color of the negatively charged, anionic chromogen of an acidic stain. Structurally, picric acid is an example of an acidic stain that produces an anionic chromogen, as illustrated in Figure P3.3.

Basic stains are cationic, because on ionization the chromogen portion exhibits a positive

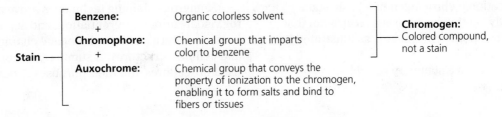

Figure P3.1 **Chemical composition of a stain**

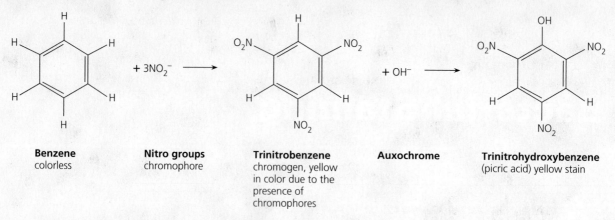

Figure P3.2 Chemical formation of picric acid

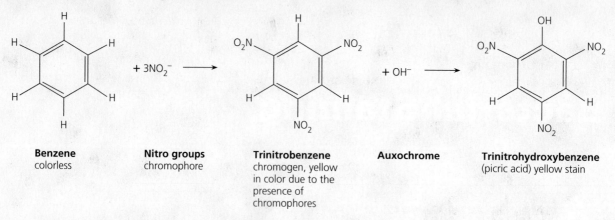

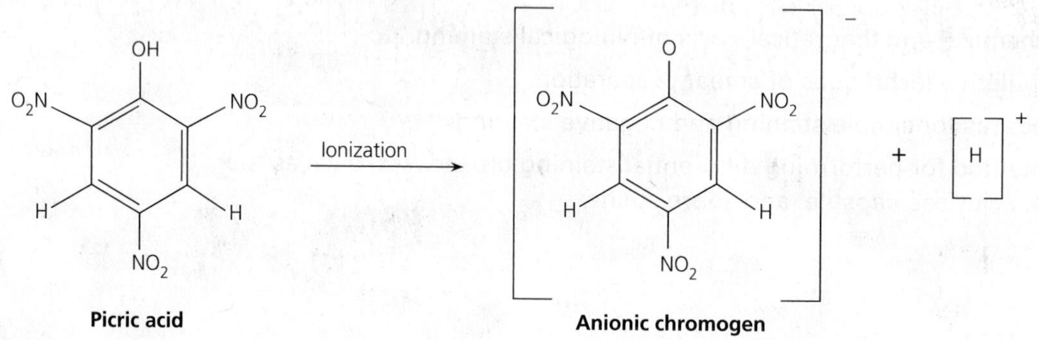

Figure P3.3 Picric acid: an acidic stain

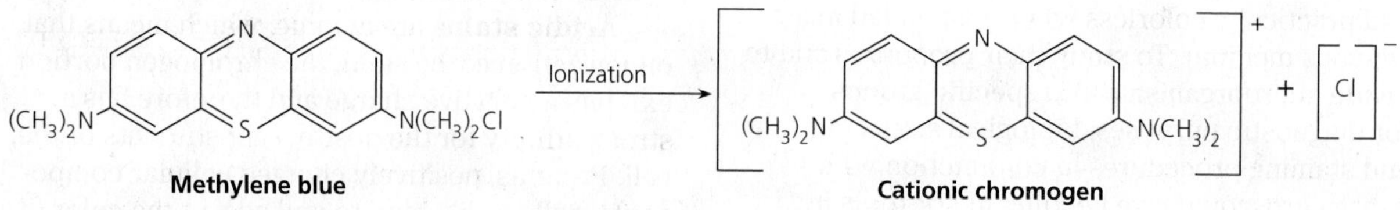

Methylene blue   **Cationic chromogen**

Figure P3.4 Methylene blue: a basic stain

charge and therefore has a strong affinity for the negative constituents of the cell. Nucleic acids, negatively charged cellular components, will readily bind to and accept the color of the positively charged, cationic chromogen of a basic stain. Structurally, methylene blue is a basic stain that produces a cationic chromogen, as illustrated in Figure P3.4.

Figure P3.5 is a summary of acidic and basic stains.

Basic stains are more commonly used for bacterial staining. The presence of a negative charge on the bacterial surface acts to repel most acidic stains and thus prevent their penetration into the cell.

Numerous staining techniques are available for visualization, differentiation, and separation of bacteria in terms of morphological characteristics and cellular structures. A summary of commonly used procedures and their purposes is outlined in Figure P3.6.

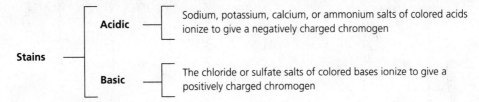

Figure P3.5 **Acidic and basic stains**

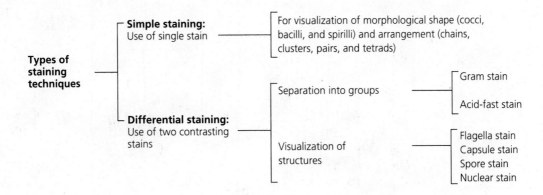

Figure P3.6 **Staining techniques**

Preparation of Bacterial Smears

LEARNING OBJECTIVE

Once you have completed this experiment, you should be able to

1. Prepare bacterial smears for the microscopic visualization of bacteria.

Principle

Bacterial smears must be prepared prior to the execution of any of the staining techniques listed in Figure P3.6 on page 49. Although not difficult, the preparation requires adequate care. Meticulously follow the rules listed below.

1. **Preparation of the glass microscope slide:** Clean slides are essential for the preparation of microbial smears. Grease or oil from the fingers on slides must be removed by washing the slides with soap and water or scouring powders such as Bon Ami®, followed by a water rinse and a rinse of 95% alcohol. After cleaning, dry the slides and place them on laboratory towels until ready for use. *Note: Remember to hold the clean slides by their edges.*

2. **Labeling of slides:** Proper labelling of the slide is essential. The initials of the organism can be written on either end of the slide with a glassware marking pencil on the surface on which the smear is to be made. Ensure that the label does not come into contact with staining reagents.

3. **Preparation of smear:** It is crucial to avoid thick, dense smears. A thick or dense smear occurs when too much of the culture is used in its preparation, which concentrates a large number of cells on the slide. This type of preparation diminishes the amount of light that can pass through and makes it difficult to visualize the morphology of single cells.

Note: Smears require only a small amount of the bacterial culture. A good smear is one that, when dried, appears as a thin whitish layer or film. The print of your textbook should be legible through the smear. Different techniques are used depending on whether the smear is made from a broth- or solid-medium culture.

a. **Broth cultures:** Resuspend the culture by tapping the tube with your finger. Depending on the size of the loop, one or two loopfuls should be applied to the center of the slide with a sterile inoculating loop and spread evenly over an area about the size of a dime. Set the smears on the laboratory table and allow to air-dry.

b. **Cultures from solid medium:** Organisms cultured in a solid medium produce thick, dense surface growth and are not amenable to direct transfer to the glass slide. These cultures must be diluted by placing one or two loopfuls of water on the center of the slide in which the cells will be emulsified. Transfer of the cells requires the use of a sterile inoculating loop or a needle, if preferred. Only the tip of the loop or needle should touch the culture to prevent the transfer of too many cells. Suspension is accomplished by spreading the cells in a circular motion in the drop of water with the loop or needle. This helps to avoid cell clumping. The finished smear should occupy an area about the size of a nickel and should appear as a translucent, or semitransparent, confluent whitish film Figure 6.1. At this point the smear should be allowed to dry completely. *Note: Do not blow on slide or wave it in the air.*

4. **Heat fixation:** Unless fixed on the glass slide, the bacterial smear will wash away during the staining procedure. This is avoided by heat fixation, during which the bacterial proteins are coagulated and fixed to the

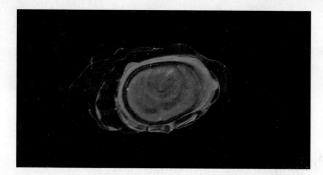

Figure 6.1 A bacterial smear following fixation

glass surface. Heat fixation is performed by the rapid passage of the air-dried smear two or three times over the flame of the Bunsen burner or in front of a microincinerator. While many texts will discuss the use of a Bunsen burner for sterilization and heat fixation, governing bodies such as the American Society for Microbiology (ASM) have changed the proscribed methods for heat fixation and bench top sterilization to utilize a microincinerator instead of a Bunsen burner to reduce the possibility of aerosolization of bacteria on the slide or loop.

The preparation of a bacterial smear is illustrated in Figure 6.2.

The preparation of a bacterial smear is illustrated in Figure 6.2.

CLINICAL APPLICATION

Proper Slide Preparation

Before any staining or visualization of a bacterial sample can take place, a proper smear must be prepared. A smear that is too thick may give a false result due to retention of dye that should have been rinsed away or because the thickness may prevent dye penetration. A smear that is too thin may have too few cells, increasing the time and energy to find the bacteria under magnification. Inconclusive results due to improperly prepared slides may have an impact on patient treatment and outcomes. Good smears are those that allow newsprint to be read through the smear.

AT THE BENCH

Materials

Cultures

Twenty-four–hour nutrient agar slant culture of *Bacillus cereus* and a 24-hour nutrient broth culture of *Staphylococcus aureus* **BSL-2** .

Equipment

Glass microscope slides, microincinerator or Bunsen burner, inoculating loop and needle, and glassware marking pencil.

Procedure

Smears from a Broth Medium

Label three clean slides with the initials of the organism, and number them 1, 2, and 3. Resuspend the sedimented cells in the broth culture by tapping the culture tube with your finger. The next four steps of this procedure are illustrated in Figure 6.2a and c:

1. With a sterile loop, place one loopful of culture on Slide 1, two loopfuls on Slide 2, and three loopfuls on Slide 3, respectively.

2. With a circular movement of the loop, spread the cell suspension into an area approximately the size of a dime.

3. Allow the slide to air-dry completely. This may be done by placing the slide on a drying tray attached to a microincinerator or by placing the slide on the bench.

4. Heat fix the preparation. *Note: Pass the air-dried slide in front of the entrance to the microincinerator or pass the slide through the outer portion of the Bunsen flame to prevent overheating, which can distort the morphology through plasmolysis of the cell wall.*

Examine each slide for the confluent, whitish film or haze and record your results in the Lab Report.

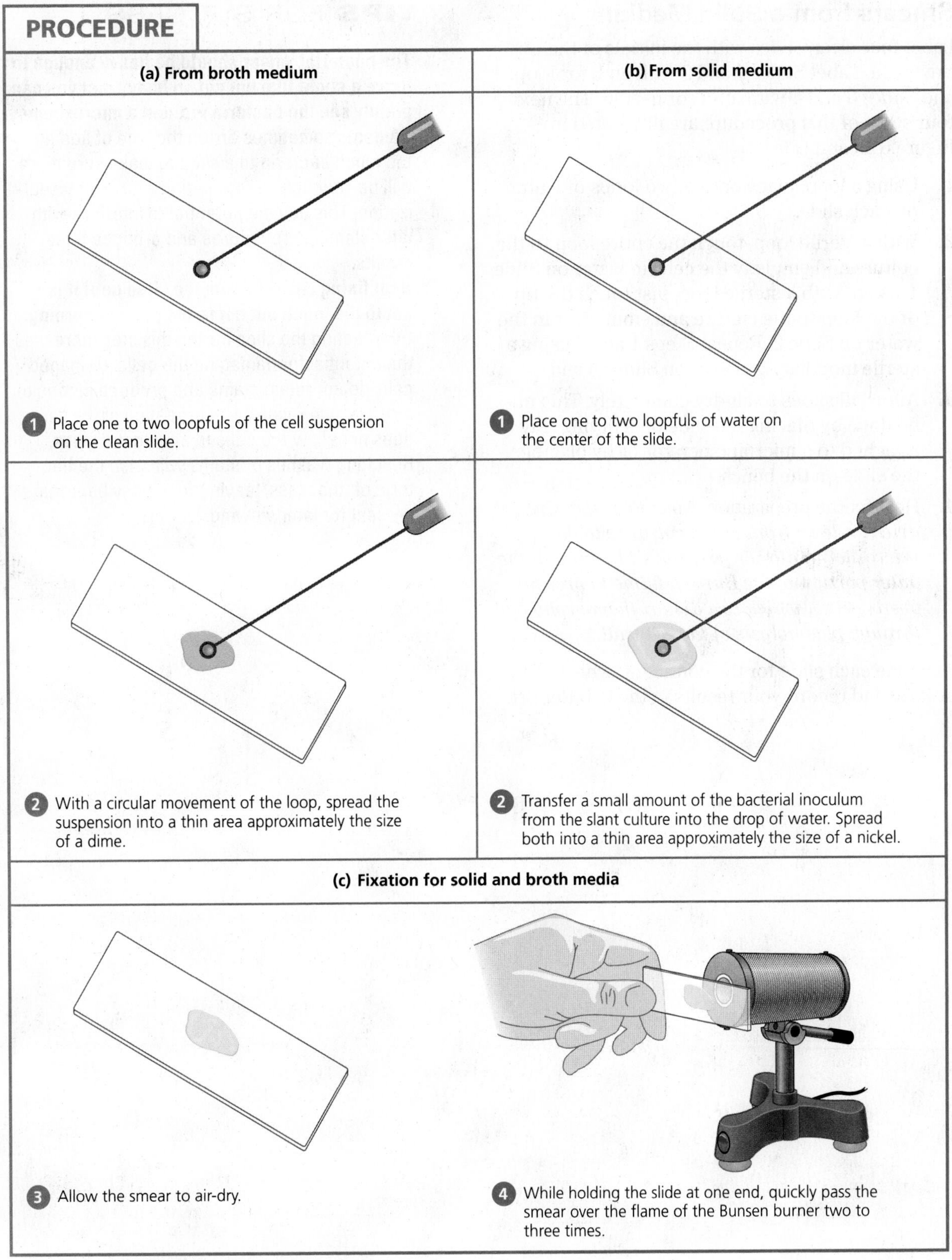

PROCEDURE

(a) From broth medium	**(b) From solid medium**

(a)
1. Place one to two loopfuls of the cell suspension on the clean slide.
2. With a circular movement of the loop, spread the suspension into a thin area approximately the size of a dime.

(b)
1. Place one to two loopfuls of water on the center of the slide.
2. Transfer a small amount of the bacterial inoculum from the slant culture into the drop of water. Spread both into a thin area approximately the size of a nickel.

(c) Fixation for solid and broth media

3. Allow the smear to air-dry.
4. While holding the slide at one end, quickly pass the smear over the flame of the Bunsen burner two to three times.

Figure 6.2 **Bacterial smear preparation**

Smears from a Solid Medium

Label four clean slides with the initials of the organism. Label Slides 1 and 2 with an L for loop, and Slides 3 and 4 with an N for needle. The next four steps of this procedure are illustrated in Figure 6.2b and c:

1. Using a loop, place one to two loops of water on each slide.

2. With a sterile loop, touch the entire loop to the culture and emulsify the cells in water on Slide 1. Then, with a sterile loop, just touch the tip of the loop to the culture and emulsify it in the water on Slide 2. Repeat Steps 1 and 2 using a sterile inoculating needle on Slides 3 and 4.

3. Allow all slides to air-dry completely. This may be done by placing the slide on a drying tray attached to a microincinerator or by placing the slide on the bench.

4. Heat fix the preparation. *Note: Pass the air-dried slide in front of the entrance to the microincinerator or pass the slide through the outer portion of the Bunsen flame to prevent overheating, which can distort the morphology through plasmolysis of the cell wall.*

Examine each slide for the confluent, whitish film or haze and record your results in the Lab Report.

TIPS FOR SUCCESS

1. **The bacterial smear should be heavy enough to leave a slight film but not so heavy that you can plainly see the bacteria without a microscope.** Students sometimes err on the side of adding too much bacteria to a slide to make sure there will be "enough" bacteria there for later visualization. This has the potential to interfere with later staining procedures and produce false results.

2. **Heat fixing should warm the slide until it is hot to the touch but not to the point of burning.** Overheating the slide during this step increases the potential for damaging the cells. Damaged cells do not retain stains and produce inconclusive staining results. Underheating of the slide does not allow the cells to affix to the glass. Resulting washes or stains will rinse the bacteria off the glass, leaving few if any bacteria present for later viewing.

Simple Staining

LEARNING OBJECTIVES

Once you have completed this experiment, you should be able to

1. Perform a simple staining procedure.
2. Compare the morphological shapes and arrangements of bacterial cells.

Principle

In **simple staining**, the bacterial smear is stained with a single reagent, which produces a distinctive contrast between the organism and its background. Basic stains with a positively charged chromogen are preferred because bacterial nucleic acids and certain cell wall components carry a negative charge that strongly attracts and binds to the cationic chromogen. The purpose of simple staining is to elucidate the morphology and arrangement of bacterial cells (**Figure 7.1**). The most commonly used basic stains are methylene blue, crystal violet, and carbol fuchsin.

CLINICAL APPLICATION

Quick and Simple Stain

Simple stains are relatively quick and useful methods of testing for the presence of, determining the shape of, or determining the numbers of bacteria present in a sample. Generally involving a single staining step, simple staining methods are not considered differential or diagnostic and will have limited uses. However, this is a quick procedure for determining whether a clinical sample has the presence of a foreign bacterial pathogen.

AT THE BENCH

Materials

Cultures

Twenty-four–hour nutrient agar slant cultures of *Escherichia coli* and *Bacillus cereus* and a 24-hour nutrient broth culture of *Staphylococcus aureus* BSL-2 . Alternatively, use the smears prepared in Experiment 6.

Cocci are spherical in shape.

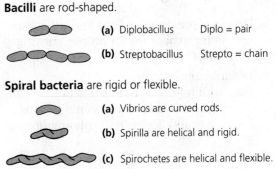

(a) Diplococcus	Diplo = pair
(b) Streptococcus	Strepto = chain
(c) Staphylococcus	Staphlyo = cluster
(d) Tetrad	Tetrad = packet of 4
(e) Sarcina	Sarcina = packet of 8

Bacilli are rod-shaped.

(a) Diplobacillus   Diplo = pair

(b) Streptobacillus   Strepto = chain

Spiral bacteria are rigid or flexible.

(a) Vibrios are curved rods.

(b) Spirilla are helical and rigid.

(c) Spirochetes are helical and flexible.

Figure 7.1 Bacterial shapes and arrangements

Reagents

Methylene blue, crystal violet, and carbol fuchsin.

Equipment

Microincinerator or Bunsen burner, inoculating loop, staining tray, microscope, lens paper, bibulous (highly absorbent) paper, and glass slides.

Procedure

1. Prepare separate bacterial smears of the organisms following the procedure described in Experiment 8. *Note: All smears must be heat fixed prior to staining.*

Simple Staining

The following steps are illustrated in Figure 7.2.

1. Place a slide on the staining tray and flood the smear with one of the indicated stains, using the appropriate exposure time for each: carbol fuchsin, 15 to 30 seconds; crystal violet, 20 to 60 seconds; methylene blue (shown in Figure 7.2), 1 to 2 minutes.

2. *Gently* wash the smear with tap water to remove excess stain. During this step, hold the slide parallel to the stream of water; in this way you can reduce the loss of organisms from the preparation.

3. Using bibulous paper, blot dry, but *do not* wipe the slide.

4. Repeat this procedure with the remaining two organisms, using a different stain for each.

5. Examine all stained slides under oil immersion.

6. In the chart provided in the Lab Report, complete the following:

   a. Draw a representative field for each organism. Refer to page xvi for proper drawing procedure.

   b. Describe the morphology of the organisms with reference to their shapes (bacilli, cocci, spirilla) and arrangements (chains, clusters, pairs). Refer to the photographs in Figure 7.3.

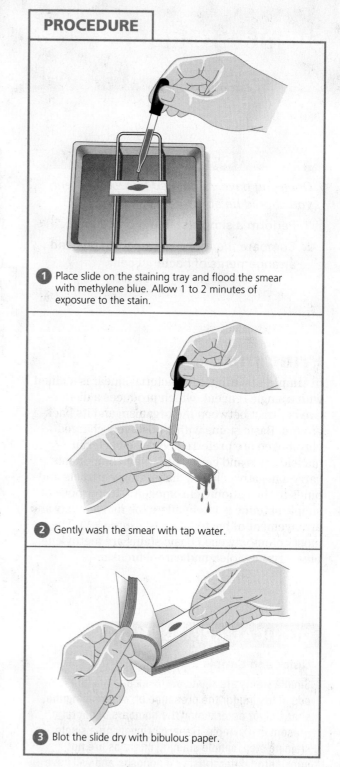

PROCEDURE

1. Place slide on the staining tray and flood the smear with methylene blue. Allow 1 to 2 minutes of exposure to the stain.

2. Gently wash the smear with tap water.

3. Blot the slide dry with bibulous paper.

Figure 7.2 **Simple staining procedure**

Diplobacilli

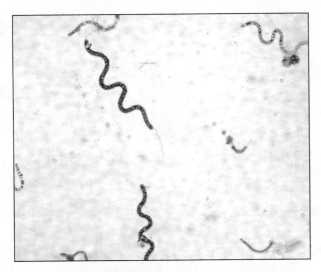

(a) Bacilli and diplobacilli (rod-shaped) bacteria

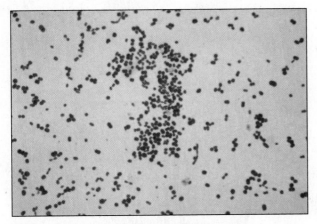

(b) Spirilla (spiral-shaped) bacteria

(c) Cocci (spherical-shaped) bacteria: *Staphylococcus*

Figure 7.3 **Micrographs showing bacteria morphology**

Negative Staining

LEARNING OBJECTIVES

Once you have completed this experiment, you should be able to

1. Perform a negative staining procedure.
2. Understand the benefit obtained from visualizing unstained microorganisms.

CLINICAL APPLICATION

Detecting Encapsulated Invaders

The principle application of negative staining is to determine if an organism possesses a capsule (a gelatinous outer layer that makes the microorganism more virulent), although it can also be used to demonstrate spore formation. The technique is frequently used in the identification of fungi such as *Cryptococcus neoformans,* an important infectious agent found in bird dropping that is linked to meningeal and lung infections in humans.

Principle

Negative staining requires the use of an acidic stain such as India ink or nigrosin. The acidic stain, with its negatively charged chromogen, will not penetrate the cells because of the negative charge on the surface of bacteria. Therefore, the unstained cells are easily discernible against the colored background.

The practical application of negative staining is twofold. First, since heat fixation is not required and the cells are not subjected to the distorting effects of chemicals and heat, their natural size and shape can be seen. Second, it is possible to observe bacteria that are difficult to stain, such as some spirilla. Because heat fixation is not done during the staining process, keep in mind that the organisms are not killed and *slides should be handled with care.* Figure 8.1 shows a negative stain of bacilli.

AT THE BENCH

Materials

Cultures

Twenty-four–hour agar slant cultures of *Micrococcus luteus, Bacillus cereus,* and *other alternate bacterial cultures.*

Reagent

Nigrosin.

Equipment

Microincinerator or Bunsen burner, inoculating loop, staining tray, glass slides, lens paper, and microscope.

Procedure

Steps 1–4 are illustrated in Figure 8.2.

1. Place a small drop of nigrosin close to one end of a clean slide.
2. Using aseptic technique, place a loopful of inoculum from the *M. luteus* culture in the drop of nigrosin and mix.

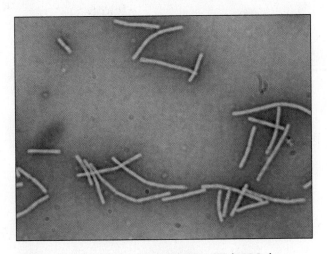

Figure 8.1 Negative staining: Bacilli (1000×)

3. Place a slide against the drop of suspended organisms at a 45° angle and allow the drop to spread along the edge of the applied slide.

4. Push the slide away from the drop of suspended organisms to form a thin smear. Air-dry. *Note: Do not heat fix the slide.*

5. Repeat Steps 1–4 for slide preparations of *the remaining cultures.*

6. Examine the slides under oil immersion, and record your observations in the Lab Report.

PROCEDURE

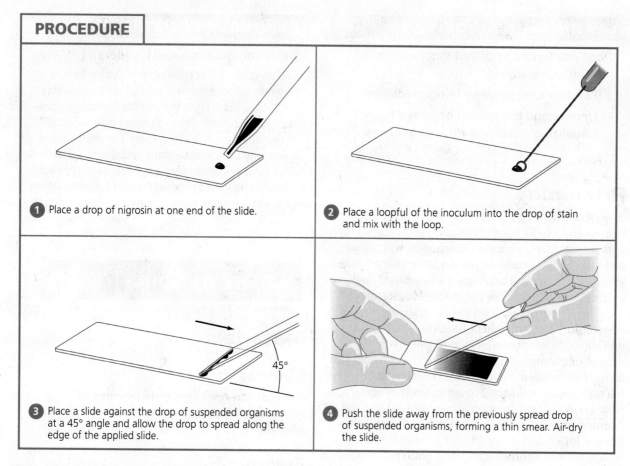

1 Place a drop of nigrosin at one end of the slide.

2 Place a loopful of the inoculum into the drop of stain and mix with the loop.

3 Place a slide against the drop of suspended organisms at a 45° angle and allow the drop to spread along the edge of the applied slide.

4 Push the slide away from the previously spread drop of suspended organisms, forming a thin smear. Air-dry the slide.

Figure 8.2 **Negative staining procedure**

Name: _____

Date: _____   Section: _____

Observations and Results

1. Draw representative fields of your microscopic observations.

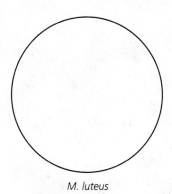

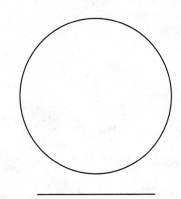

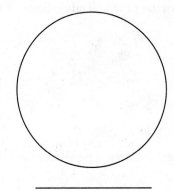

M. luteus   _____   _____

2. Describe the microscopic appearance of the different bacteria using the chart below.

Organism	M. luteus	_____	_____
Shape			
Arrangement			
Magnification			

Review Questions

1. Why can't methylene blue be used in place of nigrosin for negative staining? Explain.

2. What are the practical advantages of negative staining?

3. Why doesn't nigrosin penetrate bacterial cells?

Gram Stain

LEARNING OBJECTIVES

Once you have completed this experiment, you should understand

1. The chemical and theoretical basis for differential staining procedures.

2. The chemical basis for the Gram stain.

3. The procedure for differentiating between two principal groups of bacteria: gram positive and gram negative.

Principle

Differential staining requires the use of at least four chemical reagents that are applied sequentially to a heat-fixed smear. The first reagent is called the **primary stain**. Its function is to impart its color to all cells. The second stain is a **mordant** used to intensify the color of the primary stain. In order to establish a color contrast, the third reagent used is the decolorizing agent. Based on the chemical composition of cellular components, the **decolorizing agent** may or may not remove the primary stain from the entire cell or only from certain cell structures. The final reagent, the **counterstain**, has a contrasting color to that of the primary stain. Following decolorization, if the primary stain is not washed out, the **counterstain** cannot be absorbed and the cell or its components will retain the color of the primary stain. If the primary stain is removed, the decolorized cellular components will accept and assume the contrasting color of the counterstain. In this way, cell types or their structures can be distinguished from each other on the basis of the stain that is retained.

The most important differential stain used in bacteriology is the **Gram stain**, named after Dr. Hans Christian Gram. It divides bacterial cells into two major groups, gram positive and gram negative, which makes it an essential tool for classification and differentiation of microorganisms. **Figure 9.1** shows gram-positive and gram-negative cells. The Gram stain reaction is based on the difference in the chemical composition of bacterial cell walls. Gram-positive cells have a thick peptidoglycan layer, whereas the peptidoglycan layer in gram-negative cells is much thinner and surrounded by outer lipid-containing layers. Peptidoglycan is mainly a polysaccharide composed of two chemical subunits found only in the bacterial cell wall. These subunits are *N*-acetylglucosamine and *N*-acetylmuramic acid. With some organisms, as the adjacent layers of peptidoglycan are formed, they are cross-linked by short chains of peptides by means of a transpeptidase enzyme, resulting in the shape and rigidity of the cell wall. In the case

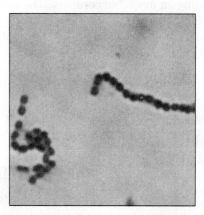

(a) Gram-positive stain of streptococci

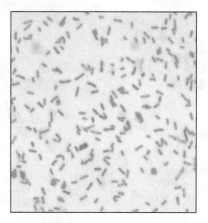

(b) Gram-negative stain of *E. coli*

Figure 9.1 Gram-stained cells

of gram-negative bacteria and several of the gram-positive such as the *Bacillus*, the cross-linking of the peptidoglycan layer is direct because the bacteria do not have short peptide tails. Early experiments have shown that a gram-positive cell denuded of its cell wall by the action of lysozyme or penicillin will stain gram-negative.

The Gram stain uses four different reagents. Descriptions of these reagents and their mechanisms of action follow. Figure 9.2 shows the microscopic appearance of cells at each step of the Gram staining procedure.

Primary Stain

Crystal Violet (Hucker's) This violet stain is used first and stains all cells purple.

Mordant

Gram's Iodine This reagent serves not only as a killing agent but also as a mordant, a substance that increases the cells' affinity for a stain. The reagent does this by binding to the primary stain, thus forming an insoluble complex. The resultant crystal-violet–iodine (CV-I) complex serves to intensify the color of the stain. At this point, all cells will appear purple-black.

Decolorizing Agent

Ethyl Alcohol, 95% This reagent serves a dual function as a protein-dehydrating agent and as a lipid solvent. Its action is determined by two factors, the concentration of lipids and the thickness of the peptidoglycan layer in bacterial cell walls. In gram-negative cells, the alcohol increases the porosity of the cell wall by dissolving the lipids in the outer layers. Thus, the CV-I complex can be more easily removed from the thinner and less highly cross-linked peptidoglycan layer. Therefore, the washing-out effect of the alcohol facilitates the release of the unbound CV-I complex, leaving the cells colorless or unstained. The much thicker peptidoglycan layer in gram-positive cells is responsible for the more stringent retention of the CV-I complex, as the pores are made smaller due to the dehydrating effect of the alcohol. Thus, the tightly bound primary stain complex is difficult to remove, and the cells remain purple. *Note: Be careful not to over-decolorize the smear with alcohol.*

Counterstain

Safranin This is the final reagent, used to stain pink those cells that have been previously decolorized. Since only gram-negative cells undergo decolorization, they may now absorb the counterstain.

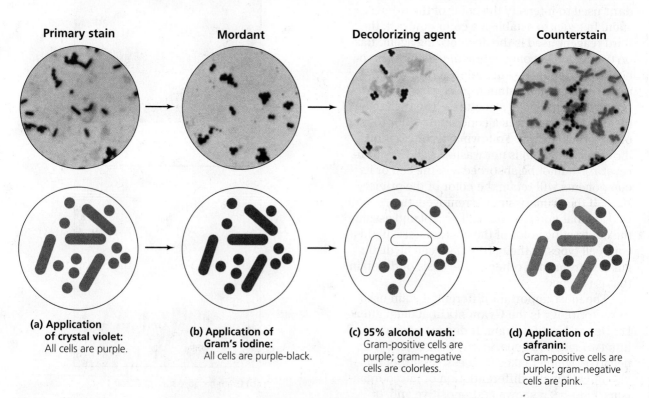

Primary stain **Mordant** **Decolorizing agent** **Counterstain**

(a) Application of crystal violet: All cells are purple.

(b) Application of Gram's iodine: All cells are purple-black.

(c) 95% alcohol wash: Gram-positive cells are purple; gram-negative cells are colorless.

(d) Application of safranin: Gram-positive cells are purple; gram-negative cells are pink.

Figure 9.2 Microscopic observation of cells following steps in the Gram staining procedure

Gram-positive cells retain the purple color of the primary stain.

The preparation of adequately stained smears requires the following precautions:

1. The most critical phase of the procedure is the decolorization step, which is based on the ease with which the CV-I complex is released from the cell. Remember that over-decolorization will result in loss of the primary stain, causing gram-positive organisms to appear gram negative. Under-decolorization, however, will not completely remove the CV-I complex, causing gram-negative organisms to appear gram positive. Strict adherence to all instructions will help remedy part of the difficulty, but individual experience and practice are the keys to correct decolorization.

2. It is imperative that, between applications of the reagents, slides need to be thoroughly washed under running water or water applied with an eyedropper. This removes excess reagent and prepares the slide for application of the subsequent reagent.

3. The best Gram-stained preparations are made with fresh cultures (i.e., not older than 24 hours). As cultures age, especially in the case of gram-positive cells, the organisms tend to lose their ability to retain the primary stain and may appear to be **gram-variable**; that is, some cells will appear purple, while others will appear pink.

CLINICAL APPLICATION

Gram Staining: The First Diagnostic Test

The Gram stain is a diagnostic staining procedure that can be done on body fluids, tissue biopsies, throat cultures, samples from abscesses when infection is suspected, and more. Clinically important results are obtained much more rapidly from staining than from culturing the specimen. The results of the Gram stain will aid a clinical lab in determining which additional tests may be required for identification of the bacterial strain in question. Once the bacterial gram type, shape, and orientation are determined, it expedites the appropriate choice of antibiotic needed to treat the patient.

AT THE BENCH

Materials

Cultures

Twenty-four–hour nutrient agar slant cultures of *Escherichia coli*, *Staphylococcus aureus* BSL-2, and *Bacillus cereus*.

Reagents

Crystal violet, Gram's iodine, 95% ethyl alcohol, and safranin.

Equipment

Microincinerator or Bunsen burner, inoculating loop or needle, staining tray, glass slides, bibulous paper, lens paper, and microscope.

Procedure

Smear Preparation

1. Obtain four clean glass slides.

2. Using aseptic technique, prepare a smear of each of the three organisms and on the remaining slide prepare a smear consisting of a mixture of *S. aureus* BSL-2, and *E. coli*. Do this by placing a drop of water on the slide, and then transferring each organism separately to the drop of water with a sterile, cooled loop. Mix and spread both organisms by means of a circular motion of the inoculating loop. (Note: If bacteria are taken from a broth culture, the drop of water is not required. Place a loop of bacterial suspension directly on the glass slide.)

3. Allow smears to air-dry and then heat fix in the usual manner.

1 Gently stain with crystal violet for 1 minute.

2 Gently wash off the stain with tap water.

3 Gently apply Gram's iodine for 1 minute.

4 Gently wash off the Gram's iodine with tap water.

5 Add 95% alcohol drop by drop until the alcohol runs almost clear.

6 Gently wash off the 95% alcohol with tap water.

7 Counterstain with safranin for 45 seconds.

8 Gently wash off the safranin with tap water.

9 Blot dry with bibulous paper.

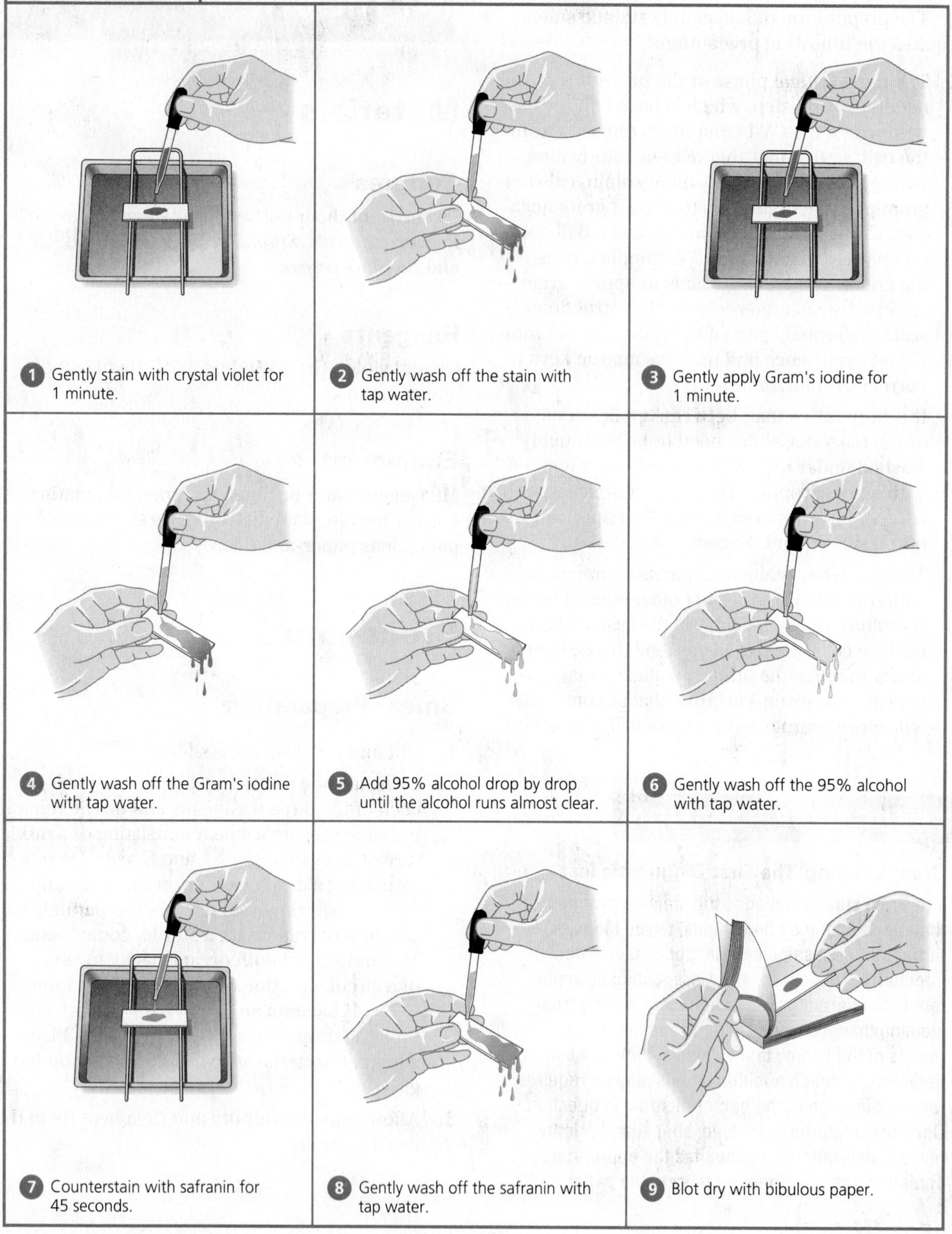

Figure 9.3 Gram staining procedure

Gram Staining

The following steps are shown in Figure 9.3.

1. *Gently* flood smears with crystal violet and let stand for 1 minute.
2. *Gently* wash with tap water.
3. *Gently* flood smears with the Gram's iodine mordant and let stand for 1 minute.
4. *Gently* wash with tap water.
5. Decolorize with 95% ethyl alcohol. *Note: Do not over-decolorize.* Add reagent drop by drop until the alcohol runs almost clear, showing only a blue tinge.
6. *Gently* wash with tap water.
7. Counterstain with safranin for 45 seconds.
8. *Gently* wash with tap water.
9. Blot dry with bibulous paper and examine under oil immersion.
10. As you observe each slide under oil immersion, complete the chart provided in the Lab Report.
    a. Draw a representative microscopic field.
    b. Describe the cells according to their morphology and arrangement.
    c. Describe the color of the stained cells.
    d. Classify the organism as to the Gram reaction: gram positive or gram negative.

TIPS FOR SUCCESS

1. **Proper slide preparation is key to successful staining.** Incorrect heat fixation will affect the number of bacteria that will be present during staining. Fixation that was not hot enough or was too short will not allow the cells to adhere to the glass slide properly, and the cells will be rinsed away during the multiple stain and rinse steps. Conversely, overheating will result in the destruction of the cells and cell debris adhered to the cell. Few, if any, cells will remain intact for the staining process.

2. Timing of the decolorizing step may be the most important step of the procedure. Over decolorizing with an incorrect alcohol solution or from allowing the slide to decolorize too long, will remove the CV-I complex by causing extensive damage to the cell membrane and cell wall, even on a gram-positive cell. Alternatively, decolorizing for too short a time period will not remove enough CV-I complexes. The safranin-stained cells will appear to be darker in color, and could be mistaken for a light purple, gram-positive stained cell.

3. **The age of the culture or colony being stained may impact the Gram stain results.** The best Gram-stained preparations are made with fresh cultures that are not older than 24 hours. As cultures age, especially in the case of gram-positive cells, the organisms tend to lose their ability to retain the primary stain and may appear to be **gram variable**; that is, some cells will appear purple, while others will appear pink.

Acid-Fast Stain

LEARNING OBJECTIVES

Once you have completed this experiment, you should understand

1. The chemical basis of the acid-fast stain.
2. The procedure for differentiating bacteria into acid-fast and non–acid-fast groups.

Principle

While the majority of bacterial organisms are stainable by either simple or Gram staining procedures, a few genera, particularly the members of the genus *Mycobacterium*, are visualized more clearly by the **acid-fast** method. Since *M. tuberculosis* and *M. leprae* represent bacteria that are pathogenic to humans, the stain is of diagnostic value in identifying these organisms.

The characteristic difference between mycobacteria and other microorganisms is the presence of a thick, waxy (lipoidal) wall that makes penetration by stains extremely difficult. Mycobacteria tend to clump together, and it is difficult to identify individual cells in stained preparations if this clumping effect occurs. To avoid or minimize this phenomenon requires careful preparation of the smear. Place a small drop of water on the slide, suspend the culture in the water, and mix the suspension thoroughly to dislodge and disperse some of the cells. Once the stain has penetrated, however, it cannot be readily removed even with the vigorous use of acid-alcohol as a decolorizing agent (unlike the 95% ethyl alcohol used in the Gram stain). Because of this property, these organisms are called acid-fast, while all other microorganisms, which are easily decolorized by acid-alcohol, are non–acid-fast.

The acid-fast stain uses three different reagents.

Primary Stain

Carbol Fuchsin Unlike cells that are easily stained by ordinary aqueous stains, most species of mycobacteria are not stainable with common dyes such as methylene blue and crystal violet. Carbol fuchsin, a dark red stain in 5% phenol that is soluble in the lipoidal materials that constitute most of the mycobacterial cell wall, does penetrate these bacteria and is retained. Penetration is further enhanced by the application of heat, which drives the carbol fuchsin through the lipoidal wall and into the cytoplasm. This application of heat is used in the **Ziehl-Neelsen method**. The **Kinyoun method**, a modification of the Ziehl-Neelsen method, circumvents the use of heat by addition of a wetting agent (Tergitol®) to this stain, which reduces surface tension between the cell wall of the mycobacteria and the stain. Following application of the primary stain, all cells will appear red.

Decolorizing Agent

Acid-Alcohol (3% HCl + 95% Ethanol) Prior to decolorization, the smear is cooled, which allows the waxy cell substances to harden. On application of acid-alcohol, acid-fast cells will be resistant to decolorization since the primary stain is more soluble in the cellular waxes than in the decolorizing agent. In this event, the primary stain is retained and the mycobacteria will stay red. This is not the case with non–acid-fast organisms, which lack cellular waxes. The primary stain is more easily removed during decolorization, leaving these cells colorless or unstained.

Counterstain

Methylene Blue This is used as the final reagent to stain previously decolorized cells. As only non–acid-fast cells undergo decolorization, they may now absorb the counterstain and take on its blue color, while acid-fast cells retain the red of the primary stain.

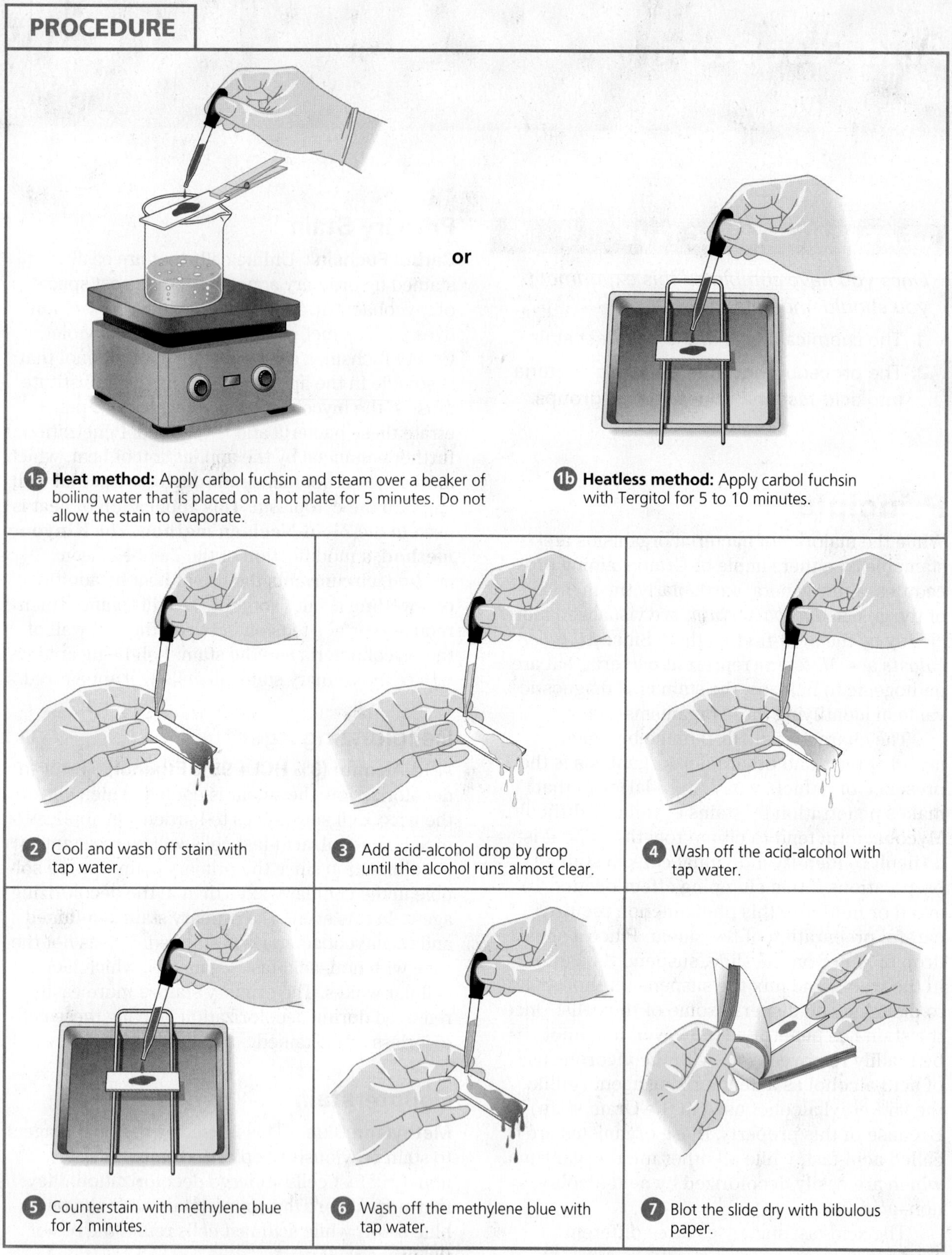

1a **Heat method:** Apply carbol fuchsin and steam over a beaker of boiling water that is placed on a hot plate for 5 minutes. Do not allow the stain to evaporate.

or

1b **Heatless method:** Apply carbol fuchsin with Tergitol for 5 to 10 minutes.

2 Cool and wash off stain with tap water.

3 Add acid-alcohol drop by drop until the alcohol runs almost clear.

4 Wash off the acid-alcohol with tap water.

5 Counterstain with methylene blue for 2 minutes.

6 Wash off the methylene blue with tap water.

7 Blot the slide dry with bibulous paper.

Figure 10.1 Acid-fast staining procedure

CLINICAL APPLICATION

Diagnosing Leprosy and Lung Infections

The cell walls of bacteria belonging to the genera *Mycobacterium* or *Nocardia* contain mycolic acid and are resistant to penetration by water-soluble stains such as the Gram stain, which can lead to a false gram-positive result. The medical importance of the acid-fast stain is for the diagnosis of the *Mycobacterium* species, which cause tuberculosis, leprosy, and other infections. The genus *Nocardia*, which is the causative agent for lung infections, will also be identified by the acid-fast staining method.

AT THE BENCH

Materials

Cultures

Seventy-two– to 96-hour Trypticase™ soy broth culture of *Mycobacterium smegmatis* and 18- to 24-hour culture of *Staphylococcus aureus* BSL-2 .

Reagents

Carbol fuchsin, acid-alcohol, and methylene blue.

Equipment

Microincinerator or Bunsen burner, hot plate, 250-ml beaker, inoculating loop, glass slides, bibulous paper, lens paper, staining tray, and microscope.

Procedure

Smear Preparation

1. Obtain three clean glass slides.
2. Using aseptic technique, prepare a bacterial smear of each organism plus a third mixed smear of *M. smegmatis* and *S. aureus* BSL-2 .
3. Allow smears to air-dry and then heat fix in the usual manner.

Acid-Fast Staining

Steps 1–7 are pictured in Figure 10.1.

1. a. Flood smears with carbol fuchsin and place over a beaker of water on a warm hot plate, allowing the preparation to steam for 5 minutes. *Note: Do not allow stain to evaporate; replenish stain as needed. Also, prevent stain from boiling by adjusting the hot-plate temperature.*

   b. For a heatless method, flood the smear with carbol fuchsin containing Tergitol® for 5 to 10 minutes.

2. Wash with tap water. Heated slides must be cooled prior to washing.

3. Decolorize with acid-alcohol, adding the reagent drop by drop until the alcohol runs almost clear with a slight red tinge.

4. Wash with tap water.

5. Counterstain with methylene blue for 2 minutes.

6. Wash smear with tap water.

7. Blot dry with bibulous paper and examine under oil immersion.

8. In the chart provided in the Lab Report, complete the following:

   a. Draw a representative microscopic field for each preparation.

   b. Describe the cells according to their shapes and arrangements.

   c. Describe the color of the stained cells.

   d. Classify the organisms as to reaction: acid-fast or non–acid-fast.

Refer to Figure 10.2 for a photograph of an acid-fast stain.

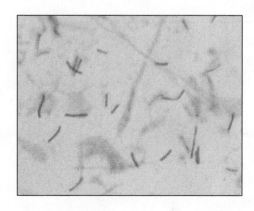

Figure 10.2 **Acid-fast stain of mycobacteria**

Name: _____

Date: _____ Section: _____

Observations and Results

	M. smegmatis	*S. aureus*	Mixture
Draw a representative field.			

Cell morphology:
    Shape
    Arrangement
Cell color
Acid-fast reaction
	_____
	_____
	_____
	_____
		_____
		_____
		_____
		_____
			_____
			_____
			_____
			_____

Review Questions

1. Why must heat or a surface-active agent be used with application of the primary stain during acid-fast staining?

2. Why is acid-alcohol rather than ethyl alcohol used as a decolorizing agent?

3. What is the specific diagnostic value of this staining procedure?

4. Why is the application of heat or a surface-active agent not required during the application of the counterstain in acid-fast staining?

5. A child presents symptoms suggestive of tuberculosis, namely a respiratory infection with a productive cough. Microscopic examination of the child's sputum reveals no acid-fast rods. However, examination of gastric washings reveals the presence of both acid-fast and non–acid-fast bacilli. Do you think the child has active tuberculosis? Explain.

Differential Staining for Visualization of Bacterial Cell Structures

LEARNING OBJECTIVES

Once you have completed this experiment, you should understand

1. The chemical basis for the spore and capsule stains.

2. The procedure for differentiation between the bacterial spore and vegetative cell forms.

3. The procedure to distinguish capsular material from the bacterial cell.

PART A Spore Stain (Schaeffer-Fulton Method)

Principle

Members of the anaerobic genera *Clostridium* and *Desulfotomaculum* and the aerobic genus *Bacillus* are examples of organisms that have the capacity to exist either as metabolically active **vegetative cells** or as highly resistant, metabolically inactive cell types called **spores**. When environmental conditions become unfavorable for continuing vegetative cellular activities, particularly with the exhaustion of a nutritional carbon source, these cells have the capacity to undergo **sporogenesis** and give rise to a new intracellular structure called the **endospore**, which is surrounded by impervious layers called spore coats. As conditions continue to worsen, the endospore is released from the degenerating vegetative cell and becomes an independent cell called a **free spore**. Because of the chemical composition of spore layers, the spore is resistant to the damaging effects of excessive heat, freezing, radiation,

desiccation, and chemical agents, as well as to the commonly employed microbiological stains. With the return of favorable environmental conditions, the free spore may revert to a metabolically active and less resistant vegetative cell through **germination** (see Figure 11.1). It should be emphasized that sporogenesis and germination are not means of reproduction but merely mechanisms that ensure cell survival under all environmental conditions.

In practice, the spore stain uses two different reagents. An alternative method known as the Dorner method is widely published and utilizes nigrosin as the counterstain which may be found online at websites similar to www.microbelibrary.org.

Primary Stain

Malachite Green Unlike most vegetative cell types that stain by common procedures, the free spore, because of its impervious coats, will not accept the primary stain easily. For further penetration, the application of heat is required. After the primary stain is applied and the smear is heated, both the vegetative cell and spore will appear green.

Decolorizing Agent

Water Once the spore accepts the malachite green, it cannot be decolorized by tap water, which removes only the excess primary stain. The spore remains green. On the other hand, the stain does not demonstrate a strong affinity for vegetative cell components; the water removes it, and these cells will be colorless.

Counterstain

Safranin This contrasting red stain is used as the second reagent to color the decolorized vegetative cells, which will absorb the counterstain

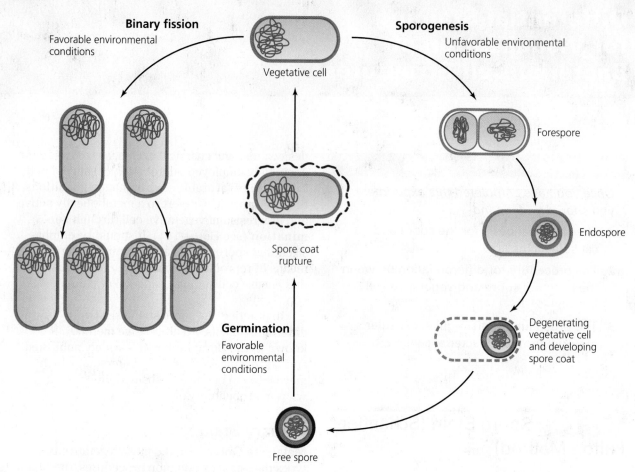

Figure 11.1 Life cycle of a spore-forming bacterium

Binary fission
Favorable environmental conditions

Vegetative cell

Sporogenesis
Unfavorable environmental conditions

Forespore

Endospore

Spore coat rupture

Germination
Favorable environmental conditions

Degenerating vegetative cell and developing spore coat

Free spore

and appear red. The spores retain the green of the primary stain. A micrograph of spore-stained cells appears in Figure 11.2.

CLINICAL APPLICATION

Identification of Dangerous Spore-Forming Bacteria

Some spore-forming bacteria can have extremely negative health effects. These bacteria include *Bacillus anthracis,* which causes anthrax, and certain *Clostridia* bacteria, which are the causative agents for tetanus, gas gangrene, food poisoning, and pseudomembranous colitis. Differential stains can stain endospores inside bacterial cells as well as free spores to identify these pathogenic bacteria.

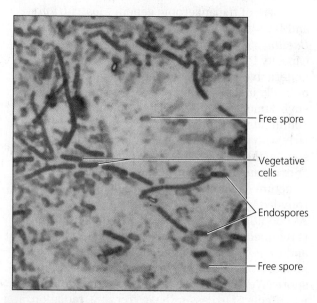

Free spore

Vegetative cells

Endospores

Free spore

Figure 11.2 Spore stain showing free spores and vegetative bacilli

Materials

Cultures

Forty–eight- to 72-hour nutrient agar slant culture of *Bacillus cereus* and thioglycollate culture of *Clostridium sporogenes*.

Reagents

Malachite green and safranin.

Equipment

Microincinerator or Bunsen burner, hot plate, staining tray, inoculating loop, glass slides, bibulous paper, lens paper, and microscope.

Procedure

Smear Preparation

1. Obtain two clean glass slides.
2. Make individual smears in the usual manner using aseptic technique.
3. Allow smear to air-dry, and heat fix in the usual manner.

Spore Staining

Steps 1–5 are illustrated in Figure 11.3.

1. Flood smears with malachite green and place on top of a beaker of water sitting on a warm hot plate, allowing the preparation to steam for 2 to 3 minutes. *Note: Do not allow stain to evaporate; replenish stain as needed.* Prevent the stain from boiling by adjusting the hot plate temperature.
2. Remove slides from hot plate, cool, and wash under running tap water.
3. Counterstain with safranin for 30 seconds.
4. Wash with tap water.
5. Blot dry with bibulous paper and examine under oil immersion.
6. In the chart provided in the Lab Report, complete the following:

   a. Draw a representative microscopic field of each preparation.

   b. Describe the location of the endospore within the vegetative cell as central, subterminal, or terminal on each preparation.

   c. Indicate the color of the spore and vegetative cell on each preparation.

PART B Capsule Stain (Anthony Method)

Principle

A **capsule** is a gelatinous outer layer that is secreted by the cell and that surrounds and adheres to the cell wall. It is not common to all organisms. Cells that have a heavy capsule are generally virulent and capable of producing disease, since the structure protects bacteria against the normal phagocytic activities of host cells. Chemically, the capsular material is composed mainly of complex polysaccharides such as levans, dextrans, and celluloses.

Capsule staining is more difficult than other types of differential staining procedures because the capsular materials are water-soluble and may be dislodged and removed with vigorous washing. Smears should not be heated because the resultant cell shrinkage may create a clear zone around the organism that is an artifact that can be mistaken for the capsule.

The capsule stain uses two reagents.

Primary Stain

Crystal Violet (1% aqueous) A violet stain is applied to a non–heat-fixed smear. At this point, the cell and the capsular material will take on the dark color.

Decolorizing Agent

Copper Sulfate (20%) Because the capsule is nonionic, unlike the bacterial cell, the primary stain adheres to the capsule but does not bind to it. In the capsule staining method, copper sulfate is used as a decolorizing agent rather than water. The copper sulfate washes the purple primary stain out of the capsular material without removing the stain bound to the cell wall. At the same time, the decolorized capsule absorbs the copper sulfate, and the capsule will now appear blue in contrast to the deep purple color of the cell.

Figure 11.4 shows the presence of a capsule as a clear zone surrounding the darker-stained cell.

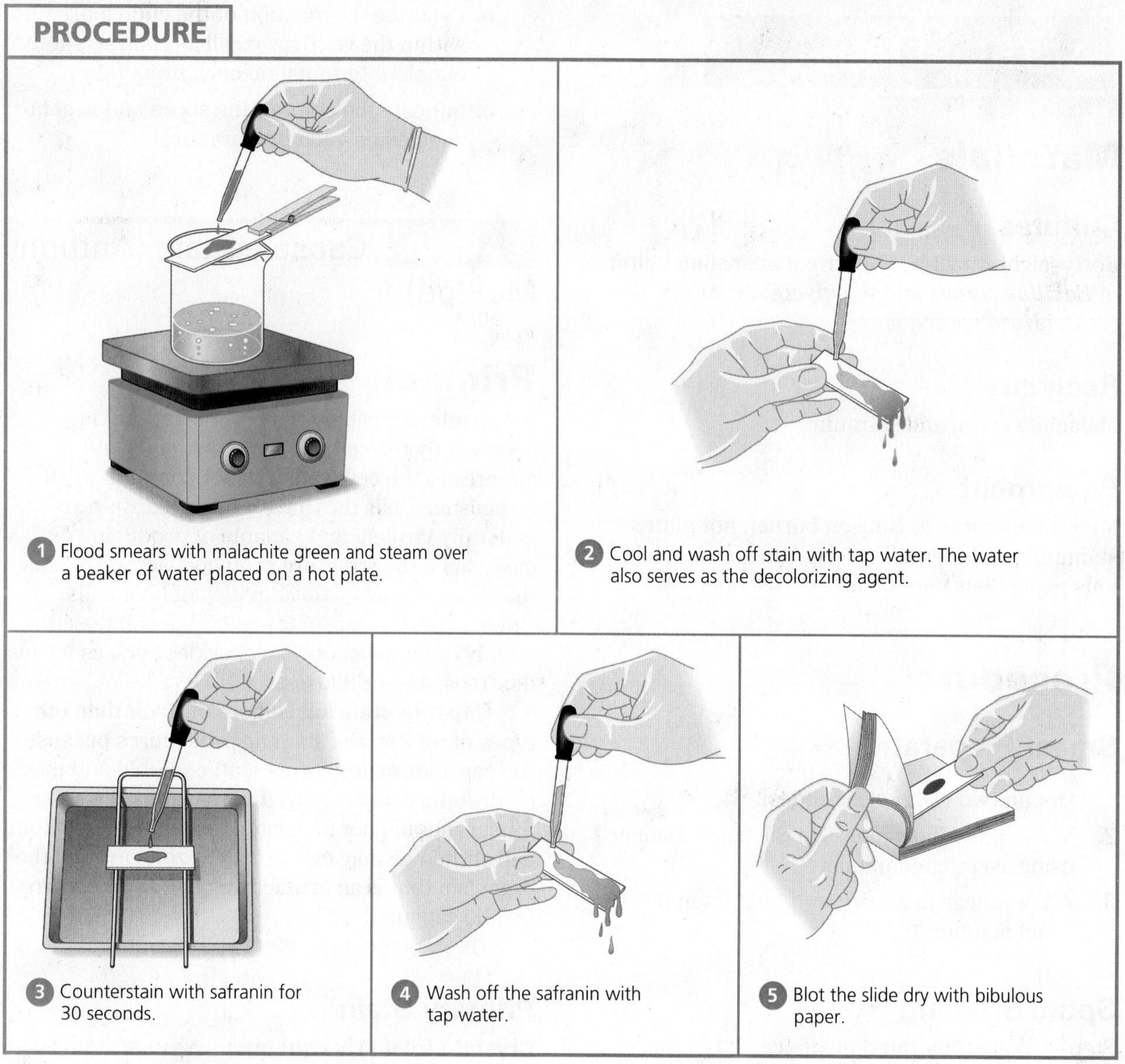

1 Flood smears with malachite green and steam over a beaker of water placed on a hot plate.

2 Cool and wash off stain with tap water. The water also serves as the decolorizing agent.

3 Counterstain with safranin for 30 seconds.

4 Wash off the safranin with tap water.

5 Blot the slide dry with bibulous paper.

Figure 11.3 **Spore-staining procedure**

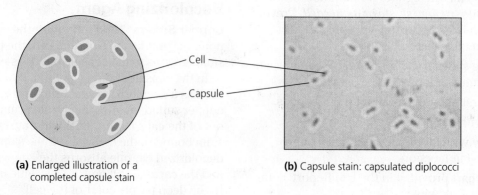

(a) Enlarged illustration of a completed capsule stain

Cell

Capsule

(b) Capsule stain: capsulated diplococci

Figure 11.4 **Capsule stain**

Encapsulated Bacterial Pneumonia

The virulence of an organism is increased by the presence of a capsule, since the capsule protects the organism from phagocytosis by white blood cells and inhibits antibody or complement fixation. The water-soluble polysaccharide and/or the polypeptide composition of the bacterial capsule makes staining this feature difficult. Gram-negative bacteria that form capsules include *Haemophilus influenzae* and *Klebsiella pneumoniae*. Gram-positive bacteria that form capsules include *Bacillus anthracis* and *Streptococcus pneumoniae*. If a bacterial infection is not being cleared or responding to antibiotic therapy as expected, staining of isolated organisms to determine the presence of a capsule may be warranted.

AT THE BENCH

Materials

Cultures

Skimmed milk cultured for 48 hours with *Alcaligenes viscolactis*, *Leuconostoc mesenteroides*, and *Enterobacter aerogenes*.

Reagents

1% crystal violet and 20% copper sulfate ($CuSO_4 \cdot 5H_2O$)

Equipment

Microincinerator or Bunsen burner, inoculating loop or needle, staining tray, bibulous paper, lens paper, glass slides, and microscope.

Procedure

Steps 1–5 are pictured in Figure 11.5.

1. Obtain one clean glass slide. Place several drops of crystal violet stain on the slide.

2. Using aseptic technique, add three loopfuls of a culture to the stain and *gently* mix with the inoculating loop.

3. With a clean glass slide, spread the mixture over the entire surface of the slide to create a very thin smear. Let stand for 5 to 7 minutes. Allow smears to air-dry. *Note: Do not heat fix.*

4. Wash smears with 20% copper sulfate solution.

5. *Gently* blot dry and examine under oil immersion.

6. Repeat Steps 1–5 for each of the remaining test cultures.

7. In the chart provided in the Lab Report, complete the following:

    a. Draw a representative microscopic field of each preparation.

    b. Record the comparative size of the capsule; that is, small, moderate, or large.

    c. Indicate the color of the capsule and of the cell on each preparation.

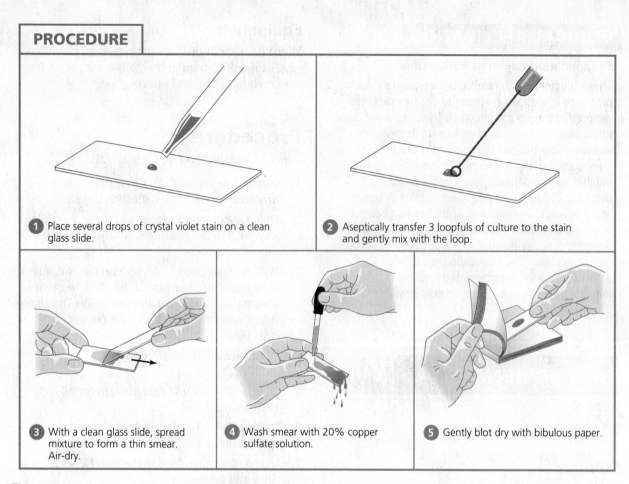

1 Place several drops of crystal violet stain on a clean glass slide.

2 Aseptically transfer 3 loopfuls of culture to the stain and gently mix with the loop.

3 With a clean glass slide, spread mixture to form a thin smear. Air-dry.

4 Wash smear with 20% copper sulfate solution.

5 Gently blot dry with bibulous paper.

Figure 11.5 **Capsule staining procedure**

c. You did not apply heat during the application of the primary stain.

4. Explain the medical significance of a capsule.

5. Explain the function of copper sulfate in this procedure.

PART 4

Cultivation of Microorganisms: Nutritional and Physical Requirements, and Enumeration of Microbial Populations

LEARNING OBJECTIVES

Once you have completed the experiments in this section, you should be familiar with

1. The basic nutritional and environmental requirements for the cellular activities of all forms of life.

2. The principles associated with the use of routine and special-purpose media for microbial cultivation.

3. The diversified physical factors essential for microbial cultivation.

4. Specialized techniques for the cultivation of anaerobic microorganisms.

5. The serial dilution–agar plate technique for enumeration of viable microorganisms.

6. The growth dynamics of bacterial populations.

Introduction

As do all living organisms, microorganisms require certain basic nutrients and physical factors for the sustenance of life. However, their particular requirements vary greatly. Understanding these needs is necessary for successful cultivation of microorganisms in the laboratory.

Nutritional Needs

Nutritional needs of microbial cells are supplied in the laboratory through a variety of media. The following list illustrates the nutritional diversity that exists among microbes.

1. **Carbon:** This is the most essential and central atom common to all cellular structures and functions. Among microbial cells, two carbon-dependent types are noted:

    a. **Autotrophs:** These organisms can be cultivated in a medium consisting solely of inorganic compounds; specifically, they use inorganic carbon in the form of carbon dioxide.

    b. **Heterotrophs:** These organisms cannot be cultivated in a medium consisting solely of inorganic compounds; they must be supplied with organic nutrients, primarily glucose.

2. **Nitrogen:** This is also an essential atom in many cellular macromolecules, particularly proteins and nucleic acids. Proteins serve as the structural molecules forming the so-called fabric of the cell and as functional molecules, enzymes, that are responsible for the metabolic activities of the cell. Nucleic acids include DNA, the genetic basis of cell life, and RNA, which plays an active role in protein synthesis within the cell. Some microbes use atmospheric nitrogen, others rely on inorganic compounds, including ammonium or nitrate salts, and still others require nitrogen-containing organic compounds such as amino acids.

3. **Nonmetallic elements:** Two major nonmetallic ions are used for cellular nutrition:

    a. **Sulfur** is integral to some amino acids and is therefore a component of proteins. Sources include organic compounds, such as sulfur-containing amino acids; inorganic compounds, such as sulfates; and elementary sulfur.

   b. **Phosphorus** is necessary for the formation of the nucleic acids DNA and RNA and also for synthesis of the high-energy organic compound adenosine triphosphate (ATP). Phosphorus is supplied in the form of phosphate salts for use by all microbial cells.

4. **Metallic elements:** Ca^{2+}, Zn^{2+}, Na^+, K^+, Cu^{2+}, Mn^{2+}, Mg^{2+}, Fe^{2+}, and Fe^{3+}, are some of the metallic ions necessary for continued efficient performance of varied cellular activities. Some of these activities are osmoregulation, regulation of enzyme activity, and electron transport during biooxidation. Remember that these ions are micronutrients and are required in trace concentrations only. Inorganic salts supply these materials.

5. **Vitamins:** These organic substances contribute to cellular growth and are essential in minute concentrations for cell activities. They are also sources of coenzymes, which are required for the formation of active enzyme systems. Some microbes require vitamins to be supplied in a preformed state for normal metabolic activities. Some possess extensive vitamin-synthesizing pathways, whereas others can synthesize only a limited number from other compounds present in the medium.

6. **Water:** All cells require distilled water in the medium so that the low-molecular-weight nutrients can cross the cell membrane.

7. **Energy:** Active transport, biosynthesis, and biodegradation of macromolecules are the metabolic activities of cellular life. These activities can be sustained only if there is a constant availability of energy within the cell. Two bioenergetic types of microorganisms exist:

   a. **Phototrophs** use radiant energy as their sole energy source.

   b. **Chemotrophs** depend on oxidation of chemical compounds as their energy source. Some microbes use organic molecules, such as glucose; others utilize inorganic compounds, such as H_2S or $NaNO_2$.

Physical Factors

Three of the most important physical factors that influence the growth and survival of cells are temperature, pH, and the gaseous environment. An understanding of the roles they play in cell metabolism is essential.

1. **Temperature** influences the rate of chemical reactions through its action on cellular enzymes. Bacteria, as a group of organisms, exist over a wide range of temperatures. However, individual species can exist only within a narrower spectrum of temperatures. Low temperatures slow down or inhibit enzyme activity, thereby slowing down or inhibiting cell metabolism and, consequently, cell growth. High temperatures cause coagulation and thus irreversibly denature thermolabile enzymes. Although enzymes differ in their degree of heat sensitivity, generally temperatures in the range of 70°C destroy most essential enzymes and cause cell death.

2. **The pH of the extracellular environment** greatly affects cells' enzymatic activities. Most commonly, the optimum pH for cell metabolism is in the neutral range of 7. An increase in the hydrogen ion concentration resulting in an acidic pH (below 7) or a decrease in the hydrogen ion concentration resulting in an alkaline pH (above 7) is often detrimental. Either increase or decrease will slow down the rate of chemical reactions because of the destruction of cellular enzymes, thereby affecting the rate of growth and, ultimately, survival.

3. **The gaseous requirement** in most cells is atmospheric oxygen, which is necessary for the biooxidative process of respiration. Atmospheric oxygen plays a vital role in ATP formation and the availability of energy in a utilizable form for cell activities. Other cell types, however, lack the enzyme systems for respiration in the presence of oxygen and therefore must use an anaerobic form of respiration or fermentation.

The following exercises will demonstrate the diversity of nutritional and environmental requirements among microorganisms.

Nutritional Requirements: Media for the Routine Cultivation of Bacteria

LEARNING OBJECTIVES

Once you have completed this experiment, you should know how to evaluate

1. The abilities of several types of media to support the growth of different bacterial species.

2. The nutritional needs of the bacteria under study.

Principle

To satisfy the diverse nutritional needs of bacteria, bacteriologists employ two major categories of media for routine cultivation.

Chemically Defined Media

These are composed of known quantities of chemically pure, specific organic and/or inorganic compounds. Their use requires knowledge of the organism's specific nutritional needs. The following two chemically defined media are used in this exercise:

1. **Inorganic synthetic broth:** This completely inorganic medium is prepared by incorporating the following salts per 1000 ml of water:

Sodium chloride (NaCl)	5.0 g
Magnesium sulfate ($MgSO_4$)	0.2 g
Ammonium dihydrogen phosphate ($NH_4H_2PO_4$)	1.0 g
Dipotassium hydrogen phosphate (K_2HPO_4)	1.0 g
Atmospheric carbon dioxide (CO_2)	

2. **Glucose salts broth:** This medium is composed of salts incorporated into the inorganic synthetic broth medium plus **glucose**, 5 g per liter, which serves as the sole organic carbon source.

Complex Media

The exact chemical composition of these media is not known. They are made of extracts of plant and animal tissue and are variable in their chemical composition. Most contain abundant amino acids, sugars, vitamins, and minerals; however, the quantities of these constituents are not known. They are capable of supporting the growth of most heterotrophs. The following two complex media are used in this exercise.

1. **Nutrient broth:** This basic complex medium is prepared by incorporating the following ingredients per 1000 ml of distilled water:

Peptone	5.0 g
Beef extract	3.0 g

   Peptone, a semidigested protein, is primarily a nitrogen source. The **beef extract**, a beef derivative, is a source of organic carbon, nitrogen, vitamins, and inorganic salts.

2. **Yeast extract broth:** This is composed of the basic artificial medium ingredients used in the nutrient broth plus **yeast extract**, 5 g per liter, which is a rich source of vitamin B and provides additional organic nitrogen and carbon compounds.

   The yeast extract broth is an example of an **enriched medium** and is used for the cultivation of **fastidious** microorganisms, organisms that have highly elaborate and specific nutritional needs. These bacteria do not grow or grow poorly on a basic artificial medium and require the addition of one or more growth-supporting substances, enrichments such as additional plant or animal extracts, vitamins, or blood.

Measuring Turbidity

In this experiment, you will evaluate (1) the abilities of media to support the growth of different species of bacteria, and (2) the nutritional needs of the bacteria. You will observe the amount of growth, measured by turbidity, present in each culture following incubation. To evaluate the amount

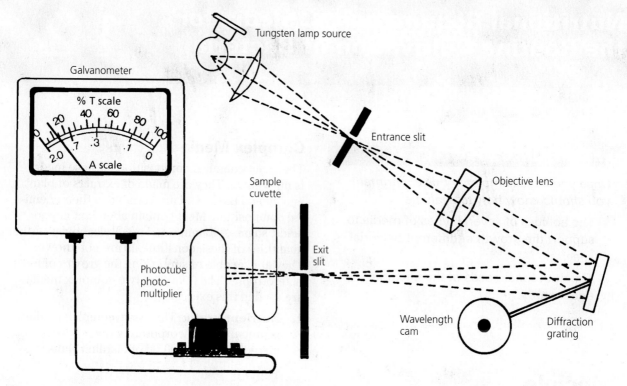

Figure 12.1 **Schematic diagram of a spectrophotometer**

of growth accurately, a Bausch & Lomb Spectronic 20 spectrophotometer will be used.

This instrument measures the amount of light transmitted (T) or absorbed (A). It transmits a beam of light at a single wavelength (monochromatic light) through a liquid culture. The cells suspended in the culture interrupt the passage of light, and the amount of light energy transmitted through the suspension is measured on a photoelectric cell and converted into electrical energy. The electrical energy is then recorded on a galvanometer as 0% to 100% T. A schematic representation of a spectrophotometer is shown in Figure 12.1.

In practice, the density of a cell suspension is expressed as absorbance (A) rather than percent T, since A is directly proportional to the concentration of cells, whereas percent T is inversely proportional to the concentration of suspended cells. Therefore, as the turbidity of a culture increases, the A increases and percent T decreases, indicating growth of the cell population in the culture. For example, in comparing three cultures with A readings of 0.10 (percent T = 78), 0.30 (percent T = 49), and 0.50 (percent T = 30), the A reading of 0.50 would be indicative of the most abundant growth, and the 0.10 reading would be indicative of the least amount of growth. Figure 12.2 shows the Bausch & Lomb Spectronic 20 spectrophotometer, as well as a Unico RS 1100 model. For the purpose

of this experiment, we will be discussing the procedure using the Busch & Lomb Spectronic 20.

You will also use a colorimetric plate reader to measure turbidity in small volume cultures. Plate readers are generally used in a laboratory setting to quantify the degree of color development at specified wavelengths in individual wells of a 96-well plate. When using a 600 nm filter, a plate reader can be used to determine the increase in turbidity in a culture with less than a 200 µL volume.

CLINICAL APPLICATION

The Purpose of Specialized Media

The successful cultivation of bacteria requires the use of culture media containing the nutritional and biochemical requirements capable of supporting growth. There is no single medium that can support the growth of all microorganisms. This challenge has been met by the development of a variety of specialized media. For example, the streptococci require media supplemented with blood in order to determine certain properties that are necessary for isolation and species identification. Another example is the thioglycollate medium, which contains thioglycolic acid that removes oxygen from the medium to encourage the growth of certain anaerobic bacteria.

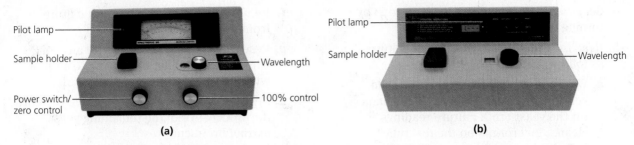

Figure 12.2 (a) The Bausch & Lomb Spectronic 20 (b) and the Unico 1100RS spectrophotometer

Current methods of using a set standard to measure growth against are being developed. Many procedures utilize sterile broth as a "blank" to set the spectrophotometer at zero and then measure the absorbance above that zero setting as a means of quantifying cell density. Commercially available standards such as the McFarland Standards utilize microscopic plastic particles to simulate cells in suspension and allow for standardization in preparation of bacterial suspensions. The 0.5 McFarland Standard has been shown to correlate to 10^8 cells per ml in *Escherichia coli* cultures.

AT THE BENCH

Materials

Cultures

Saline suspension of 24-hour Trypticase soy broth cultures, adjusted to 0.05 absorbance at a wavelength of 600 nm (or equilibrated to the 0.5 McFarland Standard), of *Escherichia coli*, *Alcaligenes faecalis*, and *Streptococcus mitis* BSL-2 .

Media

Per designated student group: three test tubes (13 × 100 mm) of each type of broth: inorganic synthetic broth, glucose salts broth, nutrient broth, and yeast extract broth.

Equipment

Microincinerator or bunsen burner, sterile 1-ml serological pipettes, mechanical pipetting device, micropipette and tips, glassware marking pencil,

test tube rack, 96-well clear plastic culture plate, Bausch & Lomb Spectronic 20 (or comparable) spectrophotometer, and a colorimetric plate reader.

Procedure Lab One

1. Using sterile tips and a micropipette, add 100 µl of the *E. coli* culture to one test tube of each of the appropriately labeled media.
2. Using a micropipette, add 100 µl of broth and 50 µl of the *E. coli* culture made in step 1 to designated wells in the 96-well plate.
3. Repeat Steps 1 and 2 for inoculation with *A. faecalis* and *S. mitis* BSL-2 .
4. Follow manufacturer's guidelines for the plate reader to measure the turbidity of the wells at 600 nm and record the preliminary readings.
5. Incubate the test cultures for 24 to 48 hours at 37°C.

Procedure Lab Two

Follow the instructions below and refer to Figure 12.2 for the use of the Bausch & Lomb Spectronic 20 spectrophotometer to obtain the absorbance readings of all your cultures. *Follow the instructions provided by the manufacturer or your instructor for all other spectrophotometers.*

1. Use the plate reader to measure change in turbidity readings at 600 nm wavelength. Subtract initial readings measured on Day One readings from readings measured on Day Two. Record the results.
2. Turn the spectrophotometer on 10 to 15 minutes prior to use.
3. Set wavelength at 600 nm.

4. Set percent transmittance to 0% (A to 2) by turning the knob on the left.

5. Read the four yeast extract broth cultures as follows:

   a. Wipe the provided test tube of sterile yeast broth that will serve as the blank for the yeast broth culture readings clean. Fingerprints on the test tube will obscure the light path of the spectrophotometer.

   b. Insert the yeast extract broth blank into the tube holder, close the cover, and set the A to 0 (percent T = 100) by turning the knob on the right.

   c. Shake lightly or tap one of the tubes of yeast extract broth culture to resuspend the bacteria, wipe the test tube clean, and allow it to sit for several seconds for the equilibration of the bacterial suspension.

   d. Remove the yeast extract broth blank from the tube holder.

   e. Insert a yeast extract broth culture into the tube holder, close the cover, and read and record the optical density reading in the chart provided in the Lab Report.

   f. Remove the yeast extract broth culture from the tube holder.

   g. Reset the spectrophotometer to an A of 2 with the tube holder empty and to an A of 0 with the yeast extract broth blank.

   h. Repeat Steps c through g to read and record the absorbance of the remaining yeast extract broth cultures.

6. Repeat Step 4 (a–h) to read and record the absorbance of the nutrient broth cultures. Use the provided nutrient broth blank to set the spectrophotometer to an A of 0.

7. Repeat Step 4 (a–h) to read and record the absorbance of the glucose salts broth cultures. Use the provided glucose salts broth blank to set the spectrophotometer to an A of 0.

8. Repeat Step 4 (a–h) to read and record the absorbance of the inorganic synthetic broth cultures. Use the provided inorganic synthetic broth blank to set the spectrophotometer to an A of 0.

9. At the end of the experiment, return all cultures to the area designated for their disposal.

10. Complete the Lab Report.

Name: _____

Date: _____ Section: _____

Observations and Results

Optical Density Readings Using the Spectrophotometer

	Yeast Extract Broth	Nutrient Broth	Glucose Broth	Inorganic Synthetic Broth
E. coli				
A. faecalis				
S. mitis				

Optical Density Readings Using a Colorimetric Plate Reader

	Yeast Extract Broth	Nutrient Broth	Glucose Broth	Inorganic Synthetic Broth
E. coli				
A. faecalis				
S. mitis				

1. On the basis of the data above, list the media in order (from best to worst) according to their ability to support the growth of bacteria.

2. List the three bacterial species in order of their increasing fastidiousness.

3. Why did the most fastidious organism grow poorly in the chemically defined medium?

Review Questions

1. Explain the advantages of using A readings rather than percent T as a means of estimating microbial growth.

2. Explain the reason for the use of different medium blanks in adjusting the spectrophotometer prior to obtaining A readings.

3. Why are complex media preferable to chemically defined media for routine cultivation of microorganisms?

4. Would you expect a heterotrophic organism to grow in an inorganic synthetic medium? Explain.

5. A soil isolate is found to grow poorly in a basic artificial medium. You suspect that a vitamin supplement is required.

   a. What supplement would you use to enrich the medium to support and maintain the growth of the organisms? Explain.

   b. Outline the procedure you would follow to determine the specific vitamins required by the organism to produce a more abundant growth.

Use of Differential, Selective, and Enriched Media

LEARNING OBJECTIVES

Once you have completed this experiment, you should be familiar with

1. The use and function of specialized media for the selection and differentiation of microorganisms.

2. How an enriched medium like blood agar can also function as both a selective and differential medium.

Principle

Numerous special-purpose media are available for functions including the following:

1. Isolation of bacterial types from a mixed population of organisms.

2. Differentiation among closely related groups of bacteria on the basis of macroscopic appearance of the colonies and biochemical reactions within the medium.

3. Enumeration of bacteria in sanitary microbiology, such as in water and sewage, and also in food and dairy products.

4. Assay of naturally occurring substances, including antibiotics, vitamins, and products of industrial fermentation.

5. Characterization and identification of bacteria by their abilities to produce chemical changes in different media.

In addition to nutrients necessary for the growth of all bacteria, special-purpose media contain both nutrients and chemical compounds important for specific metabolic pathways in different types of bacteria. In this exercise, three types of media will be studied and evaluated.

Selective Media

These media are used to select (isolate) specific groups of bacteria. They incorporate chemical substances that inhibit the growth of one type of bacteria while permitting growth of another, thus facilitating bacterial isolation.

1. **Phenylethyl alcohol agar:** This medium is used for the isolation of most gram-positive organisms. The phenylethyl alcohol is partially inhibitory to gram-negative organisms, which may form visible colonies whose size and number are much smaller than on other media.

2. **Crystal violet agar:** This medium is selective for most gram-negative microorganisms. Crystal violet dye exerts an inhibitory effect on most gram-positive organisms.

3. **7.5% sodium chloride agar:** This medium is inhibitory to most organisms other than halophilic (salt-loving) microorganisms. It is most useful in the detection of members of the genus *Staphylococcus*.

Figure 13.1 is a photo illustrating the selective effect of phenylethyl alcohol agar, which inhibits the gram-negative organism *E. coli* and selects for the gram-positive organism *S. aureus*.

Differential/Selective Media

These media can distinguish among morphologically and biochemically related groups of organisms. They incorporate chemical compounds that, following inoculation and incubation, produce a characteristic change in the appearance of bacterial growth and/or the medium surrounding the colonies, which permits differentiation.

Sometimes differential and selective characteristics are combined in a single medium. MacConkey agar is a good example of this because it contains bile salts and crystal violet, which inhibit gram-positive organisms and allow gram-negative organisms to grow. In addition, it contains the substrate lactose and the pH indicator neutral red, which differentiates the red lactose-fermenting colonies from the translucent non-fermenting colonies. The following media are examples of this type of media:

1. **Mannitol salt agar:** This medium contains a high salt concentration, 7.5% NaCl, which is inhibitory to the growth of most, but not all,

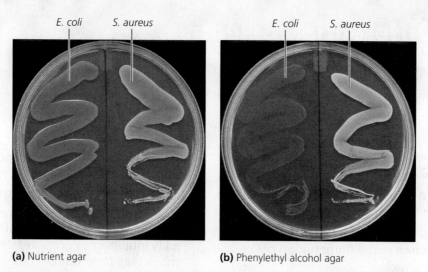

(a) Nutrient agar   (b) Phenylethyl alcohol agar

Figure 13.1 Selective effect of phenylethyl alcohol agar reduces the growth of *E. coli* and selects for *S. aureus*

bacteria other than the staphylococci. The medium also performs a differential function: It contains the carbohydrate mannitol, which some staphylococci are capable of fermenting, and phenol red, a pH indicator for detecting acid produced by mannitol-fermenting staphylococci. These staphylococci exhibit a yellow zone surrounding their growth; staphylococci that do not ferment mannitol will not produce a change in coloration.

2. **MacConkey agar:** The inhibitory action of crystal violet on the growth of gram-positive organisms allows the isolation of gram-negative bacteria. Incorporation of the carbohydrate lactose, bile salts, and the pH indicator neutral red permits differentiation of enteric bacteria on the basis of their ability to ferment lactose. On this basis, enteric bacteria are separated into two groups:

a. **Coliform bacilli** produce acid as a result of lactose fermentation. The bacteria exhibit a red coloration on their surface. *Escherichia coli* produce greater quantities of acid from lactose than other coliform species. When this occurs, the medium surrounding the growth also becomes pink because of the action of the acid that precipitates the bile salts, followed by absorption of the neutral red.

b. **Dysentery, typhoid, and paratyphoid bacilli** are not lactose fermenters and therefore do not produce acid. The colonies appear tan and frequently transparent when grown in MacConkey agar.

3. **Eosin–methylene blue agar (Levine):** Lactose and the dyes eosin and methylene blue permit differentiation between enteric lactose fermenters and nonfermenters as well as identification of the colon bacillus, *E. coli*. The *E. coli* colonies are blue-black with a metallic green sheen caused by the large quantity of acid that is produced and that precipitates the dyes onto the growth's surface. Other coliform bacteria, such as *Enterobacter aerogenes*, produce thick, mucoid, pink colonies on this medium. Enteric bacteria that do not ferment lactose produce colorless colonies, which, because of their transparency, appear to take on the purple color of the medium. This medium is also partially inhibitory to the growth of gram-positive organisms, and thus gram-negative growth is more abundant.

A photographic representation of the effects of selective/differential media is presented in Figure 13.2.

Enriched Media

Enriched media are media that have been supplemented with highly nutritious materials, such as blood, serum, or yeast extract, for the purpose of cultivating fastidious organisms.

For example, in **blood agar**, the blood incorporated into the medium is an enrichment ingredient for the cultivation of fastidious organisms, such as the *Streptococcus* spp. The blood also permits demonstration of the hemolytic properties of some microorganisms, particularly

Fermenter   Nonfermenter

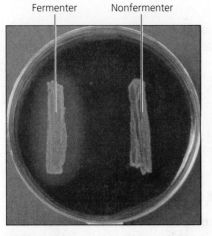

Fermenter   Nonfermenter

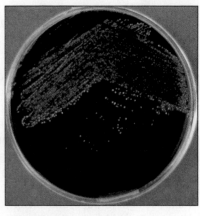

(a) Mannitol salt agar   **(b)** MacConkey agar   **(c)** Eosin–methylene blue agar

Figure 13.2 Effects of selective/differential media

the streptococci, whose hemolytic activities are classified as follows:

1. **Gamma hemolysis:** No lysis of red blood cells results in no significant change in the appearance of the medium surrounding the colonies.

2. **Alpha hemolysis:** Incomplete lysis of red blood cells, with reduction of hemoglobin to methemoglobin, results in a greenish halo around the bacterial growth.

3. **Beta hemolysis:** Lysis of red blood cells with complete destruction and use of hemoglobin by the organism results in a clear zone surrounding the colonies. This hemolysis is produced by two types of beta hemolysins,

namely **streptolysin O**, an antigenic, oxygen-labile enzyme, and **streptolysin S**, a nonantigenic, oxygen-stable lysin. The hemolytic reaction is enhanced when blood agar plates are streaked and simultaneously stabbed to show subsurface hemolysis by streptolysin O in an environment with reduced oxygen tension. Based on the hemolytic patterns on blood agar, the pathogenic beta-hemolytic streptococci may be differentiated from other streptococci.

Figure 13.3 shows the different types of hemolysis exhibited by different species of the genus *Streptococcus* on blood agar.

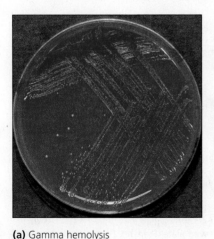

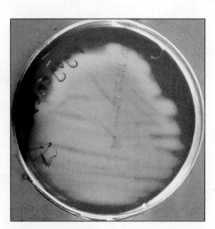

(a) Gamma hemolysis   **(b)** Alpha hemolysis   **(c)** Beta hemolysis

Figure 13.3 Types of hemolysis exhibited on a blood agar plate

CLINICAL APPLICATION

First Steps in Infected Wound Diagnosis

Wounds that have become infected may be swabbed or surgically processed to remove tissue. Once stained samples have revealed infectious agents, cultures are typically made on (1) blood agar for isolation of staphylococci and streptrococci bacteria, (2) MacConkey agar for gram-negative rods, and (3) enriched media that can support aerobes or anaerobes, such as thioglycollate broth. Additional media may be used, depending on what was observed microscopically, including Sabouraud dextrose agar for fungi and Löwenstein-Jensen medium for acid-fast rods. Once the microbes are isolated, further tests (which you will learn soon!) would likely be needed for complete identification.

AT THE BENCH

Materials

Cultures

24- to 48-hour Trypticase soy broth cultures of *Enterobacter aerogenes*, *Escherichia coli*, *Streptococcus* var. Lancefield Group E, *Streptococcus mitis* BSL-2 , *Enterococcus faecalis* BSL-2 , *Staphylococcus aureus* BSL-2 , *Staphylococcus epidermidis*, and *Salmonella typhimurium* BSL-2 .

Media

Per designated student group: one each of phenyl-ethyl alcohol agar, crystal violet agar, 7.5% sodium chloride agar, mannitol salt agar, MacConkey agar, eosin–methylene blue agar, and blood agar.

Equipment

Microincinerator or Bunsen burner, inoculating loop, and glassware marking pencil.

Procedure Lab One

1. Using the bacterial organisms listed in Step 2, prepare and inoculate each of the plates in the following manner:

   a. Label the cover of each plate appropriately, as indicated in the Laboratory Protocol section on page xv.

   b. Divide each of the Petri dishes into the required number of sections (one section for each different organism) by marking the *bottom of the dish*. Label each section with the name of the organism to be inoculated, as illustrated in **Figure 13.4a**.

   c. Using aseptic technique, inoculate all plates, except the blood agar plate, with the designated organisms by making a single line of inoculation of each organism in its appropriate section (**Figure 13.4b**). Be sure to close the Petri dish and flame the inoculating needle between inoculations of the different organisms.

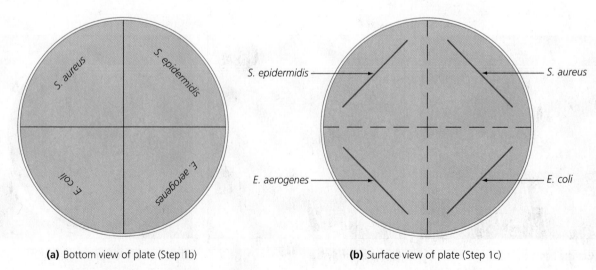

(a) Bottom view of plate (Step 1b)   **(b)** Surface view of plate (Step 1c)

Figure 13.4 Mannitol salt agar plate preparation and inoculation procedure

d. Using aseptic technique, inoculate the blood agar plate as described in Step 1c. On completion of each single line of inoculation, use the inoculating loop and make three or four stabs at a 45° angle across the streak.

2. Inoculate each of the different media with the following:

a. Phenylethyl alcohol agar: *E. coli*, *S. aureus* `BSL-2`, and *E. faecalis* `BSL-2`.

b. Crystal violet agar: *E. coli*, *S. aureus* `BSL-2`, and *E. faecalis* `BSL-2`.

c. 7.5% sodium chloride agar: *S. aureus* `BSL-2`, *S. epidermidis*, and *E. coli*.

d. Mannitol salt agar: *S. aureus* `BSL-2`, *S. epidermidis*, *E. aerogenes*, and *E. coli*.

e. MacConkey agar: *E. coli*, *E. aerogenes*, *S. typhimurium* `BSL-2`, and *S. aureus* `BSL-2`.

f. Eosin–methylene blue agar: *E. coli*, *E. aerogenes*, *S. typhimurium* `BSL-2`, and *S. aureus* `BSL-2`.

g. Blood agar: *E. faecalis* `BSL-2`, *S. mitis* `BSL-2`, and *Streptococcus* var. Lancefield Group E.

3. Incubate the phenylethyl alcohol agar plate in an inverted position for 48 to 72 hours at 37°C. Incubate the remaining plates in an inverted position for 24 to 48 hours at 37°C.

Procedure Lab Two

1. Carefully examine each of the plates. Note and record the following in the chart provided in the Lab Report:

a. Amount of growth along line of inoculation as follows: 0 = none; 1+ = scant; and 2+ = moderate to abundant.

b. Appearance of the growth: coloration, transparency.

c. Change in the appearance of the medium surrounding the growth: coloration, transparency indicative of hemolysis.

Physical Factors: Temperature

LEARNING OBJECTIVES

Once you have completed this experiment, you should know

1. The diverse growth temperature requirements of bacteria.

2. How to determine whether the optimum growth temperature is also the ideal temperature for enzyme-regulated cell activities, such as pigment production and carbohydrate fermentation.

Principle

Microbial growth is directly dependent on how temperature affects cellular enzymes. With increasing temperatures, enzyme activity increases until the three-dimensional configuration of these molecules is lost because of denaturation of their protein structure. As the temperature is lowered toward the freezing point, enzyme inactivation occurs and cellular metabolism gradually diminishes. At 0°C, biochemical reactions cease in most cells.

Bacteria, as a group of living organisms, are capable of growth within an overall temperature range of minus 5°C to 80°C. Each species, however, requires a narrower range that is determined by the heat sensitivity of its enzyme systems. Specific temperature ranges consist of the following **cardinal (significant) temperature points** (Figure 14.1):

1. **Minimum growth temperature:** The lowest temperature at which growth will occur. Below this temperature, enzyme activity is inhibited and the cells are metabolically inactive so that growth is negligible or absent.

2. **Maximum growth temperature:** The highest temperature at which growth will occur. Above this temperature, most cell enzymes are destroyed and the organism dies.

3. **Optimum growth temperature:** The temperature at which the rate of reproduction is most rapid; however, it is not necessarily optimum or ideal for all enzymatic activities of the cell.

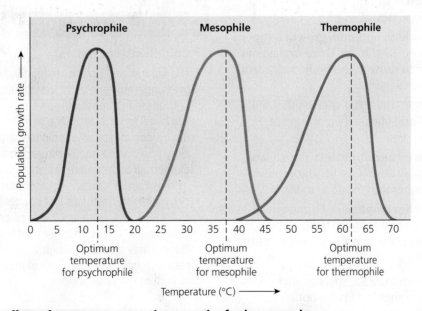

Figure 14.1 **The effect of temperature on the growth of microorganisms**

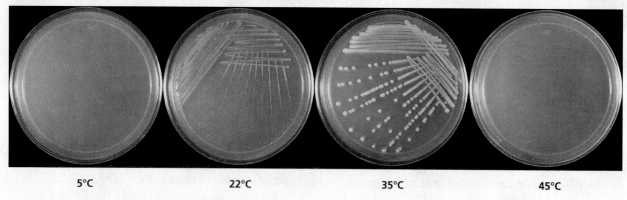

| 5°C | 22°C | 35°C | 45°C |

Figure 14.2 Effect of temperature on bacterial growth and pigmentation

Figure 14.2 shows the effects of temperature on bacterial growth and pigment production.

All bacteria can be classified into one of three major groups, depending on their temperature requirements:

1. **Psychrophiles:** Bacterial species that will grow within a temperature range of –5°C to 20°C. The distinguishing characteristic of all psychrophiles is that they will grow between 0° and 5°C.

2. **Mesophiles:** Bacterial species that will grow within a temperature range of 20°C to 45°C. The distinguishing characteristics of all mesophiles are their ability to grow at human body temperature (37°C) and their inability to grow at temperatures above 45°C. Included among the mesophiles are two distinct groups:

    a. Mesophiles with optimum growth temperature between 20°C and 30°C are plant saprophytes.

    b. Mesophiles with optimum growth temperature between 35°C to 40°C are organisms that prefer to grow in the bodies of warm-blooded hosts.

3. **Thermophiles:** Bacterial species that will grow at 35°C and above. Two groups of thermophiles exist:

    a. **Facultative thermophiles:** Organisms that will grow at 37°C, with an optimum growth temperature of 45°C to 60°C.

    b. **Obligate thermophiles:** Organisms that will grow only at temperatures above 50°C, with optimum growth temperatures above 60°C.

The ideal temperature for specific enzymatic activities may not coincide with the optimum growth temperature for a given organism. To understand this concept, you will investigate pigment production and carbohydrate fermentation by selected organisms at a variety of incubation temperatures.

1. The production of an endogenous red or magenta pigment by *Serratia marcescens* is determined by the presence of an orange to deep red coloration on the surface of the colonial growth.

2. Carbohydrate fermentation by *Saccharomyces cerevisiae* is indicated by the presence of gas, one of the end products of this fermentative process. Detection of this accumulated gas may be noted as an air pocket, of varying size, in an inverted inner vial (Durham tube) within the culture tube. Refer to Experiment 21 for a more extensive discussion of carbohydrate fermentation.

CLINICAL APPLICATION

Cold-Resistant Killers

The field of food science is highly concerned with the temperature-related growth patterns of bacteria. Refrigeration temperatures below 4.4°C are generally considered safe for the short-term storage of food, since most pathogenic bacteria grow very slowly below that temperature. However, some dangerous bacteria are resistant to cold. *Listeria monocytogenes,* which causes a flu-like illness and can be deadly, is capable of doubling its population every 36 hours, even at 4.2°C, and can still attain slow growth below 2°C. The cold tolerance of *Listeria* may be due to adaptive genes, prompting research into novel methods of controlling its growth at low temperatures.

Materials

Cultures

24- to 48-hour nutrient broth cultures of *Escherichia coli*, *Bacillus stearothermophilus*, *Pseudomonas savastanoi*, *Serratia marcescens*, and Sabouraud broth culture of *Saccharomyces cerevisiae*.

Media

Per designated student group: four Trypticase soy agar plates and four Sabouraud broth tubes containing inverted Durham tubes.

Equipment

Microincinerator or Bunsen burner, inoculating loop, refrigerator set at 4°C, two incubators set at 37°C and 60°C, sterile Pasteur pipette, test tube rack, and glassware marking pencil.

Procedure Lab One

1. Score the underside of all plates into four quadrants with a glassware marker. Label each section with the name of the test organism to be inoculated. When labeling the cover of each plate, include the temperature of incubation (4°C, 20°C, 37°C, or 60°C).

2. Aseptically inoculate each of the plates with *E. coli*, *B. stearothermophilus*, *P. savastanoi*, and *S. marcescens* by means of a single line of inoculation of each organism in its appropriately labeled section.

3. Appropriately label the four Sabouraud broth tubes, including the temperatures of incubation as indicated above.

4. Gently shake the *S. cerevisiae* culture to suspend the organisms. Using a sterile Pasteur pipette, aseptically add one drop of the culture into each of the four tubes of broth media.

5. Incubate all plates in an inverted position and the broth cultures at each of the four experimental temperatures (4°C, 20°C, 37°C, and 60°C) for 24 to 48 hours.

Procedure Lab Two

1. In the chart provided in the Lab Report, complete the following:

   a. Observe all the cultures for the presence of growth. Record your observations: (1+) for scant growth; (2+) for moderate growth; (3+) for abundant growth; and (−) for the absence of growth. Evaluate the amount of growth in the *S. cerevisiae* cultures by noting the degree of developed turbidity.

   b. Observe the *S. marcescens* growth on all the plate cultures for the presence or absence of orange to deep red pigmentation. Record the presence of pigment on a scale of 1+ to 3+, and enter (−) for the absence of pigmentation.

   c. Observe the *S. cerevisiae* cultures for the presence of a gas pocket in the Durham tube, which is indicative of carbohydrate fermentation. Record your observations using the following designations: (1+) for a minimal amount of gas; (2+) for a moderate amount of gas; (3+) for a large amount of gas; and (−) for the absence of gas.

   d. Record and classify the cultures as psychrophiles, mesophiles, facultative thermophiles, or obligate thermophiles.

Name: _____

Date: _____   Section: _____

Observations and Results

Temperature	Serratia marcescens		Pseudomonas savastanoi	Escherichia coli	Bacillus stearothermophilus	Saccharomyces cerevisiae	
	Pigment	Growth	Growth	Growth	Growth	Growth	Gas
4°C (refrigerator)							
20°C (room temp.)							
37°C (body temp.)							
60°C							
Classification							

Based on your observations of the *S. marcescens* and *S. cerevisiae* cultures, is the optimum growth temperature the ideal temperature for all cell activities? Explain.

Review Questions

1. In the following chart, indicate the types of organisms that would grow pref-erentially in or on various environments, and indicate the optimum tempera-ture for their growth.

Environment	Type of Organism	Optimum Temperature
Ocean bottom near shore		
Ocean bottom near hot vent		
Hot sulfur spring		
Compost pile (middle)		
High mountain lake		
Center of an abscess		
Antarctic ice		

2. Explain the effects of temperatures above the maximum and below the minimum growth temperatures on cellular enzymes.

3. If an organism grew at 20°C, explain how you would determine experimentally whether the organism was a psychrophile or a mesophile.

4. Is it possible for thermophilic organisms to induce infections in warm-blooded animals? Explain.

Physical Factors: pH of the Extracellular Environment

LEARNING OBJECTIVE

Once you have completed this experiment, you should be familiar with

1. The pH requirements of microorganisms.

Principle

Growth and survival of microorganisms are greatly influenced by the pH of the environment, and all bacteria and other microorganisms differ as to their requirements. Based on their optimal pH, microorganisms may be classified as acidophiles, neutrophiles, or alkalophiles (Figure 15.1). Each species has the ability to grow within a specific pH range; the range may be broad or limited, with the most rapid growth occurring within a narrow optimum range. These specific pH needs reflect the organisms' adaptations to their natural environment. For example, enteric bacteria are capable of survival within a broad pH range, which is characteristic of their natural habitat, the digestive system. Bacterial blood parasites, on the other hand, can tolerate only a narrow range; the pH of the circulatory system remains fairly constant at approximately 7.4.

Despite this diversity and the fact that certain organisms can grow at extremes of the pH scale, generalities can be made. The specific range for bacteria is between 4 and 9, with the optimum being 6.5 to 7.5. Fungi (molds and yeasts) prefer an acidic environment, with optimum activities at a pH of 4 to 6.

Because a neutral or nearly neutral environment is generally advantageous to the growth of microorganisms, the pH of the laboratory medium is frequently adjusted to approximately 7. Metabolic activities of the microorganism will result in the production of wastes, such as acids from carbohydrate degradation and alkali from protein breakdown, and these will cause shifts in pH that can be detrimental to growth.

To retard this shift, chemical substances that act as **buffers** are frequently incorporated when the medium is prepared. A commonly used **buffering system** involves the addition of equimolar concentrations of K_2HPO_4, a salt of a weak base, and KH_2PO_4, a salt of a weak acid.

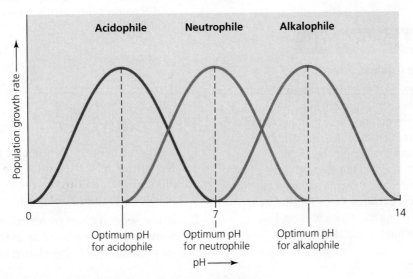

Figure 15.1 The effect of pH on the growth of microorganisms

In a medium that has become acidic, the K_2HPO_4 absorbs excess H^+ to form a weakly acidic salt and a potassium salt with the anion of the strong acid.

$$K_2HPO_4 \quad + \quad HCL \quad \rightarrow \quad KH_2PO_4 \quad + \quad KCL$$

Salt of a weak base	**Strong acid**	**Salt of a weak acid**	**Potassium chloride salt**

In a medium that has become alkaline, KH_2PO_4 releases H^+, which combines with the excess OH^- to form water, and the remaining anionic portion of the weakly acidic salt combines with the cation of the alkali.

$$KH_2PO_4 \quad + \quad KOH \quad \rightarrow \quad K_2HPO_4 \quad + \quad H_2O$$

Salt of a weak acid	**Strong base**	**Salt of a weak base**	**water**

Most media contain amino acids, peptones, and proteins, which can act as natural buffers because of their amphoteric nature. For example, amino acids are zwitterions, molecules in which the amino group and the carboxyl group ionize to form dipolar ions. These behave in the following manner:

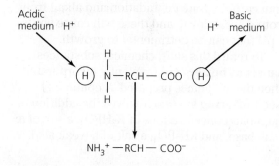

CLINICAL APPLICATION

pH as a Defense Against Infection

Most bacteria grow best at a pH between 6.5 and 7.5, and fungi show optimal growth between a pH of 4 and 6. Many microorganisms are not able to cause stomach infections because the pH of the stomach is 2.0, resembling that of hydrochloric acid. In this way, the acid of the stomach acts as a defense against infection. By the same token, the pH of the skin varies between 4 and 7, with lower ranges (around 5) being the most common, helping prevent many infections of the skin.

AT THE BENCH

Materials

Cultures

Saline suspensions of 24-hour nutrient broth cultures, adjusted to an absorbance (A) of 0.05 or equilibrated to a 0.5 McFarland Standard at a wavelength of 600 nm, of *Alcaligenes faecalis*, *Escherichia coli*, and *Saccharomyces cerevisiae*.

Media

Per designated student group: 12 Trypticase soy broth (TSB) tubes, three at each of the following pH designations: 3, 6, 7, and 9. The pH has been adjusted with 1N sodium hydroxide or 1N hydrochloric acid.

Equipment

Microincinerator or Bunsen burner, sterile 1-ml pipettes, mechanical pipetting device, Bausch & Lomb Spectronic 20 spectrophotometer, test tube rack, and glassware marking pencil.

Procedure Lab One

1. Using a sterile pipette, inoculate a series of the appropriately labeled TSB tubes of media, pH values of 3, 6, 7, and 9, with *E. coli* by adding 0.1 ml of the saline culture to each.

2. Repeat Step 1 for the inoculation of *A. faecalis* and *S. cerevisiae*, using a new sterile pipette each time.

3. Incubate the *A. faecalis* and *E. coli* cultures for 24 to 48 hours at 37°C and the *S. cerevisiae* cultures for 48 to 72 hours at 25°C.

Procedure Lab Two

1. Using the spectrophotometer as described in Experiment 14, determine the absorbance of all cultures. Record the readings in the chart provided in the Lab Report.

2. In the second chart provided in the Lab Report, summarize your results as to the overall range and optimum pH of each organism studied.

Name: _____

Date: _____   Section: _____

Observations and Results

Absorbance Readings

Microbial Species	ABSORBANCE READINGS			
	pH 3	pH 6	pH 7	pH 9
E. coli				
A. faecalis				
S. cerevisiae				

pH Summary

Microbial Species	pH Range	Optimum pH
E. coli		
A. faecalis		
S. cerevisiae		

Review Questions

1. Explain the mechanism by which buffers prevent radical shifts in pH.

2. Explain why it is necessary to incorporate buffers into media in which microorganisms are grown.

3. Why are proteins and amino acids considered to be natural buffers?

4. Explain why microorganisms differ in their pH requirements.

5. Will all microorganisms grow optimally at a neutral pH? Explain.

6. You are instructed to grow *E. coli* in a chemically defined medium containing glucose and NH_4Cl as the carbon and nitrogen sources and also in nutrient broth that contains beef extract and peptone. Both media are adjusted to a pH of 7. With turbidity as an index for the amount of growth in each of the cultures, the following spectrophotometric readings are obtained following incubation:

	ABSORBANCE READINGS	
Time (hours)	Chemically Defined Medium	Nutrient Broth Medium
6	0.100	0.100
12	0.300	0.500
18	0.275	0.900
24	0.125	1.500

Based on the above data, explain why *E. coli* ceased growing in the chemically defined medium but continued to grow in the nutrient broth.

Physical Factors: Atmospheric Oxygen Requirements

LEARNING OBJECTIVE

Once you have completed this experiment, you should be familiar with

1. The diverse atmospheric oxygen requirements of microorganisms.

Principle

Microorganisms exhibit great diversity in their ability to use free oxygen (O_2) for cellular respiration. These variations in O_2 requirements reflect the differences in biooxidative enzyme systems present in the various species. Microorganisms can be classified into one of five major groups according to their O_2 needs:

1. **Aerobes** require the presence of atmospheric oxygen for growth. Their enzyme system necessitates use of O_2 as the final hydrogen (electron) acceptor in the complete oxidative degradation of high-energy molecules, such as glucose.

2. **Microaerophiles** require limited amounts of atmospheric oxygen for growth. Oxygen in excess of the required amount appears to block the activities of their oxidative enzymes and results in death.

3. **Obligate anaerobes** require the absence of free oxygen for growth because their oxidative enzyme system requires the presence of molecules other than O_2 to act as the final hydrogen (electron) acceptor. In these organisms, as in aerobes, the presence of atmospheric oxygen results in the formation of toxic metabolic end products, such as superoxide, O_2^-, a free radical of oxygen. However, these organisms lack the enzymes superoxide dismutase

and catalase, whose function is to degrade the superoxide to water and oxygen as follows:

$$2O_2^- + 2H^+ \xrightarrow{\text{Superoxide dismutase}} H_2O_2 + O_2$$

$$2H_2O_2 \xrightarrow{\text{Catalase}} 2H_2O + O_2$$

In the absence of these enzymes, small amounts of atmospheric oxygen are lethal, and these organisms are justifiably called obligate anaerobes.

4. **Aerotolerant anaerobes** are fermentative organisms, and therefore they do not use O_2 as a final electron acceptor. Unlike the obligate anaerobes, they produce catalase and/or superoxide dismutase, and thus they are not killed by the presence of O_2. Hence, these organisms are anaerobes that are termed aerotolerant.

5. **Facultative anaerobes** can grow in the presence or absence of free oxygen. They preferentially use oxygen for aerobic respiration. However, in an oxygen-poor environment, cellular respiration may occur anaerobically, utilizing such compounds as nitrates (NO_3^-) or sulfates (SO_4^{2-}) as final hydrogen acceptors, or via a fermentative pathway (refer to Experiment 23).

The oxygen needs of microorganisms can be determined by noting their growth distributions following a **shake-tube inoculation**. This procedure requires introduction of the inoculum into a melted agar medium, shaking of the test tube to disperse the microorganisms throughout the agar, and rapid solidification of the medium to ensure that the cells remain dispersed. Following incubation, the growth distribution indicates the organisms' oxygen requirements. Aerobes exhibit surface growth, whereas anaerobic growth is limited to the bottom of the deep tube. Facultative anaerobes, because of their indifference to the presence or absence of oxygen, exhibit growth throughout the medium. Microaerophiles grow in a zone slightly below the surface. Figure 16.1 illustrates the shake-tube inoculation procedure and the distribution of growth following an appropriate incubation period.

CLINICAL APPLICATION

Differentiating Aerobes and Anaerobes

Samples suspected of containing anaerobes need to be handled carefully and transported promptly to a lab, where they are typically inoculated onto anaerobic blood agar plates and anaerobic broth, as well as onto MacConkey agar and an aerobic blood plate. Growth on aerobic or anaerobic agars will determine oxygen requirements, while comparable growth on both aerobic and anaerobic media suggests a facultative anaerobe.

AT THE BENCH

Materials

Cultures

24- to 48-hour nutrient broth cultures of *Staphylococcus aureus* BSL-2 , *Corynebacterium xerosis*, and *Enterococcus faecalis* BSL-2 ; 48- to 72-hour Sabouraud broth cultures of *Saccharomyces cerevisiae* and *Aspergillus niger*; and a 48-hour thioglycollate broth culture of *Clostridium sporogenes*.

Media

Six brain heart infusion agar deep tubes per designated student group.

Equipment

Microincinerator or Bunsen burner, waterbath, iced waterbath, thermometer, sterile Pasteur pipettes, test tube rack, and glassware marking pencil.

Procedure Lab One

1. Liquefy the sterile brain heart infusion agar by boiling in a waterbath at 100°C.
2. Cool molten agar to 45°C; check temperature with a thermometer inserted into the waterbath.

Determining Oxygen Requirements

1. Using aseptic technique, inoculate each experimental organism by introducing two drops of the culture from a sterile Pasteur pipette into the appropriately labeled tubes of molten agar.
2. Vigorously rotate the freshly inoculated molten infusion agar between the palms of the hands to distribute the organisms.
3. Place inoculated test tubes in an upright position in the iced waterbath to solidify the medium rapidly.
4. Incubate the *S. aureus* BSL-2 , *C. xerosis*, *E. faecalis* BSL-2 , and *C. sporogenes* cultures for 24 to 48 hours at 37°C, and the *A. niger* and *S. cerevisiae* cultures for 48 to 72 hours at 25°C.

Procedure Lab Two

1. Observe each of the experimental cultures for the distribution of growth in each tube.
2. Record your observations and your determination of the oxygen requirements for each of the experimental species in the chart provided in the Lab Report.

PROCEDURE

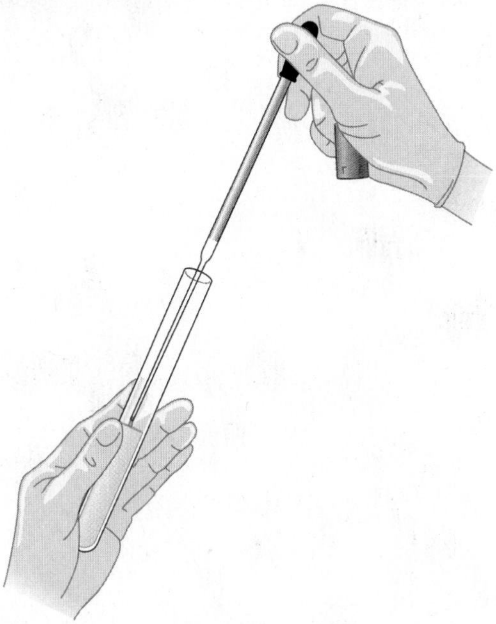

1 Transfer two drops of inoculum from the test culture into a melted agar deep tube.

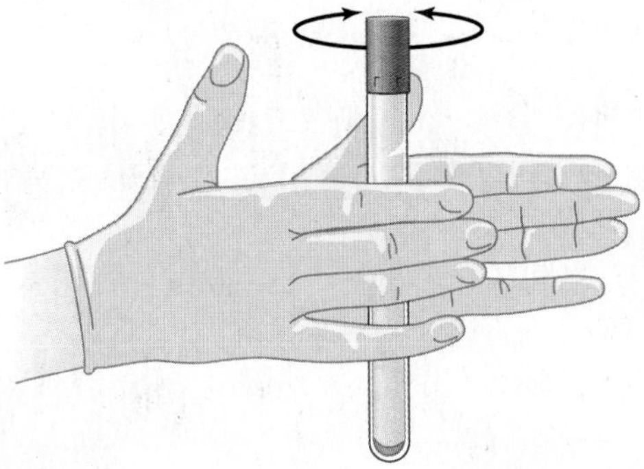

2 Disperse the organisms throughout the molten agar medium by rapidly rotating the tube between the palms of your hands.

3 Cool rapidly by immersion in an iced waterbath.

4 Incubate at 37°C.

Distribution of Growth

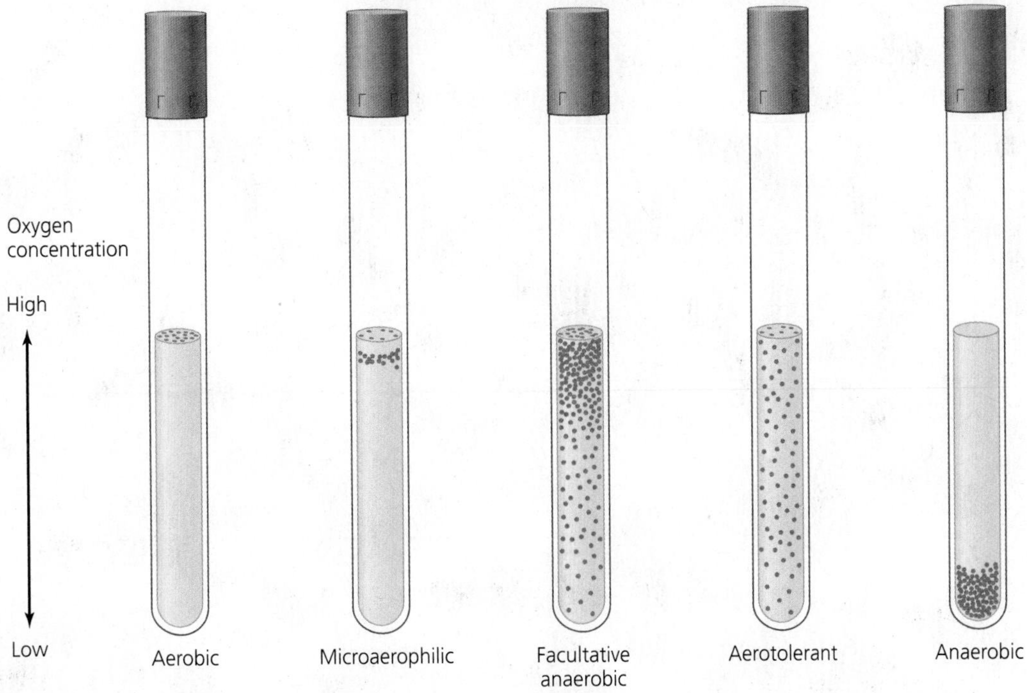

Oxygen concentration

High

Low

| Aerobic | Microaerophilic | Facultative anaerobic | Aerotolerant | Anaerobic |

Figure 16.1 **Procedure for determination of oxygen requirements**

Name: _____

Date: _____   Section: _____

Observations and Results

Species	Distribution of Growth	Classification According to Oxygen Requirement
S. aureus		
C. xerosis		
E. faecalis		
A. niger		
S. cerevisiae		
C. sporogenes		

Review Questions

1. Why is it necessary to place the inoculated molten agar cultures in an iced waterbath for rapid solidification?

2. As indicated by its oxygen requirements, which group of microorganisms has the most extensive bioenergetic enzyme system? Explain.

3. Account for the inability of aerobes to grow in the absence of O_2.

4. Account for the subsurface growth of microaerophiles in a shake-tube culture.

5. Consider the culture type in which growth was distributed throughout the entire medium and explain why the growth was more abundant toward the surface of the medium in some cultures, whereas other cultures showed an equal distribution of growth throughout the tubes.

6. Account for the fact that the *C. sporogenes* culture showed a separation within the medium or an elevation of the medium from the bottom of the test tube.

7. Your instructor asks you to explain why the *Streptococcus* species that are catalase negative are capable of growth in the presence of oxygen. How would you respond?

Techniques for the Cultivation of Anaerobic Microorganisms

LEARNING OBJECTIVE

Once you have completed this experiment, you should be familiar with

1. The methods for cultivation of anaerobic organisms.

Principle

Microorganisms differ in their abilities to use oxygen for cellular respiration. **Respiration** involves the oxidation of substrates for energy necessary to life. A substrate is **oxidized** when it loses a hydrogen ion and its electron (H^+e^-). Since the H^+e^- cannot remain free in the cell, it must immediately be picked up by an electron acceptor, which becomes reduced. Therefore, reduction means gaining the H^+e^-. These are termed **oxidation-reduction (redox)** reactions. Some microorganisms have enzyme systems in which oxygen can serve as an electron acceptor, thereby being reduced to water. These cells have high oxidation-reduction potentials; others have low potentials and must use other substances as electron acceptors.

The enzymatic differences in microorganisms are explained more fully in the section dealing with metabolism (see Part 5). This discussion is limited to cultivation of the strict anaerobes, which cannot be cultivated in the presence of atmospheric oxygen (**Figure 17.1**). The procedure is somewhat more difficult because it involves sophisticated equipment and media enriched with substances that lower the redox potential. **Figure 17.2** shows some of the methods available for anaerobic cultivation.

The following experiment uses fluid thioglycollate medium and the GasPak™ anaerobic system.

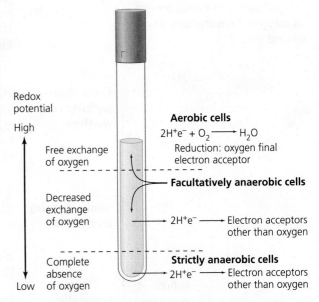

Figure 17.1 Illusvotentials in an agar deep tube

Redox potential

High

Free exchange of oxygen

Decreased exchange of oxygen

Complete absence
Low   of oxygen

Aerobic cells
$2H^+e^- + O_2 \longrightarrow H_2O$
Reduction: oxygen final electron acceptor

Facultatively anaerobic cells

$2H^+e^- \longrightarrow$ Electron acceptors other than oxygen

Strictly anaerobic cells
$2H^+e^- \longrightarrow$ Electron acceptors other than oxygen

CLINICAL APPLICATION

Oxygen as a Treatment?

The causative agent of gas gangrene, *Clostridium perfringens,* is an anaerobic bacterium that thrives in wounds deprived of circulation and oxygen and can cause limb loss and death. Treatment may involve amputation or surgical removal of infected tissue. Doctors may also prescribe therapy using enriched oxygen delivered to the patient in a hyperbaric chamber. This allows the blood to carry more oxygen to the wounds, slowing the growth of anaerobic microbes. Patients typically undergo five 90-minute sessions lying in a chamber pressurized to 2.5 atmospheres, possibly alleviating the need for surgery.

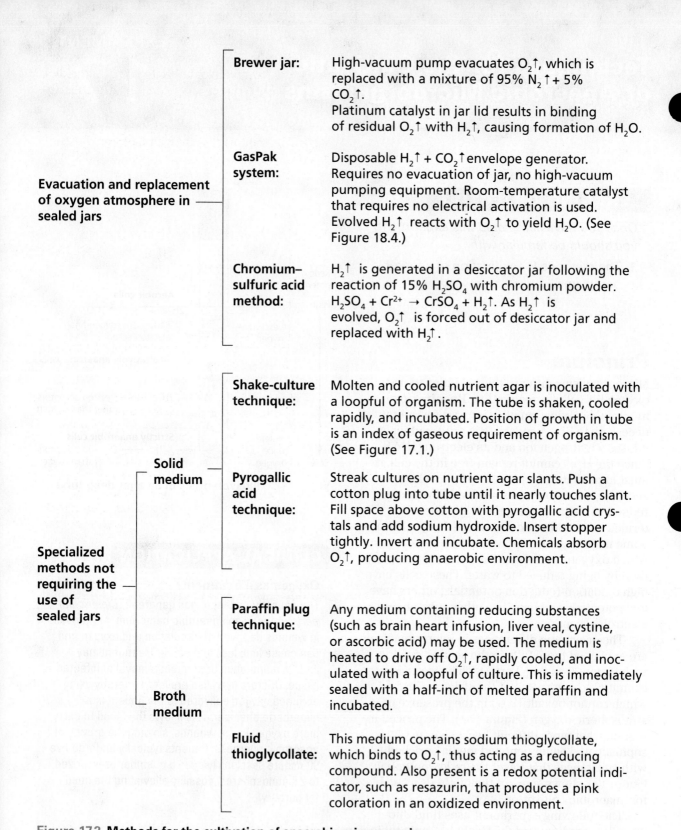

Evacuation and replacement of oxygen atmosphere in sealed jars

Brewer jar: High-vacuum pump evacuates $O_2\uparrow$, which is replaced with a mixture of 95% $N_2\uparrow$ + 5% $CO_2\uparrow$.
Platinum catalyst in jar lid results in binding of residual $O_2\uparrow$ with $H_2\uparrow$, causing formation of H_2O.

GasPak system: Disposable $H_2\uparrow$ + $CO_2\uparrow$ envelope generator. Requires no evacuation of jar, no high-vacuum pumping equipment. Room-temperature catalyst that requires no electrical activation is used. Evolved $H_2\uparrow$ reacts with $O_2\uparrow$ to yield H_2O. (See Figure 18.4.)

Chromium–sulfuric acid method: $H_2\uparrow$ is generated in a desiccator jar following the reaction of 15% H_2SO_4 with chromium powder. $H_2SO_4 + Cr^{2+} \rightarrow CrSO_4 + H_2\uparrow$. As $H_2\uparrow$ is evolved, $O_2\uparrow$ is forced out of desiccator jar and replaced with $H_2\uparrow$.

Specialized methods not requiring the use of sealed jars

Solid medium

Shake-culture technique: Molten and cooled nutrient agar is inoculated with a loopful of organism. The tube is shaken, cooled rapidly, and incubated. Position of growth in tube is an index of gaseous requirement of organism. (See Figure 17.1.)

Pyrogallic acid technique: Streak cultures on nutrient agar slants. Push a cotton plug into tube until it nearly touches slant. Fill space above cotton with pyrogallic acid crystals and add sodium hydroxide. Insert stopper tightly. Invert and incubate. Chemicals absorb $O_2\uparrow$, producing anaerobic environment.

Broth medium

Paraffin plug technique: Any medium containing reducing substances (such as brain heart infusion, liver veal, cystine, or ascorbic acid) may be used. The medium is heated to drive off $O_2\uparrow$, rapidly cooled, and inoculated with a loopful of culture. This is immediately sealed with a half-inch of melted paraffin and incubated.

Fluid thioglycollate: This medium contains sodium thioglycollate, which binds to $O_2\uparrow$, thus acting as a reducing compound. Also present is a redox potential indicator, such as resazurin, that produces a pink coloration in an oxidized environment.

Figure 17.2 **Methods for the cultivation of anaerobic microorganisms**

Materials

Cultures

24- to 48-hour nutrient broth cultures of *Bacillus cereus*, *Escherichia coli*, and *Micrococcus luteus;* and 48-hour thioglycollate broth culture of *Clostridium sporogenes*.

Media

Per designated student group: four screw-cap tubes of fluid thioglycollate medium and four nutrient agar plates.

Equipment

Microincinerator or Bunsen burner, inoculating loop, GasPak anaerobic system, test tube rack, and glassware marking pencil.

Procedure Lab One

Fluid Thioglycollate Medium

1. For the performance of this procedure, the fluid thioglycollate medium must be fresh. Freshness is indicated by the absence of a pink color in the upper one-third of the medium. If this coloration is present, loosen the screw caps and place the tubes in a boiling water bath for 10 minutes to drive off the dissolved O_2 from the medium. Cool the tubes to 45°C before inoculation.

2. Aseptically inoculate the appropriately labeled tubes of thioglycollate with their respective test organisms by means of loop inoculations *to the depths of the media.*

3. Incubate the cultures for 24 to 48 hours at 37°C.

The appearance of the growth of organisms according to their gaseous requirements in thioglycollate medium is shown in Figure 17.3.

GasPak Anaerobic Technique

The GasPak system, shown in Figure 17.4, is a contemporary method for the exclusion of oxygen from a sealed jar used for incubation of anaerobic cultures in a nonreducing medium. This system uses a GasPak generator that consists of a foil package that generates hydrogen and carbon dioxide upon the addition of water. A palladium catalyst in the lid of the jar combines the evolved hydrogen with residual oxygen to form water, thereby creating a carbon dioxide environment within the jar that is conducive for anaerobic growth. The establishment of anaerobic conditions is verified by the color change of a methylene blue indicator strip in the jar. This blue indicator becomes colorless in the absence of oxygen.

1. With a glassware marking pencil, divide the bottom of each nutrient agar plate into two sections.

(a)   (b)   (c)   (d)   (e)

Figure 17.3 Bacterial growth patterns in thioglycollate broth tubes.
(a) Uninoculated control. **(b, c)** Uniform growth indicates facultative anaerobic bacteria. **(d)** Bubbles indicate gas-producing bacteria. **(e)** Bottom growth indicates anaerobic bacteria.

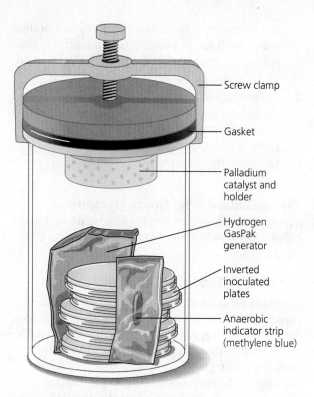

Figure 17.4 **GasPak system**

2. Label each section on two plates with the name of the organism to be inoculated.

3. Repeat Step 2 to prepare a duplicate set of cultures.

4. Using aseptic technique, make a single-line streak inoculation of each test organism in its respectively labeled section on both sets of plates.

Screw clamp

Gasket

Palladium catalyst and holder

Hydrogen GasPak generator

Inverted inoculated plates

Anaerobic indicator strip (methylene blue)

5. Tear off the corner of the hydrogen and carbon dioxide gas generator and insert this inside the GasPak jar.

6. Place one set of plate cultures in an inverted position inside the GasPak chamber.

7. Expose the anaerobic indicator strip and place it inside the anaerobic jar so that the wick is visible from the outside.

8. With a pipette, add the required 10 ml of water to the gas generator and quickly seal the chamber with its lid.

9. Place the sealed jar in an incubator at 37°C for 24 to 48 hours. After several hours of incubation, observe the indicator strip for a color change to colorless, which is indicative of anaerobic conditions.

10. Incubate the duplicate set of plates in an inverted position for 24 to 48 hours at 37°C under aerobic conditions.

Procedure Lab Two

1. Observe the fluid thioglycollate cultures, GasPak system, and aerobically incubated plate cultures for the presence of growth. Record your results in the chart provided in the Lab Report.

2. Based on your observation, record the oxygen requirement classification of each test organism as anaerobe, facultative anaerobe, or aerobe.

Serial Dilution—Agar Plate Procedure to Quantitate Viable Cells

LEARNING OBJECTIVES

Once you have completed this experiment, you should understand

1. The diverse methods used to determine the number of cells in a bacterial culture.

2. How to determine quantitatively the number of viable cells in a bacterial culture.

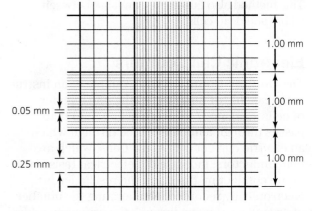

(a) Petroff-Hausser counting chamber

Principle

Studies involving the analysis of materials, including food, water, milk, and, in some cases, air require quantitative enumeration of microorganisms in the substances. Many methods have been devised to accomplish this, including direct microscopic counts, use of an electronic cell counter such as the Coulter Counter®, chemical methods for estimating cell mass or cellular constituents, turbidimetric measurements for increases in cell mass, and the serial dilution–agar plate method.

Direct Microscopic Counts

Direct microscopic counts require the use of a specialized slide called the **Petroff-Hausser counting chamber**, in which an aliquot of a eukaryotic cell suspension is counted and the total number of cells is determined mathematically. The Petroff-Hausser counting chamber is a thick glass microscope slide with a chamber 0.02 mm (1/50 mm) deep in the center. The chamber contains an etched grid and has improved Neubauer rulings (1/400 square mm). The slide and the counting chamber are illustrated in **Figure 18.1**.

The rulings cover 9 mm². The boundary lines (Neubauer rulings) are the center lines of the groups of three. The center square millimeter is ruled into groups of 16 small squares, and each group is separated by triple lines, the middle one of which is the boundary. The ruled surface is 0.02 mm below the cover glass, which makes

(b) Petroff-Hausser counting chamber grid

Figure 18.1 The Petroff-Hausser counting chamber

the volume over a square millimeter 0.02 mm³ (cubic mm). All cells are counted in this square millimeter.

The number of cells counted is calculated as follows:

number of cells per mm = number of cells counted × dilution × 50,000

[The factor of 50,000 is used in order to determine the cell count for 1 ml: 1 ml = 1000 mm³ = (50 times the chamber depth of 0.02 mm) × 1000.] Although rapid, a direct count has the disadvantages that both living and dead cells are counted and that it is not sensitive to populations of fewer than 1 million cells.

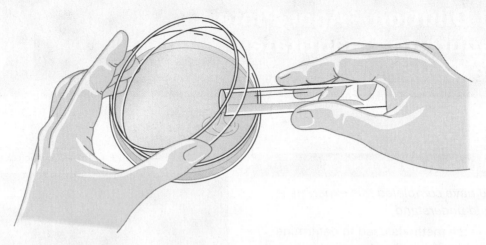

Figure 18.2 **Pour-plate technique**

Breed smears are used mainly to quantitate bacterial cells in milk. Using stained smears confined to a 1-square-millimeter ruled area of the slide, the total population is determined mathematically. This method also fails to discriminate between viable and dead cells.

Electronic Cell Counters

The **Coulter Counter** is an example of an instrument capable of rapidly counting the number of cells suspended in a conducting fluid that passes through a minute orifice through which an electric current is flowing. Cells, which are nonconductors, increase the electrical resistance of the conducting fluid, and the resistance is electronically recorded, enumerating the number of organisms flowing through the orifice. In addition to its inability to distinguish between living and dead cells, the apparatus is also unable to differentiate inert particulate matter from cellular material.

Chemical Methods

While not considered means of direct quantitative analysis, chemical methods may be used to indirectly measure increases both in protein concentration and in DNA production. In addition, cell mass can be estimated by dry weight determination of a specific aliquot of the culture. Measurement of certain metabolic parameters may also be used to quantitate bacterial populations. The amount of oxygen consumed (oxygen uptake) is directly proportional to the increasing number of vigorously growing aerobic cells, and the rate of carbon dioxide production is related to increased growth of anaerobic organisms.

Spectrophotometric Analysis

Increased turbidity in a culture is another index of growth. With turbidimetric instruments, the amount of transmitted light decreases as the cell population increases, and the decrease in radiant energy is converted to electrical energy and indicated on a galvanometer. This method is rapid but limited because sensitivity is restricted to microbial suspensions of 10 million cells or greater.

Serial Dilution–Agar Plate Analysis

While all these methods may be used to enumerate the number of cells in a bacterial culture, the major disadvantage common to all is that the total count includes dead as well as living cells. Sanitary and medical microbiology at times require determination of viable cells. To accomplish this, the serial dilution–agar plate technique is used. Briefly, this method involves serial dilution of a bacterial suspension in sterile water blanks, which serve as a diluent of known volume. Once diluted, the suspensions are placed on suitable nutrient media. The **pour-plate technique**, illustrated in Figure 18.2, is the procedure usually employed. Molten agar, cooled to 45°C, is poured into a Petri dish containing a specified amount of the diluted sample. Following addition of the molten-then-cooled agar, the cover is replaced, and the plate is gently rotated in a circular motion to achieve uniform distribution of microorganisms. This procedure is repeated for all dilutions to be plated. Dilutions should be plated in duplicate for greater accuracy, incubated overnight, and counted on a **Quebec colony counter** either by hand or by an electronically modified version of this instrument. Figure 18.3 is an illustration of this apparatus for counting colonies.

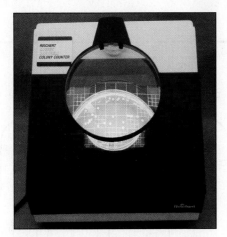

Figure 18.3 Quebec colony counter for the enumeration of bacterial colonies

Plates suitable for counting must contain not fewer than 30 nor more than 300 colonies. See Figure 18.4. The total count of the suspension is obtained by multiplying the number of cells per plate by the dilution factor, which is the reciprocal of the dilution.

Advantages of the serial dilution–agar plate technique are as follows:

1. Only viable cells are counted.

2. It allows isolation of discrete colonies that can be subcultured into pure cultures, which may then be easily studied and identified.

Disadvantages of this method are as follows:

1. Overnight incubation is necessary before colonies develop on the agar surface.

2. More glassware is used in this procedure.

3. The need for greater manipulation may result in erroneous counts due to errors in dilution or plating.

The following experiment uses the pour-plate technique for plating serially diluted culture samples.

CLINICAL APPLICATION

The Multiple Uses of Cell Counts

Determining how many cells are present in a sample is of particular importance in the food and dairy industries, which monitor the number and types of bacteria in their products. Elevated bacteria counts can indicate a sick animal, inadequate sanitation, or improper storage. Viable cell counts are also used in water treatment facilities as well as in wineries and breweries, where the number of yeast cells is monitored. In medical laboratories, sometimes the number of cells and growth rates are used to determine antimicrobial sensitivity as well as the course of infection.

AT THE BENCH

Materials

Culture

24- to 48-hour nutrient broth culture of *Escherichia coli*.

Media

Per designated student group: six 20-ml nutrient agar deep tubes and seven sterile 0.9-ml water blanks.

Equipment

Hot plate, waterbath, thermometer, test tube rack, microincinerator or Bunsen burner, vortex mixer, micropipette tips, mechanical pipetting device, sterile Petri dishes, Quebec colony counter, manual hand counter, disinfectant solution in a 500-ml beaker, glassware marking pencil, turntable, bent glass rod, and beaker with 95% alcohol.

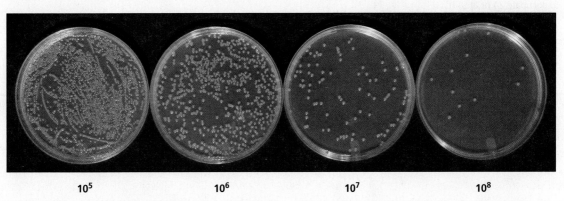

| 10^5 | 10^6 | 10^7 | 10^8 |

Figure 18.4 Agar plating method for viable cell counts using dilutions 1×10^5, 1×10^6, 1×10^7, and 1×10^8

Transfer 1 ml with pipette 1. Transfer 1 ml with pipette 2. Transfer 1 ml with pipette 3. Transfer 1 ml with pipette 4. Transfer 1 ml with pipette 5. Transfer 1 ml with pipette 6. Transfer 1 ml with pipette 7.

Tube 1 — E. coli culure
Tube 2 — H_2O 9 ml
Tube 3 — H_2O 9 ml
Tube 4 — H_2O 9 ml
Tube 5 — H_2O 9 ml
Tube 6 — H_2O 9 ml
Tube 7 — H_2O 9 ml
Tube 8 — H_2O 9 ml

* 1. Dilutions

10^{-1} 10^{-2} 10^{-3} 10^{-4} 10^{-5} 10^{-6} 10^{-7}

Transfer 0.1 ml with pipette 5. Transfer 1.0 ml with pipette 6. Transfer 0.1 ml with pipette 6. Transfer 1.0 ml with pipette 7. Transfer 0.1 ml with pipette 7. Transfer 1.0 ml with pipette 8.

2. Addition of sample of suspension to plates

3. Pour nutrient agar, 45°C, into each plate. Mix by rotation of plate for the serial dilution–agar plate method or use the spread-plate method.

1A — 0.1 ml — 10^5
1B — 1.0 ml — 10^5
2A — 0.1 ml — 10^6
2B — 1.0 ml — 10^6
3A — 0.1 ml — 10^7
3B — 1.0 ml — 10^7

** 4. Dilution factor

5. Incubate 24 hr at 37°C.

6. Enumerate using Quebec colony counter.

* **Dilution** refers to varying the concentration of the substance.
** **Dilution factor** is expressed mathematically as the reciprocal of the dilution. For example, a dilution of 10^{-3} has a dilution factor of 10^3.

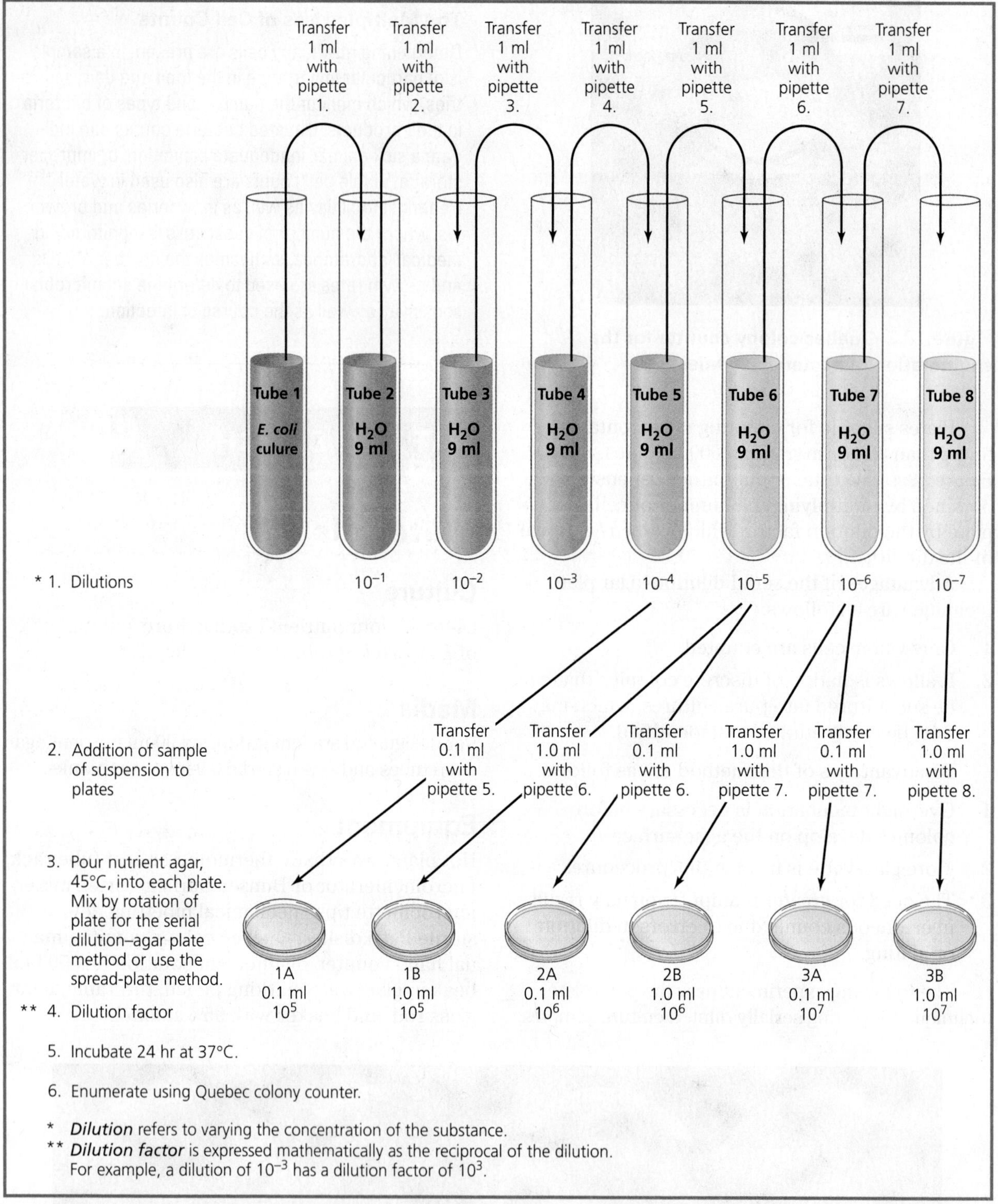

Figure 18.5 **Serial dilution–agar plate procedure**

Procedure Lab One

The following **pour-plate** procedure is illustrated in Figure 18.5. A photograph of the dilutions is shown in Figure 18.6.

1. Liquefy six agar deep tubes in an autoclave or by boiling. Cool the molten agar tubes and maintain in a water bath at 45°C.

2. Label the *E. coli* culture tube with the number 1 and the seven 9-ml water blanks as numbers 2 through 8. Place the labeled tubes in a test tube rack. Label the Petri dishes 1A, 1B, 2A, 2B, 3A, and 3B.

3. Mix the *E. coli* culture (Tube 1) by rolling the tube between the palms of your hands to ensure even dispersal of cells in the culture.

4. With a sterile pipette, aseptically transfer 1 ml from the bacterial suspension, Tube 1, to water blank Tube 2. Discard the pipette in the beaker of disinfectant. The culture has been diluted 10 times to 10^{-1}.

5. Mix Tube 2 and, with a fresh pipette, transfer 1 ml from Tube 2 to Tube 3. Discard the pipette. The culture has been diluted 100 times to 10^{-2}.

6. Mix Tube 3 and, with a fresh pipette, transfer 1 ml from Tube 3 to Tube 4. Discard the pipette. The culture has been diluted 1000 times to 10^{-3}.

7. Mix Tube 4 and, with a fresh pipette, transfer 1 ml from Tube 4 to Tube 5. Discard the pipette. The culture has been diluted 10,000 times to 10^{-4}.

8. Mix Tube 5 and, with a fresh pipette, transfer 0.1 ml of this suspension from Tube 5 to Plate 1A. Return the pipette to Tube 5 and transfer 1 ml from Tube 5 to Tube 6. Discard the pipette. The culture has been diluted 100,000 times to 10^{-5}.

9. Mix Tube 6 and, with a fresh pipette, transfer 1 ml of this suspension from Tube 6 to Plate 1B. Return the pipette to Tube 6 and transfer 0.1 ml from Tube 6 to Plate 2A. Return the pipette to Tube 6 and transfer 1 ml from Tube 6 to Tube 7. Discard the pipette. The culture has been diluted 1,000,000 times to 10^{-6}.

10. Mix Tube 7 and, with a fresh pipette, transfer 1 ml of this suspension from Tube 7 to Plate 2B. Return the pipette to Tube 7 and transfer 0.1 ml to from Tube 7 Plate 3A. Return the pipette to Tube 7 and transfer 1 ml from Tube 7 to Tube 8. Discard the pipette. The culture has been diluted 10,000,000 times to 10^{-7}.

11. Mix Tube 8 and, with a fresh pipette, transfer 1 ml of this suspension from Tube 8 to Plate 3B. Discard the pipette. The dilution procedure is now complete.

12. Check the temperature of the molten agar medium to be sure the temperature is 45°C. Remove a tube from the waterbath and wipe the outside surface dry with a paper towel. Using the pour-plate technique, pour the agar into Plate 1A as shown in Figure 18.2 and rotate the plate gently to ensure uniform distribution of the cells in the medium.

13. Repeat Step 12 for the addition of molten nutrient agar to Plates 1B, 2A, 2B, 3A, and 3B.

14. Once the agar has solidified, incubate the plates in an inverted position for 24 hours at 37°C.

The **spread-plate** technique requires that a previously diluted mixture of microorganisms be used. During inoculation, the cells are spread over the surface of a solid agar medium with a sterile, L-shaped bent glass rod while the Petri dish is

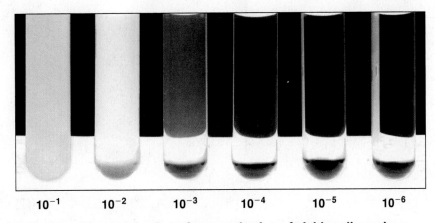

Figure 18.6 Serial dilution of bacterial culture for quantitation of viable cell numbers

spun on a lazy Susan/turntable. The step-by-step procedure for this technique is as follows:

1. Prepare bacterial suspensions as described above and label agar plates accordingly.

2. Place the bent glass rod into a beaker and add a sufficient amount of 95% ethyl alcohol to cover the lower, bent portion.

3. Place an appropriately labeled nutrient agar plate on the turntable. With a sterile pipette, place 0.1 ml of bacterial suspension on the center of the plate.

4. Remove the glass rod from the beaker, and pass it through the Bunsen burner flame with the bent portion of the rod pointing downward to prevent the burning alcohol from running down your arm. Allow the alcohol to burn off the rod completely. Cool the rod for 10 to 15 seconds.

5. Remove the Petri dish cover and spin the turntable.

6. While the turntable is spinning, lightly touch the sterile bent rod to the surface of the agar and move it back and forth. This will spread the culture over the agar surface.

7. When the turntable comes to a stop, replace the cover. Immerse the rod in alcohol and reflame.

8. In the absence of a turntable, turn the Petri dish manually and spread the culture with the sterile bent glass rod.

Procedure Lab Two

1. Using a Quebec colony counter and a mechanical hand counter, observe all colonies on plates. Statistically valid plate counts are only obtained from bacterial cell dilutions that yield between 30 and 300 colonies. Plates with more than 300 colonies cannot be counted and are designated as **too numerous to count— TNTC**; plates with fewer than 30 colonies are designated as **too few to count—TFTC**. Count only plates containing between 30 and 300 colonies. Remember to count all subsurface as well as surface colonies.

2. The number of organisms per ml of original culture is calculated by multiplying the number of colonies counted by the dilution factor:

$$\text{number of cells per ml} = \text{number of colonies} \times \text{dilution factor}$$

Examples:

a. Colonies per plate = 50

 Dilution factor = $1{:}1 \times 10^6$ (1:1,000,000)

 Volume of dilution added to plate = 1 ml

 $50 \times 1{,}000{,}000 = 50{,}000{,}000$ or

 (5×10^7) CFUs/ml

 (colony − forming units)

b. Colonies per plate = 50

 Dilution factor = $1{:}1 \times 10^5$ (1:1,00,000)

 Volume of dilution added to plate = 0.1 ml

 $50 \times 100{,}000 = 50{,}000{,}000$ (5×10^6)

 cells/0.1 ml

 $5{,}000{,}000 \times 10 = 50{,}000{,}000$

 (5×10^7) CFUs/ml

3. Record your observations and calculated bacterial counts per ml of sample in the Lab Report.

4. Since the dilutions plated are replicates of each other, determine the average of the duplicate bacterial counts per ml of sample and record in the chart provided in the Lab Report.

Name: _____

Date: _____   Section: _____

Observations and Results

Plate	Dilution Factor	ml of Dilution Plated	Final Dilution on Plate	Number of Colonies	Bacterial Count per ml of Sample (CFU/ml)	Average Count per ml of Sample (CFU/ml)
1A						
1B						
2A						
2B						
3A						
3B						

Review Questions

1. What is the major disadvantage of microbial counts performed by methods other than the serial dilution–agar plate procedure?

2. Distinguish between dilution and dilution factor.

3. What are the advantages and disadvantages of the serial dilution–agar plate procedure?

4. If 0.1 ml of a 1×10^{-6} dilution plate contains 56 colonies, calculate the number of cells per ml of the original culture.

5. How would you record your observation of a plate containing 305 colonies? A plate with 15 colonies?

6. Explain the chemical methods for measuring cell growth.

7. Your instructor asks you to determine the number of organisms in a water sample. Observation of your dilution plates reveals the presence of spreading colonial forms on some of the culture plates. What is the rationale for the elimination of these plate counts from your experimental data?

The Bacterial Growth Curve

LEARNING OBJECTIVES

Once you have completed this experiment, you should be able to

1. Understand the population growth dynamics of bacterial cultures.

2. Plot a bacterial growth curve.

3. Determine the generation time of a bacterial culture from the bacterial growth curve.

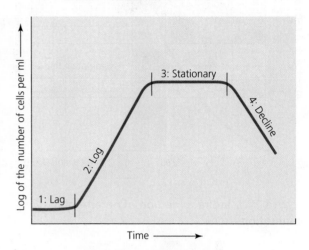

Figure 19.1 **Population growth curve**

Principle

Bacterial population growth studies require inoculation of viable cells into a sterile broth medium and incubation of the culture under optimum temperature, pH, and gaseous conditions. Under these conditions, the cells will reproduce rapidly and the dynamics of the microbial growth can be charted in a population growth curve, which is constructed by plotting the increase in cell numbers versus time of incubation. The curve can be used to delineate stages of the growth cycle. It also facilitates measurement of cell numbers and the rate of growth of a particular organism under standardized conditions as expressed by its **generation time**, the time required for a microbial population to double. The stages of a typical growth curve (**Figure 19.1**) are as follows:

1. **Lag phase:** During this stage, the cells are adjusting to their new environment. Cellular metabolism is accelerated, resulting in rapid biosynthesis of cellular macromolecules, primarily enzymes, in preparation for the next phase of the cycle. Although the cells are increasing in size, there is no cell division and therefore no increase in numbers.

2. **Logarithmic (log) phase:** Under optimum nutritional and physical conditions, the physiologically robust cells reproduce at a uniform and rapid rate by binary fission. Thus, there

is a rapid exponential increase in population, which doubles regularly until a maximum number of cells is reached. The time required for the population to double is the generation time. The length of the log phase varies, depending on the organisms and the composition of the medium. The average may be estimated to last 6 to 12 hours.

3. **Stationary phase:** During this stage, the number of cells undergoing division is equal to the number of cells that are dying. Therefore, there is no further increase in cell number, and the population is maintained at its maximum level for a period of time. The primary factors responsible for this phase are the depletion of some essential metabolites and the accumulation of toxic acidic or alkaline end products in the medium.

4. **Decline, or death, phase:** Because of the continuing depletion of nutrients and buildup of metabolic wastes, the microorganisms die at a rapid and uniform rate. The decrease in population closely parallels its increase during the log phase. Theoretically, the entire population should die during a time interval equal to that of the log phase. This does not occur, however, since a small number of highly resistant organisms persist for an indeterminate length of time.

139

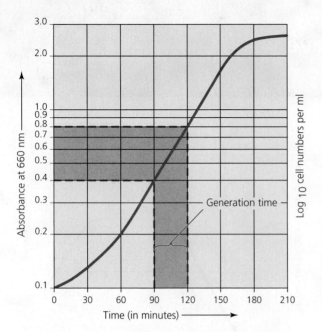

Figure 19.2 Indirect method of determining generation time

Construction of a complete bacterial growth curve requires that aliquots of a 24-hour shake-flask culture be measured for population size at intervals during the incubation period. Such a procedure does not lend itself to a regular laboratory session. Therefore, this experiment follows a modified procedure designed to demonstrate only the lag and log phases. The curve will be plotted on semilog paper by using two values for the measurement of growth. The direct method requires enumeration of viable cells in serially diluted samples of the test culture taken at 30-minute intervals as described in Experiment 18. The indirect method uses spectrophotometric measurement of the developing turbidity at the same 30-minute intervals, as an index of increasing cellular mass.

You will determine generation time with indirect and direct methods by using data on the growth curve. Indirect determination is made by simple extrapolation from the log phase as illustrated in Figure 19.2. Select two points on the absorbance scale that represent a doubling of turbidity, such as 0.2 and 0.4. Using a ruler, extrapolate by drawing a line between each of the selected absorbances on the ordinate (y-axis) and the plotted line of the growth curve. Then draw perpendicular lines from these endpoints on the plotted line of the growth curve to their respective time intervals on the abscissa (x-axis). With this information, determine the generation time (GT) as follows:

$$GT = t_{(A\ 0.4)} - t_{(A\ 0.2)}$$
$$GT = 90 \text{ minutes} - 60 \text{ minutes} = 30 \text{ minutes}$$

The generation time may be calculated directly using the log of cell numbers scale on a growth curve. The following example uses information from a hypothetical growth curve to calculate the generation time directly.

C_O = number of cells at time zero

C_E = number of cells at end of a specified time (t)

N = number of generations (doublings)

To describe logarithmic growth, the following equation is used:

$$N = (\log C_E - \log C_O)/\log 2$$

Using this formula, the logarithmic tables to the base 10, and the following supplied information, we may now solve for the generation time:

C_E = 52,000,000 cells	$\log C_E$ = 7.7218
C_O = 25,000 cells	$\log C_O$ = 4.4048
	$\log 2$ = 0.301

$$N = (7.7218 - 4.4048)/0.301 = 11 \text{ generations}$$

$$\text{generation time (GT)} = \frac{\text{the specified time (t)}}{\text{number of generations (N)}}$$

$$t = 180 \text{ minutes}$$
$$GT = 180/11 = 16 \text{ minutes}$$

CLINICAL APPLICATION

Using Growth Curves to Determine Antimicrobial Resistance

In medical laboratories, growth curves are being mathematically modeled to quickly determine antimicrobial susceptibility. By monitoring turbidity in a series of wells, each containing a test bacterium and a dilution of an antimicrobial agent, the entire growth curve of the bacterium can be determined from early measurements, greatly speeding up the testing process for drugs as well as assessing newly resistant bacteria.

AT THE BENCH

Materials

Cultures

5- to 10-hour (log phase) brain heart infusion broth culture of *Escherichia coli* with A of 0.08 to 0.10 or equilibrated to a 0.5 McFarland Standard at 600 nm.

Media

Per designated student group: 100 ml of brain heart infusion in a 250-ml Erlenmeyer flask; 42 sterile 9-ml water blanks, and 24 nutrient agar plates.

Equipment

37°C waterbath shaker incubator, Bausch & Lomb Spectronic 20 spectrophotometer, 13×100-mm cuvettes, Quebec colony counter, 1-ml sterile pipettes, mechanical pipetting device, glassware marking pencil, 1000-ml beaker, L-shaped bent glass rod, 95% ethyl alcohol, and microincinerator or Bunsen burner.

Procedure Lab One

1. Separate the 42 sterile 9-ml water blanks into six sets of seven water blanks each. Label each set as to time of inoculation (t_0, t_{30}, t_{60}, t_{90}, t_{120}, t_{150}) and the dilution to be effected in each water blank (10^{-1}, 10^{-2}, 10^{-3}, 10^{-4}, 10^{-5}, $10^{-6, 10-7}$).

2. Label six sets of four nutrient agar plates as to time of inoculation and dilution to be plated (10^{-4}, 10^{-5}, 10^{-6}, 10^{-7}).

3. With a sterile pipette, add approximately 5 ml of the log phase *E. coli* culture to the flask containing 100 ml of brain heart infusion broth. The approximate initial A (t_0) should be 0.08 to 0.1 at 600 nm. Refer to Experiment 12 for proper use of the spectrophotometer.

4. After the t_0 A has been determined, shake the culture flask and aseptically transfer 1 ml to the 9-ml water blank labeled t_0 10^{-1} and continue to dilute serially to 10^{-2} through 10^{-7}.

Note: A new pipette must be used for each subsequent dilution.

5. Place the culture flask in a waterbath shaker set at 120 rpm at 37°C, and time for the required 30-minute intervals.

6. Place 1 ml of bacterial suspension from tubes labeled 10-4, 10-5, 10-6,and 10-7 on appropriately labeled nutrient agar plates. Use sterile L-shaped rod to spread bacteria and allow plates to dry, covered with lid, for 15 minutes. Refer to Experiment 18 for proper spread plate techniques.

7. Thereafter, at each 30-minute interval, shake and aseptically transfer a 5-ml aliquot of the culture to a cuvette and determine its absorbance. Also, aseptically transfer a 1-ml aliquot of the culture into the 10^{-1} water blank of the set labeled with the appropriate time, complete the serial dilution, and plate in the respectively labeled Petri dishes as shown in Figure 19.3. *Note: A new pipette must be used for each subsequent dilution.*

8. Incubate plates in an inverted position for 24 hours at 37°C.

Procedure Lab Two

1. Perform cell counts on all plates as described in Experiment 18. Cell counts are often referred to as colony-forming units (CFUs) because each single cell in the plate becomes visible as a colony, which can then be counted.

2. Record your results in the Lab Report.

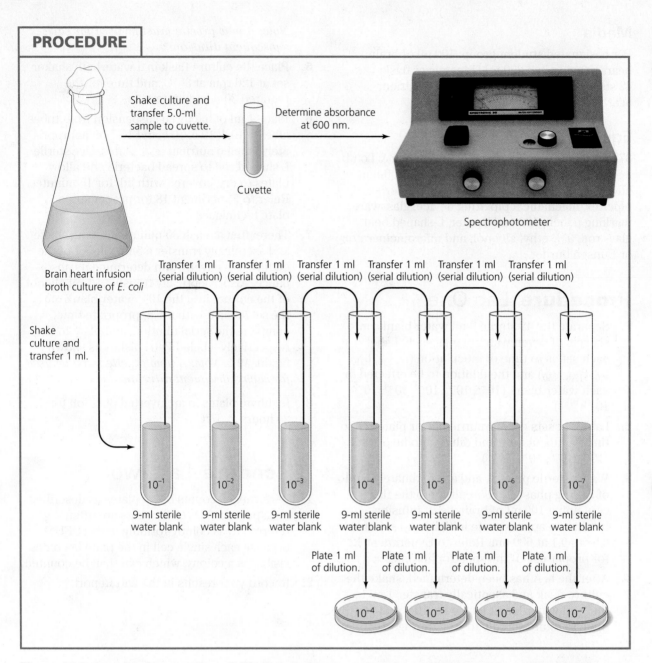

Shake culture and transfer 5.0-ml sample to cuvette.

Determine absorbance at 600 nm.

Cuvette

Spectrophotometer

Brain heart infusion broth culture of *E. coli*

Shake culture and transfer 1 ml.

Transfer 1 ml (serial dilution)

Transfer 1 ml (serial dilution)

Transfer 1 ml (serial dilution)

Transfer 1 ml (serial dilution)

Transfer 1 ml (serial dilution)

Transfer 1 ml (serial dilution)

10^{-1}

10^{-2}

10^{-3}

10^{-4}

10^{-5}

10^{-6}

10^{-7}

9-ml sterile water blank

9-ml sterile water blank

9-ml sterile water blank

9-ml sterile water blank

9-ml sterile water blank

9-ml sterile water blank

9-ml sterile water blank

Plate 1 ml of dilution.

Plate 1 ml of dilution.

Plate 1 ml of dilution.

Plate 1 ml of dilution.

10^{-4}

10^{-5}

10^{-6}

10^{-7}

Figure 19.3 **Spectrophotometric and dilution-plating procedure for use in bacterial growth curves**

Name: _____

Date: _____   Section: _____

Observations and Results

1. Record the absorbances and corresponding cell counts in the chart below.

Incubation Time (minutes)	Absorbance at 600 nm	Plate Counts (CFU/ml)	Log of CFU/ml
0			
30			
60			
90			
120			
150			

2. On the semilog paper provided on pages 145 and 146:

    a. Plot a curve relating the absorbances on the ordinate versus incubation time on the abscissa as shown in Figure 19.2.

    b. Plot a population curve with the log of the viable cells/ml on the ordinate and the incubation time on the abscissa. On both graphs, use a ruler to draw the best line connecting the plotted points. The straight-line portion of the curve represents the log phase.

3. Calculate the generation time for this culture by the direct method (using the mathematical formula) and by the indirect method (extrapolating from the A scale on the plotted curve). Show calculations, and record the generation time.

    a. Direct method:

    b. Indirect method:

Review Questions

1. Does the term *growth* convey the same meaning when applied to bacteria and to multicellular organisms? Explain.

2. Why do variations in generation time exist:

    a. Among different species of microorganisms?

    b. Within a single microbial species?

3. The generation time and growth rate of an organism grown in the laboratory can be easily determined by constructing a typical growth curve.

    a. Would you expect the growth rate of the infectious organisms found in an abscess that developed from a wound to mimic the growth curve obtained in the laboratory? Explain.

    b. Would you expect antibiotic therapy to be effective without any other concurrent treatment of the abscess?

4. Is generation time a useful parameter to indicate the types of media best suited to support the growth of a specific organism? Explain.

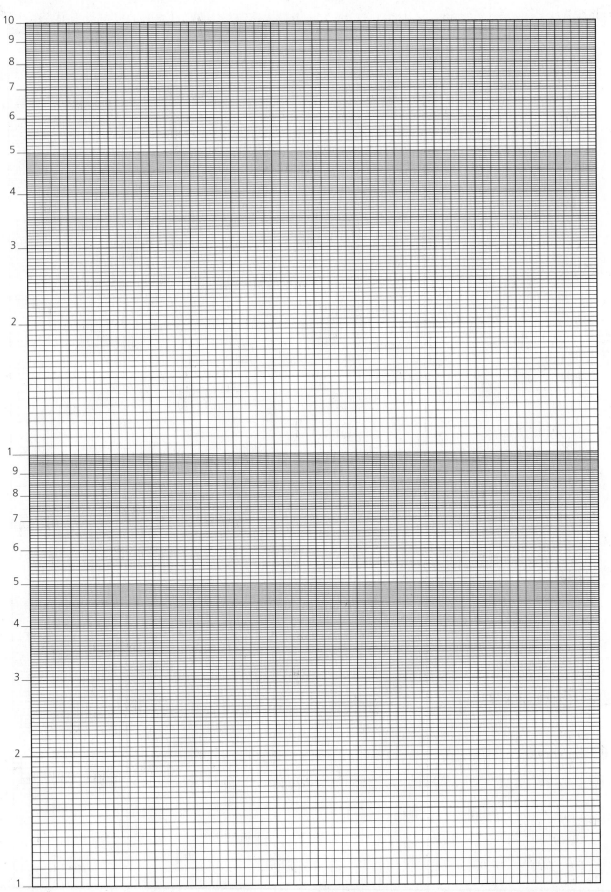

Absorbance versus Incubation Time

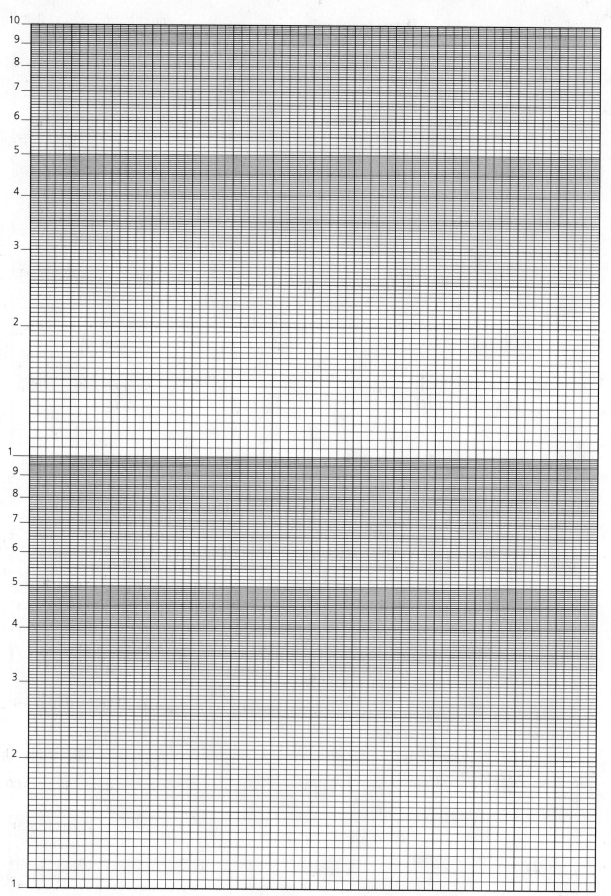

Colony-Forming Units (CFUs) versus Incubation Time

PART 5

Biochemical Activities of Microorganisms

When you have completed the experiments in this section, you should understand

1. The nature and activities of exoenzymes and endoenzymes.
2. Experimental procedures for differentiation of enteric microorganisms.
3. Biochemical test procedures for identification of microorganisms.

Introduction

Microorganisms must be separated and identified for a wide variety of reasons, including

1. Determination of pathogens responsible for infectious diseases.
2. Selection and isolation of strains of fermentative microorganisms necessary for the industrial production of alcohols, solvents, vitamins, organic acids, antibiotics, and industrial enzymes.
3. Isolation and development of suitable microbial strains necessary for the manufacture and the enhancement of quality and flavor in certain food materials, including yogurt, cheeses, and other milk products.
4. Comparison of biochemical activities for taxonomic purposes.

To accomplish these tasks, the microbiologist utilized the fact that microorganisms all have their own identifying biochemical characteristics. These so-called biochemical fingerprints are the properties controlled by the cells' enzymatic activity, and they are responsible for bioenergetics, biosynthesis, and biodegradation.

The sum of all these chemical reactions is defined as **cellular metabolism**, and the biochemical transformations that occur both outside and inside the cell are governed by biological catalysts called **enzymes**.

Extracellular Enzymes (Exoenzymes)

Exoenzymes act on substances outside of the cell. Most high-molecular-weight substances are not able to pass through cell membranes, and therefore these raw materials—food-related substances, including polysaccharides, lipids, and proteins—must be degraded to low-molecular-weight materials—nutrients—before they can be transported into the cell. Because of the reactions involved, exoenzymes are mainly **hydrolytic enzymes** that reduce high-molecular-weight materials into their building blocks by introducing water into the molecule. This liberates smaller molecules, which may then be transported into the cell and assimilated.

Intracellular Enzymes (Endoenzymes)

Endoenzymes function inside the cell and are mainly responsible for synthesis of new protoplasmic requirements and production of cellular energy from assimilated materials. The ability of cells to act on nutritional substrates permeating cell membranes indicates the presence of many endoenzymes capable of transforming the chemically specific substrates into essential materials.

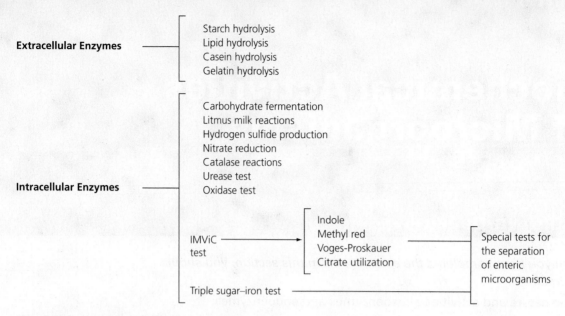

Figure P5.1 Biochemical activities of microorganisms

This transformation is necessary for cellular survival and function, and it is the basis of cellular metabolism. As a result of these metabolic processes, metabolic products are formed and excreted by the cell into the environment. Assay of these end products not only aids in identification of specific enzyme systems but also serves to identify, separate, and classify microorganisms. Figure P5.1 represents a simplified schema of experimental procedures used to acquaint students with the intracellular and extracellular enzymatic activities of microorganisms.

The experiments you will carry out in this section can be performed in either of two ways. A short version uses a limited number of organisms to illustrate the possible end product(s) that may result from enzyme action on a substrate. The organisms for this version are designated in the individual exercises.

The alternative, or long, version involves the use of 13 microorganisms. This version provides a complete overview of the biochemical fingerprints of the organisms and supplies the format for their separation and identification. These organisms were chosen to serve as a basis for identification of an unknown microorganism in Experiment 31.

If this alternative version is selected, the following organisms are recommended for use:

Escherichia coli

Enterobacter aerogenes

Klebsiella pneumoniae `BSL-2`

Shigella dysenteriae `BSL-2`

Salmonella typhimurium `BSL-2`

Proteus vulgaris

Pseudomonas aeruginosa `BSL-2`

Alcaligenes faecalis

Micrococcus luteus

Lactococcus lactis

Staphylococcus aureus `BSL-2`

Bacillus cereus

Corynebacterium xerosis

Bacteria have been designated as "BSL-2" level organisms using guidelines published by the American Society for Microbiology (www.asm.org) and the Center for Disease Control and Prevention (www.cdc.gov) covering organisms to be utilized in an undergraduate teaching laboratory.

Extracellular Enzymatic Activities of Microorganisms

LEARNING OBJECTIVES

Once you have completed this experiment, you should be able to

1. Understand the function of microbial extracellular enzymes.

2. Determine the ability of microorganisms to excrete hydrolytic extracellular enzymes capable of degrading the polysaccharide starch, the lipid tributyrin, and the proteins casein and gelatin.

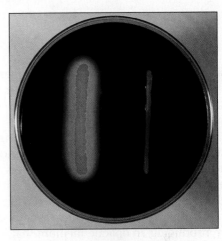

Figure 20.1 Starch agar plate. Starch hydrolysis on left; no starch hydrolysis on right.

Principle

Because of their large sizes, high-molecular-weight nutrients such as polysaccharides, lipids, and proteins are not capable of permeating the cell membrane. These macromolecules must first be hydrolyzed by specific extracellular enzymes into their respective basic building blocks. These low-molecular-weight substances can then be transported into the cells and used for the synthesis of protoplasmic requirements and energy production. The following procedures are designed to investigate the exoenzymatic activities of different microorganisms.

Starch Hydrolysis

Starch is a high-molecular-weight, branching polymer composed of **glucose** molecules linked together by **glycosidic bonds**. The degradation of this macromolecule first requires the presence of the extracellular enzyme **amylase** for its hydrolysis into shorter polysaccharides, namely **dextrins**, and ultimately into **maltose** molecules. The final hydrolysis of this disaccharide, which is catalyzed by **maltase**, yields low-molecular-weight, soluble **glucose** molecules that can be transported into the cell and used for energy production through the process of glycolysis.

In this experimental procedure, starch agar is used to demonstrate the hydrolytic activities

of these exoenzymes. The medium is composed of nutrient agar supplemented with starch, which serves as the polysaccharide substrate. The detection of the hydrolytic activity following the growth period is made by performing the starch test to determine the presence or absence of starch in the medium. Starch in the presence of iodine will impart a blue-black color to the medium, indicating the absence of starch-splitting enzymes and representing a negative result. If the starch has been hydrolyzed, a clear zone of hydrolysis will surround the growth of the organism. This is a positive result. Positive and negative results are shown in **Figure 20.1**.

Lipid Hydrolysis

Lipids are high-molecular-weight compounds possessing large amounts of energy. The degradation of lipids such as **triglycerides** is accomplished by extracellular hydrolyzing enzymes, called **lipases** (esterases), that cleave the **ester bonds** in this molecule by the addition of water to form the building blocks **glycerol** (an alcohol) and **fatty acids**. **Figure 20.2** shows this reaction. Once assimilated into the cell, these basic components can be further metabolized through aerobic respiration to produce cellular energy,

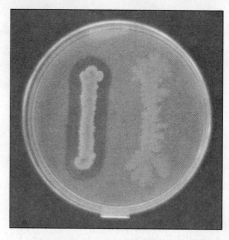

Figure 20.2 **Lipid hydrolysis**

adenosine triphosphate (ATP). The components may also enter other metabolic pathways for the synthesis of other cellular protoplasmic requirements.

In this experimental procedure, tributyrin agar is used to demonstrate the hydrolytic activities of the exoenzyme lipase. The medium is composed of nutrient agar supplemented with the triglyceride tributyrin as the lipid substrate. Tributyrin forms an emulsion when dispersed in the agar, producing an opaque medium that is necessary for observing exoenzymatic activity.

Following inoculation and incubation of the agar plate cultures, organisms excreting lipase will show a zone of **lipolysis**, which is demonstrated by a clear area surrounding the bacterial growth. This loss of opacity is the result of the hydrolytic reaction yielding soluble glycerol and fatty acids and represents a positive reaction for lipid hydrolysis. In the absence of lipolytic enzymes, the medium retains its opacity. This is a negative reaction. Positive and negative results are shown in Figure 20.3.

Casein Hydrolysis

Casein, the major milk protein, is a macromolecule composed of **amino acid** subunits linked together by **peptide bonds** (CO—NH). Before their assimilation into the cell, proteins must undergo step-by-step degradation into **peptones**, **polypeptides**, **dipeptides**, and ultimately into their building blocks, **amino acids**. This process is called peptonization, or **proteolysis**, and it is mediated by extracellular enzymes called **proteases**. The function of these proteases is to cleave the peptide bond CO–NH by introducing water into the molecule. The reaction then liberates the amino acids, as illustrated in Figure 20.4.

The low-molecular-weight soluble amino acids can now be transported through the cell membrane into the intracellular amino acid pool for use in the synthesis of structural and functional cellular proteins.

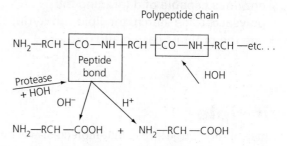

Figure 20.3 **Tributyrin agar plate.** Lipid hydrolysis on left; no lipid hydrolysis on right.

Figure 20.4 **Protein hydrolysis**

In this experimental procedure, milk agar is used to demonstrate the hydrolytic activity of these exoenzymes. The medium is composed of nutrient agar supplemented with milk that contains the protein substrate casein. Similar to other proteins, milk protein is a colloidal suspension that gives the medium its color and opacity because it deflects light rays rather than transmitting them.

Following inoculation and incubation of the agar plate cultures, organisms secreting proteases will exhibit a zone of proteolysis, which is demonstrated by a clear area surrounding the bacterial growth. This loss of opacity is the result of a hydrolytic reaction yielding soluble, noncolloidal amino acids, and it represents a positive reaction. In the absence of protease activity, the medium surrounding the growth of the organism remains opaque, which is a negative reaction.

Gelatin Hydrolysis

Although the value of gelatin as a nutritional source is questionable (it is an incomplete protein, lacking the essential amino acid tryptophan), its value in identifying bacterial species is well established. Gelatin is a protein produced by hydrolysis of collagen, a major component of connective tissue and tendons in humans and other animals. Below temperatures of 25°C, gelatin will maintain its gel properties and exist as a solid; at temperatures above 25°C, gelatin is liquid. **Figure 20.5** shows gelatin hydrolysis.

Liquefaction is accomplished by some microorganisms capable of producing a proteolytic extracellular enzyme called **gelatinase**, which acts to hydrolyze this protein to **amino acids**. Once this degradation occurs, even very low temperatures of 4°C will not restore the gel characteristic.

In this experimental procedure, you will use nutrient gelatin deep tubes to demonstrate the hydrolytic activity of gelatinase. The medium consists of nutrient broth supplemented with 12% gelatin. This high gelatin concentration results in a stiff medium and also serves as the substrate for the activity of gelatinase.

Following inoculation and incubation for 48 hours, the cultures are placed in a refrigerator at 4°C for 30 minutes. Cultures that remain liquefied produce gelatinase and demonstrate *rapid* gelatin hydrolysis. Re-incubate all solidified cultures for an additional 5 days. Refrigerate for 30 minutes and observe for liquefaction. Cultures that remain liquefied are indicative of *slow* gelatin hydrolysis.

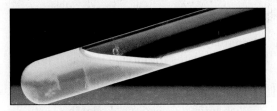

(a) Positive for gelatin liquefaction

(b) Negative for gelatin liquefaction

Figure 20.5 Nutrient gelatin hydrolysis

for the short version. 24- to 48-hour brain heart infusion broth cultures of the 13 organisms listed on page 148 for the long version.

Media

Short version: Two plates each of starch agar, tributyrin agar, and milk agar, and three nutrient gelatin deep tubes per designated student group. Long version: Four plates each of starch agar, tributyrin agar, and milk agar, and 14 nutrient gelatin deep tubes per designated student group.

Reagent

Gram's iodine solution.

Equipment

Microincinerator or Bunsen burner, inoculating loop and needle, glassware marking pencil, test tube rack, and refrigerator.

Controls

Test	Positive Control	Negative Control
Starch Hydrolysis	B. cereus	E. coli
Lipis Hydrolysis	S. aureus	E. coli
Casein Hydrolysis	B. cereus	E. coli
Gelatin Hydrolysis	B. cereus	E. coli

CLINICAL APPLICATION

Pathogens and Extracellular Enzymes

Bacteria use enzymes to alter their environments and to gain new sources of nutrients. When known bacterial pathogens are causing symptoms or damage not normally associated with that species, laboratories may test for newly acquired extracellular enzymes. Most known pathogens have been characterized by their abilities to digest proteins (fibronectin and collagen) as well as lipids and starches (glycolipids and glycoproteins).

AT THE BENCH

Materials

Cultures

24- to 48-hour trypticase soy broth cultures of *Escherichia coli*, *Bacillus cereus*, *Pseudomonas aeruginosa* BSL-2 , and *Staphylococcus aureus* BSL-2

Procedure Lab One

1. Prepare the starch agar, tributyrin agar, and milk agar plates for inoculation as follows:

   a. Short procedure: Using two plates per medium, divide the bottom of each Petri dish into two sections. Label the sections as *E. coli*, *B. cereus*, *P. aeruginosa*, and *S. aureus* BSL-2, respectively.

   b. Long procedure: Repeat Step 1a, dividing three plate bottoms into three sections and one plate bottom into four sections for each of the required media, to accommodate the 13 test organisms.

2. Using aseptic technique, make a single-line streak inoculation of each test organism on the agar surface of its appropriately labeled section on the agar plates.

3. Using aseptic technique, inoculate each experimental organism in its appropriately labeled gelatin deep tube by means of a stab inoculation.

4. Incubate all plates in an inverted position for 24 to 48 hours at 37°C. Incubate the gelatin deep tube cultures for 48 hours. Re-incubate all negative cultures for an additional 5 days.

Procedure Lab Two

Starch Hydrolysis

1. Flood the starch agar plate cultures with Gram's iodine solution, allow the iodine to remain in contact with the medium for 30 seconds, and pour off the excess.

2. Examine the cultures for the presence or absence of a blue-black color surrounding the growth of each test organism. Record your results in the chart provided in the Lab Report.

3. Based on your observations, determine and record the organisms that were capable of hydrolyzing the starch.

Lipid Hydrolysis

1. Examine the tributyrin agar plate cultures for the presence or absence of a clear area, or zone of lipolysis, surrounding the growth of each of the organisms. Record your results in the chart provided in the Lab Report.

2. Based on your observations, determine and record which organisms were capable of hydrolyzing the lipid.

Casein Hydrolysis

1. Examine the milk agar plate cultures for the presence or absence of a clear area, or zone of proteolysis, surrounding the growth of each of the bacterial test organisms. Record your results in the chart provided in the Lab Report.

2. Based on your observations, determine and record which of the organisms were capable of hydrolyzing the milk protein casein.

Gelatin Hydrolysis

1. Place all gelatin deep tube cultures into a refrigerator at 4°C for 30 minutes.

2. Examine all the cultures to determine whether the medium is solid or liquid. Record your results in the chart provided in the Lab Report.

3. Based on your observations following the 2-day and 7-day incubation periods, determine and record in the Lab Report (a) which organisms were capable of hydrolyzing gelatin and (b) the rate of hydrolysis.

Carbohydrate Fermentation

LEARNING OBJECTIVES

Once you have completed this experiment, you should

1. Understand the difference between cellular respiration and fermentation.

2. Be able to determine the ability of microorganisms to degrade and ferment carbohydrates with the production of acid and gas.

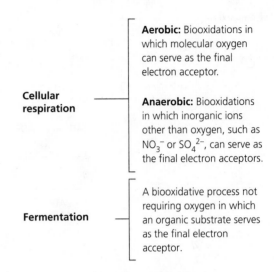

Figure 21.1 **Biooxidative pathways**

Principle

Most microorganisms obtain their energy through a series of orderly and integrated enzymatic reactions leading to the biooxidation of a substrate, frequently a carbohydrate. The major pathways by which this is accomplished are shown in Figure 21.1.

Organisms use carbohydrates differently depending on their enzyme complement. Some organisms are capable of fermenting sugars such as glucose anaerobically, while others use the aerobic pathway. Still others, facultative anaerobes, are enzymatically competent to use both aerobic and anaerobic pathways, and some organisms lack the ability to oxidize glucose by either. In this exercise, the fermentative pathways are of prime concern.

In fermentation, substrates such as carbohydrates and alcohols undergo anaerobic dissimilation and produce an organic acid (for example, lactic, formic, or acetic acid) that may be accompanied by gases such as hydrogen or carbon dioxide. Facultative anaerobes are usually the so-called fermenters of carbohydrates. Fermentation is best described by considering the degradation of glucose by way of the **Embden-Meyerhof pathway**, also known as the **glycolytic pathway**, illustrated in Figure 21.2.

As the diagram shows, one mole of glucose is converted into two moles of pyruvic acid, which is the major intermediate compound produced by glucose degradation. Subsequent metabolism

of pyruvate is not the same for all organisms, and a variety of end products result that define their different fermentative capabilities. This can be seen in Figure 21.3.

Fermentative degradation under anaerobic conditions is carried out in a fermentation broth tube containing a Durham tube, an inverted inner vial for the detection of gas production as illustrated in Figure 21.4. A typical carbohydrate fermentation medium contains

1. Nutrient broth ingredients for the support of the growth of all organisms.

2. A specific carbohydrate that serves as the substrate for determining the organism's fermentative capabilities.

3. The pH indicator phenol red, which is red at a neutral pH (7) and changes to yellow at a slightly acidic pH of 6.8, indicating that slight amounts of acid will cause a color change.

The critical nature of the fermentation reaction and the activity of the indicator make it imperative that all cultures should be observed within 48 hours. Extended incubation may mask acid-producing reactions by production of alkali because of enzymatic action on substrates other than the carbohydrate.

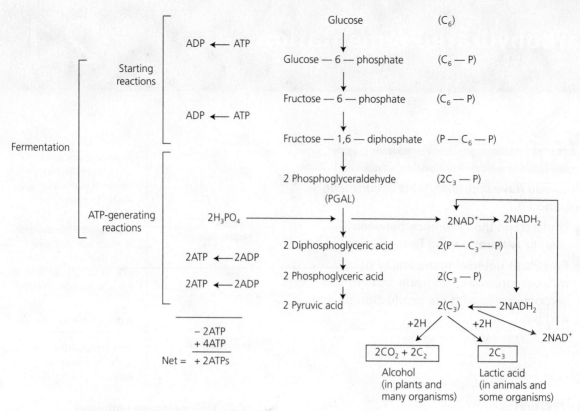

Figure 21.2 **The Embden-Meyerhof pathway**

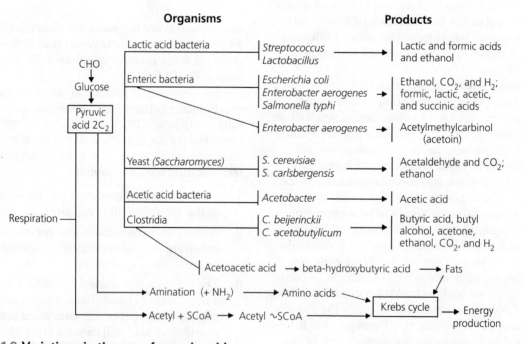

Figure 21.3 **Variations in the use of pyruvic acid**

Following incubation, carbohydrates that have been fermented with the production of acidic wastes will cause the phenol red (**Figure 21.5a**) to turn yellow, thereby indicating a positive reaction (**Figures 21.5b** and **c**). In some cases, acid production is accompanied by the evolution of a gas (CO_2) that will be visible as a bubble in the inverted tube (Figure 21.5b). Cultures that are not capable of fermenting a carbohydrate substrate will not change the indicator, and the tubes will appear

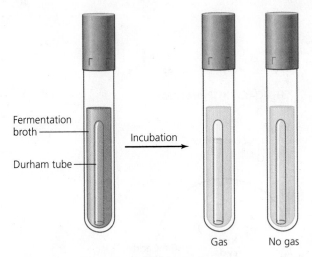

Figure 21.4 **Detection of gas production**

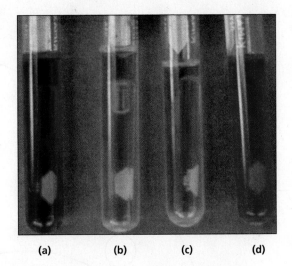

Figure 21.5 **Carbohydrate fermentation test.**
(a) Uninoculated, **(b)** acid and gas, **(c)** acid, and
(d) negative.

red; there will not be a concomitant evolution of gas. This is a negative reaction (Figure 21.5d).

The lack of carbohydrate fermentation by some organisms should not be construed as absence of growth. The organisms use other nutrients in the medium as energy sources. Among these nutrients are peptones present in nutrient broth. Peptones can be degraded by microbial enzymes to amino acids that are in turn enzymatically converted by oxidative deamination to ketoamino acids. These are then metabolized through the Krebs cycle for energy production. These reactions liberate ammonia, which accumulates in the medium, forming ammonium hydroxide (NH_4OH) and producing an alkaline environment. When this occurs, the phenol red turns to a deep red in the now basic medium. This alternative pathway of aerobic respiration is illustrated in Figure 21.6.

CLINICAL APPLICATION

Using Fermentation Products to Identify Bacteria

The fermentation of carbohydrates assists in the identification of some bacteria by determining what nutrients they are using and what products they produce. The pattern of sugars fermented may be unique to a particular genus, species, or strain. Lactose fermentation is one test that distinguishes between enteric and non-enteric bacteria. Dextrose fermentation allows for the differentiation between the oxidase (+) *Vibrio* and *Pseudomonads* species in patients suffering from septicemia after eating contaminated fish.

AT THE BENCH

Materials

Cultures

24- to 48-hour Trypticase soy broth cultures of *Escherichia coli*, *Alcaligenes faecalis*, *Salmonella typhimurium* BSL-2 , and *Staphylococcus aureus* BSL-2 for the short version. 24- to 48-hour Trypticase soy broth cultures of the 13 organisms listed on page 148 for the long version.

Media

Per designated student group: phenol red lactose, dextrose (glucose), and sucrose broths: 5 of each for the short version, 14 of each for the long version.

Equipment

Microincinerator or Bunsen burner, inoculating loop, and glassware marking pencil.

Controls

	REACTION	
Sugar	*Acid*	*Acid w/Gas*
Dextrose	S. aureus	E. coli
Sucrose	S. aureus	K. pneumoniae
Lactose	S. aureus	E. coli

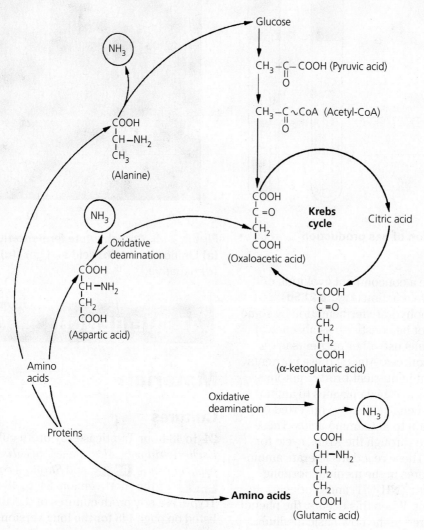

Figure 21.6 Proteins as energy sources for microbes

Procedure Lab One

1. Using aseptic technique, inoculate each experimental organism into its appropriately labeled medium by means of loop inoculation. *Note: Take care during this step not to shake the fermentation tube; shaking the tube may accidentally force a bubble of air into the inverted gas vial, displacing the medium and possibly rendering a false-positive result. The last tube will serve as a control.*

2. Incubate all tubes for 24 hours at 37°C.

Procedure Lab Two

1. Examine all carbohydrate broth cultures for color and the presence or absence of a gas bubble. Record your results in the chart provided in the Lab Report.

2. Based on your observations, determine and record whether or not each organism was capable of fermenting the carbohydrate substrate with the production of acid or acid and gas.

Triple Sugar–Iron Agar Test

LEARNING OBJECTIVES

Once you have completed this experiment, you should understand a rapid screening procedure that will

1. Differentiate among members of the Enterobacteriaceae.

2. Distinguish between the Enterobacteriaceae and other groups of intestinal bacilli.

Principle

The **triple sugar–iron (TSI) agar test** is designed to differentiate among the different groups or genera of the Enterobacteriaceae, which are all gram-negative bacilli capable of fermenting glucose with the production of acid, and to distinguish Enterobacteriaceae from other gram-negative intestinal bacilli. This differentiation is made on the basis of differences in carbohydrate fermentation patterns and hydrogen sulfide production by the various groups of intestinal organisms.

To facilitate observation of carbohydrate utilization patterns, the TSI agar slants contain lactose and sucrose in 1% concentrations and glucose (dextrose) in a concentration of 0.1%, which permits detection of the utilization of this substrate only. The acid-base indicator phenol red is also incorporated to detect carbohydrate fermentation that is indicated by a change in color of the medium from orange-red to yellow in the presence of acids. The slant is inoculated by means of a stab-and-streak procedure. This requires the insertion of a sterile, straight needle from the base of the slant into the butt. Upon withdrawal of the needle, the slanted surface of the medium is streaked. Following incubation, you will determine the fermentative activities of the organisms as described below.

1. **Alkaline slant (red) and acid butt (yellow) with or without gas production (breaks in the agar butt).** Only glucose fermentation has occurred. The organisms preferentially degrade glucose first. Since this substrate is present in minimal concentration, the small amount of acid produced on the slant surface is oxidized rapidly. The peptones in the medium are also used in the production of alkali. In the butt the acid reaction is maintained because of reduced oxygen tension and slower growth of the organisms.

2. **Acid slant (yellow) and acid butt (yellow) with or without gas production.** Lactose and/or sucrose fermentation has occurred. Since these substances are present in higher concentrations, they serve as substrates for continued fermentative activities with maintenance of an acid reaction in both slant and butt.

3. **Alkaline slant (red) and alkaline butt (red) or no change (orange-red) butt.** No carbohydrate fermentation has occurred. Instead, peptones are catabolized under anaerobic and/or aerobic conditions, resulting in an alkaline pH due to production of ammonia. If only aerobic degradation of peptones occurs, the alkaline reaction is evidenced only on the slant surface. If there is aerobic and anaerobic utilization of peptone, the alkaline reaction is present on the slant and the butt.

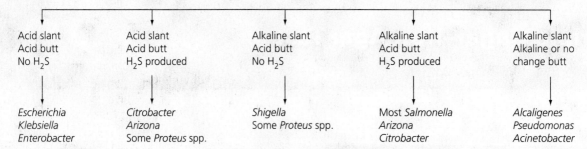

Acid slant	Acid slant	Alkaline slant	Alkaline slant	Alkaline slant
Acid butt	Acid butt	Acid butt	Acid butt	Alkaline or no
No H_2S	H_2S produced	No H_2S	H_2S produced	change butt

Escherichia	*Citrobacter*	*Shigella*	Most *Salmonella*	*Alcaligenes*
Klebsiella	*Arizona*	Some *Proteus* spp.	*Arizona*	*Pseudomonas*
Enterobacter	Some *Proteus* spp.		*Citrobacter*	*Acinetobacter*

Figure 22.1 **TSI reactions for differentiation of enteric microorganisms**

For you to obtain accurate results, it is absolutely essential to observe the cultures within 18 to 24 hours following incubation. Doing so will ensure that the carbohydrate substrates have not been depleted and that degradation of peptones yielding alkaline end products has not taken place.

The TSI agar medium also contains sodium thiosulfate, a substrate for hydrogen sulfide (H_2S) production, and ferrous sulfate for detection of this colorless end product. Following incubation, only cultures of organisms capable of producing H_2S will show an extensive blackening in the butt because of the precipitation of the insoluble ferrous sulfide. (Refer to Experiment 00 for a more detailed biochemical explanation of H_2S production.)

Figure 22.1 is a schema for the differentiation of intestinal bacilli on the basis of the TSI agar reactions.

CLINICAL APPLICATION

Differentiating Between *Proteus* Species
The TSI test can differentiate enteric organisms based on their abilities to reduce sulfur and ferment carbohydrates. It can be used to separate the three species of *Proteus, P. vulgaris, P. mirabilis,* and *P. penneri,* all of which are human opportunistic pathogens. *P. mirabilis* causes urinary tract infections and is sensitive to treatment with ampicillin and cephalosporins. *P. vulgaris,* a less common cause of urinary tract infections, is not sensitive to these antibiotics and is found as a nosocomial infectious agent among immunocompromised patients.

AT THE BENCH

Materials

Cultures

24-hour Trypticase soy broth cultures of *Alcaligenes faecalis, Escherichia coli, Pseudomonas aeruginosa* `BSL-2`, *Salmonella typhimurium* `BSL-2`, *Shigella dysenteriae* `BSL-2`, *Proteus vulgaris,* for the short version. 24-hour Trypticase soy broth cultures of the 13 organisms listed on page 148 for the long version.

Media

Per designated student group: triple sugar–iron agar slants: 7 for the short version, 14 for the long version.

Equipment

Microincinerator or Bunsen burner, inoculating needle, test tube rack, and glassware marking pencil.

Controls

Refer to Figure 22.1 for a description of positive controls for the different results exhibited when using a triple sugar-iron agar slant.

Procedure Lab One

1. Using aseptic technique, inoculate each experimental organism into its appropriately labeled tube by means of a stab-and-streak inoculation. *Note: Do not fully tighten screw cap.* The last tube will serve as a control.
2. Incubate for 18 to 24 hours at 37°C.

Procedure Lab Two

1. Examine the color of both the butt and slant of all agar slant cultures (Figure 22.2). Based on your observations, determine the type of reaction that has taken place (acid, alkaline, or none) and the carbohydrate that has been fermented (dextrose, lactose, sucrose, all, or none) in each culture. Record your observations and results in the chart provided in the Lab Report.
2. Examine all cultures for the presence or absence of blackening within the medium. Based on your observations, determine whether or not each organism was capable of H_2S production. Record your observations and results in the chart provided in the Lab Report.

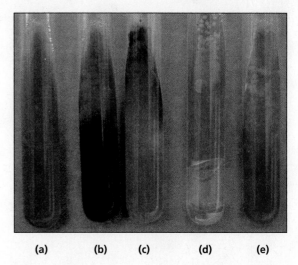

(a)   (b)   (c)   (d)   (e)

Figure 22.2 **Reactions in triple sugar–iron agar.** **(a)** Uninoculated; **(b)** alkaline slant/acid butt, H_2S; **(c)** alkaline slant/acid butt; **(d)** acid slant/acid butt, gas; and **(e)** acid slant/acid butt.

Name: _____

Date: _____ Section: _____

Observations and Results

Bacterial Species	CARBOHYDRATE FERMENTATION			H₂S PRODUCTION	
	Butt Color and Reaction	Slant Color and Reaction	Carbohydrate Fermented	Blackening	H₂S (+) or (−)
E. coli					
A. faecalis					
P. aeruginosa					
S. dysenteriae					
S. typhimurium					
P. vulgaris					
K. pneumoniae					
E. aerogenes					
M. luteus					
L. lactis					
S. aureus					
B. cereus					
C. xerosis					
Alternate organism					
Control					

Review Questions

1. What is the purpose of the TSI test?

2. Explain why the TSI medium contains a lower concentration of glucose than of lactose and sucrose.

3. Explain the purpose of the phenol red in the medium.

4. Explain the purpose of thiosulfate in the medium.

5. Explain why the test observations must be made between 18 and 24 hours after inoculation.

Identification of enteric bacilli is of prime importance in controlling intestinal infections by preventing contamination of food and water supplies. The groups of bacteria that can be found in the intestinal tract of humans and lower mammals are classified as members of the family **Enterobacteriaceae**. They are short, gram-negative, non–spore-forming bacilli. Included in this family are:

1. **Pathogens**, such as members of the genera *Salmonella* and *Shigella*.

2. **Occasional pathogens**, such as members of the genera *Proteus* and *Klebsiella*.

3. **Normal intestinal flora**, such as members of the genera *Escherichia* and *Enterobacter*, which are saprophytic inhabitants of the intestinal tract.

Differentiation of the principal groups of Enterobacteriaceae can be accomplished on the basis of their biochemical properties and enzymatic reactions in the presence of specific substrates. The **IMViC** series of tests (**indole**, **methyl red**, **Voges-Proskauer**, and **citrate utilization**) can be used.

Figure 23.11 on page 173 shows the biochemical reactions that occur during the IMViC tests. It is designed to assist you in the execution and interpretation of each test.

The following experiments are designed for either a short or long version. The short version uses selected members of the enteric family. The long procedure makes use of bacterial species that do not belong solely to the Enterobacteriaceae. Nonenteric forms are included to acquaint you with the biochemical activities of other organisms grown in these media and to enable you to use these data for further comparisons of both types of bacteria. Selected organisms to be used

in the long-version procedures are listed below. The enteric organisms are subdivided as lactose fermenters and non–lactose fermenters.

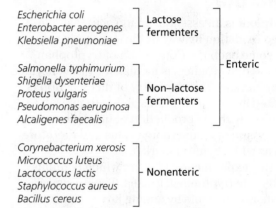

CLINICAL APPLICATION

Identification of Enteric Bacteria

The IMViC test is used to identify members of the *Enterobacteriaceae,* some of which are powerful pathogens such as members of the genera *Shigella* and *Salmonella,* which cause intestinal infections. Identification of the causative agent may lead to the source of the infection, such as raw food (*Salmonella*) or fecal contamination of food (*Shigella*). This will aid in determining the possible number of individuals that have been exposed and who may require medical attention. This test uses the organisms' biochemical properties and enzymatic reactions on specific substrates as a means of identification.

Once you have completed this test, you should be able to

1. Determine the ability of microorganisms to degrade the amino acid tryptophan.

Cultures producing a red reagent layer following addition of Kovac's reagent are indole-positive; an example of this is *E. coli*. The absence of red coloration demonstrates that the substrate tryptophan was not hydrolyzed and indicates an indole-negative reaction.

Principle

Tryptophan is an essential amino acid that can undergo oxidation by way of the enzymatic activities of some bacteria. Conversion of tryptophan into metabolic products is mediated by the enzyme **tryptophanase**. The chemistry of this reaction is illustrated in Figure 23.1. This ability to hydrolyze tryptophan with the production of indole is not a characteristic of all microorganisms and therefore serves as a biochemical marker.

In this experiment, SIM agar, which contains the substrate tryptophan, is used. The presence of indole is detectable by adding Kovac's reagent, which produces a cherry red reagent layer. This color is produced by the reagent, which is composed of *p*-dimethylaminobenzaldehyde, butanol, and hydrochloric acid. Indole is extracted from the medium into the reagent layer by the acidified butyl alcohol component and forms a complex with the *p*-dimethylaminobenzaldehyde, yielding the cherry red color. The chemistry of this reaction is illustrated in Figure 23.2.

Materials

Cultures

24- to 48-hour Trypticase soy broth cultures of *E. coli*, *P. vulgaris*, and *E. aerogenes* for the short version. 24- to 48-hour Trypticase soy broth cultures of the 13 organisms listed on page 148 for the long version.

Media

SIM agar deep tubes per designated student group: 4 for the short version, 14 for the long version.

Reagent

Kovac's reagent.

Equipment

Microincinerator or Bunsen burner, inoculating needle, test tube rack, and glassware marking pencil.

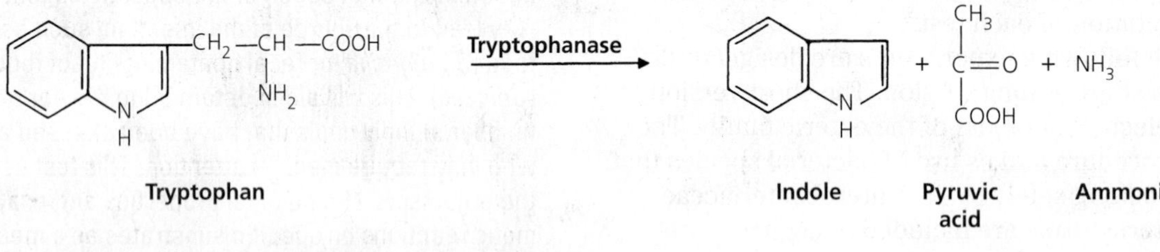

Tryptophan —— Tryptophanase ——> Indole + Pyruvic acid + Ammonia

Figure 23.1 Enzymatic degradation of tryptophan

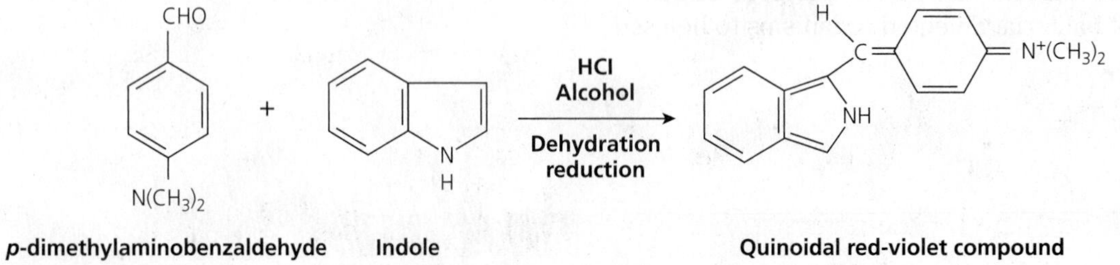

p-dimethylaminobenzaldehyde + Indole —— HCl Alcohol / Dehydration reduction ——> Quinoidal red-violet compound

Figure 23.2 Indole reaction with Kovac's reagent

Figure 23.3 Indole production test.
(a) Uninoculated, **(b)** negative, and **(c)** positive.

Procedure Lab One

1. Using aseptic technique, inoculate each experimental organism into its appropriately labeled deep tube by means of a stab inoculation. The last tube will serve as a control.

2. Incubate tubes for 24 to 48 hours at 37°C.

Procedure Lab Two

1. Add 10 drops of Kovac's reagent to all deep tube cultures and agitate the cultures gently.

2. Examine the color of the reagent layer in each culture (refer to Figure 23.3). Record your results in the chart in the Lab Report.

3. Based on your observations, determine and record whether or not each organism was capable of hydrolyzing the tryptophan.

PART B   Methyl Red Test

LEARNING OBJECTIVES

Once you have completed this experiment, you should be able to

1. Determine the ability of microorganisms to ferment glucose with the production and stabilization of high concentrations of acid end products.

2. Differentiate between all glucose-fermenting enteric organisms, particularly *E. coli* and *E. aerogenes*.

Principle

The hexose monosaccharide **glucose** is the major substrate utilized by all enteric organisms for energy production. The end products of this process will vary depending on the specific enzymatic pathways present in the bacteria. In this test, the pH indicator methyl red detects the presence of large concentrations of acid end products. Although most enteric microorganisms ferment glucose with the production of organic acids, this test is of value in the separation of *E. coli* and *E. aerogenes*.

Both of these organisms initially produce organic acid end products during the early incubation period. The low acidic pH (4) is stabilized and maintained by *E. coli* at the end of incubation. During the later incubation period, *E. aerogenes* enzymatically converts these acids to nonacidic end products, such as 2,3-butanediol and acetoin (acetylmethylcarbinol), resulting in an elevated pH of approximately 6. The glucose fermentation reaction generated by *E. coli* is illustrated in Figure 23.4.

As shown, at a pH of 4.4 or lower, the methyl red indicator in the pH range of 4 will turn red, which is indicative of a positive test. At a pH of 6.2 or higher, still indicating the presence of acid but with a lower hydrogen ion concentration, the indicator turns yellow and is a negative test. Production and detection of the nonacidic end products from glucose fermentation by *E. aerogenes* is amplified in Part C of this exercise, the Voges-Proskauer test, which is performed simultaneously with the methyl red test.

AT THE BENCH

Materials

Cultures

24- to 48-hour Trypticase soy broth cultures of *E. coli*, *E. aerogenes*, and *K. pneumoniae* BSL-2 for the short version. 24- to 48-hour Trypticase soy broth cultures of the 13 organisms listed on page 148 for the long version. In Lab Two, aliquots of these experimental cultures must be set aside for the Voges-Proskauer test.

Media

MR-VP broth per designated student group: 4 for the short version, 14 for the long version.

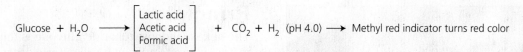

$$\text{Glucose} + H_2O \longrightarrow \begin{bmatrix} \text{Lactic acid} \\ \text{Acetic acid} \\ \text{Formic acid} \end{bmatrix} + CO_2 + H_2 \text{ (pH 4.0)} \longrightarrow \text{Methyl red indicator turns red color}$$

Figure 23.4 Glucose fermentation reaction with methyl red pH reagent

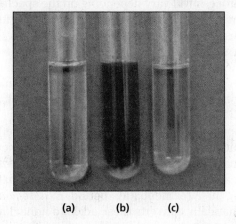

(a)   (b)   (c)

Figure 23.5 Methyl red test. (a) Uninoculated, **(b)** positive, and **(c)** negative.

Reagent

Methyl red indicator.

Equipment

Microincinerator or Bunsen burner, inoculating loop, test tubes, and glassware marking pencil.

Procedure Lab One

1. Using aseptic technique, inoculate each experimental organism into its appropriately labeled tube of medium by means of a loop inoculation. The last tube will serve as a control.
2. Incubate all cultures for 24 to 48 hours at 37°C.

Procedure Lab Two

1. Transfer approximately one-third of each culture into an empty test tube and set these tubes aside for the Voges-Proskauer test.
2. Add five drops of the methyl red indicator to the remaining aliquot of each culture.
3. Examine the color of all cultures (refer to Figure 23.5). Record the results in the chart in the Lab Report.
4. Based on your observations, determine and record whether or not each organism was capable of fermenting glucose with the production and maintenance of a high concentration of acid.

LEARNING OBJECTIVE

Once you have completed this experiment, you should be able to

1. Differentiate further among enteric organisms such as *E. coli, E. aerogenes,* and *K. pneumoniae.*

Principle

The Voges-Proskauer test determines the capability of some organisms to produce nonacidic or neutral end products, such as acetylmethylcarbinol, from the organic acids that result from glucose metabolism. This glucose fermentation, which is characteristic of *E. aerogenes*, is illustrated in Figure 23.6.

The reagent used in this test, Barritt's reagent, consists of a mixture of alcoholic α-naphthol and 40% potassium hydroxide solution. Detection of acetylmethylcarbinol requires this end product to be oxidized to a diacetyl compound. This reaction will occur in the presence of the α-naphthol catalyst and a guanidine group that is present in the peptone of the MR-VP medium. As a result, a pink complex is formed, imparting a rose color to the medium. The chemistry of this reaction is illustrated in Figure 23.7.

Development of a deep rose color in the culture 15 minutes following the addition of Barritt's reagent is indicative of the presence of acetylmethylcarbinol and represents a positive result. The absence of rose coloration is a negative result.

$$\text{Glucose} + O_2 \longrightarrow \text{Acetic} \atop \text{acid} \longrightarrow \left[{2,3\text{-butanediol} \atop \text{acetylmethylcarbinol}} \right] + CO_2 + H_2 \text{ (pH 6.0)}$$

Figure 23.6 Glucose fermentation by E. aerogenes

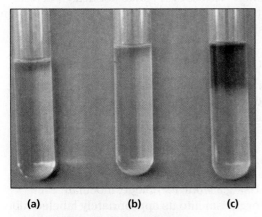

Figure 23.7 Acetylmethylcarbinol reaction with Barritt's reagent

(a)   **(b)**   **(c)**

Figure 23.8 Voges-Proskauer test
(a) Uninoculated, **(b)** negative, and **(c)** positive

AT THE BENCH

Materials

Cultures
24- to 48-hour Trypticase soy broth cultures of *E. coli*, *E. aerogenes*, and *K. pneumoniae* **BSL-2** for the short version. 24- to 48-hour Trypticase soy broth cultures of the 13 organisms listed on page 148 for the long version. *Note: Aliquots of these experimental cultures must be set aside from the methyl red test.*

Reagent
Barritt's reagents A and B.

Equipment
Microincinerator or Bunsen burner, inoculating loop, and glassware marking pencil.

Procedure Lab One
Refer to the methyl red test in Part B of this exercise.

Procedure Lab Two

1. To the aliquots of each broth culture separated during the methyl red test, add 10 drops of Barritt's reagent A and shake the cultures. Immediately add 10 drops of Barritt's reagent B and shake. Reshake the cultures every 3 to 4 minutes.

2. Examine the color of the cultures 15 minutes after the addition of Barritt's reagent. Refer to Figure 23.8. Record your results in the Lab Report.

3. Based on your observations, determine and record whether or not each organism was capable of fermenting glucose with ultimate production of acetylmethylcarbinol.

PART D   Citrate Utilization Test

LEARNING OBJECTIVE

Once you have completed this experiment, you should be able to

1. Differentiate among enteric organisms on the basis of their ability to ferment citrate as a sole source of carbon.

1.

$$
\underset{\text{Citrate}}{
\begin{array}{c}
\text{COOH} \\
| \\
\text{CH}_2 \\
| \\
\text{HO} - \text{C} - \text{COOH} \\
| \\
\text{CH}_2 \\
| \\
\text{COOH}
\end{array}
}
\xrightarrow{\text{Citrase}}
\underset{\substack{\text{Oxaloacetic}\\\text{acid}}}{
\begin{array}{c}
\text{COOH} \\
| \\
\text{C}=\text{O} \\
| \\
\text{CH}_2 \\
| \\
\text{COOH}
\end{array}
}
\longrightarrow
\underset{\substack{\text{Pyruvic}\\\text{acid}}}{
\begin{array}{c}
\text{COOH} \\
| \\
\text{C}=\text{O} \\
| \\
\text{CH}_3
\end{array}
}
+
\underset{\substack{\text{Acetic acid}}}{
\begin{array}{c}
\text{CH}_3 \\
| \\
\text{COOH}
\end{array}
}
+
\underset{\substack{\text{Excess}\\\text{carbon}\\\text{dioxide}}}{\text{CO}_2}
$$

2. $CO_2 + 2Na^+ + H_2O \longrightarrow Na_2CO_3 \longrightarrow$ Alkaline pH $\longrightarrow$ Color change from green to blue

Figure 23.9 **Enzymatic degradation of citrate**

Principle

In the absence of fermentable glucose or lactose, some microorganisms are capable of using **citrate** as a carbon source for their energy. This ability depends on the presence of a **citrate permease** that facilitates the transport of citrate in the cell. Citrate is the first major intermediate in the Krebs cycle and is produced by the condensation of active acetyl with oxaloacetic acid. Citrate is acted on by the enzyme **citrase**, which produces oxaloacetic acid and acetate. These products are then enzymatically converted to pyruvic acid and carbon dioxide. During this reaction, the medium becomes alkaline—the carbon dioxide that is generated combines with sodium and water to form sodium carbonate, an alkaline product. The presence of sodium carbonate changes the bromthymol blue indicator incorporated into the medium from green to deep Prussian blue. The chemistry of this reaction is illustrated in Figure 23.9.

Following incubation, citrate-positive cultures are identified by the presence of growth on the surface of the slant, which is accompanied by blue coloration, as *seen with E. aerogenes*. Citrate-negative cultures will show no growth, and the medium will remain green.

AT THE BENCH

Materials

Cultures

24- to 48-hour Trypticase soy broth cultures of *E. coli*, *E. aerogenes*, and *K. pneumoniae* BSL-2 for the short version. 24- to 48-hour Trypticase soy broth cultures of the 13 organisms listed on page 148 for the long version.

Media

Simmons citrate agar slants per designated student group: 4 for the short version, 14 for the long version.

Equipment

Microincinerator or Bunsen burner, inoculating needle, test tube rack, and glassware marking pencil.

Procedure Lab One

1. Using aseptic technique, inoculate each organism into its appropriately labeled tube by means of streak inoculation. The last tube will serve as a control.

2. Incubate all cultures for 24 to 48 hours at 37°C.

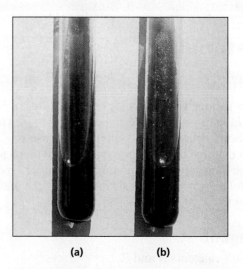

(a)   (b)

Figure 23.10 **Citrate utilization test. (a)** Tube is negative, showing no growth on slant surface. **(b)** Tube is positive, showing growth on slant surface.

Procedure Lab Two

1. Examine all agar slant cultures for the presence or absence of growth and coloration of the medium. Refer to **Figure 23.10**. Record your results in the chart in the Lab Report.

2. Based on your observations, determine and record whether or not each organism was capable of using citrate as its sole source of carbon.

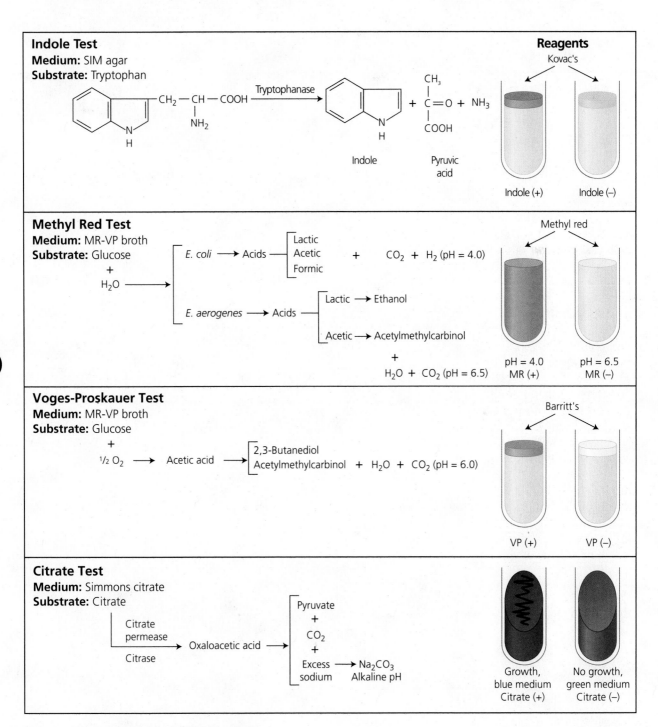

Indole Test
Medium: SIM agar
Substrate: Tryptophan

Tryptophanase

Indole   Pyruvic acid

Reagents
Kovac's

Indole (+)   Indole (–)

Methyl Red Test
Medium: MR-VP broth
Substrate: Glucose
+
H_2O

E. coli → Acids → Lactic / Acetic / Formic   + CO_2 + H_2 (pH = 4.0)

E. aerogenes → Acids → Lactic → Ethanol / Acetic → Acetylmethylcarbinol
+
H_2O + CO_2 (pH = 6.5)

Methyl red

pH = 4.0   pH = 6.5
MR (+)   MR (–)

Voges-Proskauer Test
Medium: MR-VP broth
Substrate: Glucose
+
½ O_2 → Acetic acid → 2,3-Butanediol / Acetylmethylcarbinol + H_2O + CO_2 (pH = 6.0)

Barritt's

VP (+)   VP (–)

Citrate Test
Medium: Simmons citrate
Substrate: Citrate

Citrate permease / Citrase → Oxaloacetic acid → Pyruvate + CO_2 + Excess sodium → Na_2CO_3 Alkaline pH

Growth, blue medium
Citrate (+)

No growth, green medium
Citrate (–)

Figure 23.11 **Summary of IMViC reactions**

Hydrogen Sulfide Test

LEARNING OBJECTIVES

Once you have completed this experiment, you will be able to determine

1. The ability of microorganisms to produce hydrogen sulfide from sulfur-containing amino acids or inorganic sulfur compounds.

2. Mobility of microorganisms in SIM agar.

Principle

There are two major fermentative pathways by which some microorganisms are able to produce hydrogen sulfide (H_2S).

Pathway 1: Gaseous H_2S may be produced by the reduction (hydrogenation) of organic sulfur present in the amino acid cysteine, which is a component of peptones contained in the medium. These peptones are degraded by microbial enzymes to amino acids, including the sulfur-containing amino acid cysteine. This amino acid in the presence of a **cysteine desulfurase** loses the sulfur atom, which is then reduced by the addition of hydrogen from water to form bubbles of hydrogen sulfide gas ($H_2S \uparrow$) as illustrated:

$$
\begin{array}{c}
CH_2-SH \\
| \\
CH-NH_2 \\
| \\
COOH
\end{array}
\xrightarrow[\text{desulfurase}]{\text{Cysteine}}
\begin{array}{c}
CH_3 \\
| \\
C=O \\
| \\
COOH
\end{array}
+ \; H_2S \uparrow \; + \; NH_3
$$

Cysteine   **Pyruvic**   **Hydrogen**   **Ammonia**
                      acid   **sulfide**
                            gas

Pathway 2: Gaseous H_2S may also be produced by the reduction of inorganic sulfur compounds such as the thiosulfates ($S_2O_3^{2-}$), sulfates (SO_4^{2-}), or sulfites (SO_3^{2-}). The medium contains sodium thiosulfate, which certain microorganisms are capable of reducing to sulfite with the liberation of hydrogen sulfide. The sulfur atoms act as hydrogen acceptors during oxidation of the inorganic compound as illustrated in the following:

$$
2S_2O_3^{2-} + 4H^+ + 4e^- \xrightarrow[\text{reductase}]{\text{Thiosulfate}} 2SO_3^{2-} + 2H_2S \uparrow
$$

Thiosulfate   **Sulfite**   **Hydrogen**
                                           Sulfide
                                           gas

In this experiment the SIM medium contains peptone and sodium thiosulfate as the sulfur substrates; ferrous sulfate ($FeSO_4$), which behaves as the H_2S indicator; and sufficient agar to make the medium semisolid and thus enhance anaerobic respiration. Regardless of which pathway is used, the hydrogen sulfide gas is colorless and therefore not visible. Ferrous ammonium sulfate in the medium serves as an indicator by combining with the gas, forming an insoluble black ferrous sulfide precipitate that is seen along the line of the stab inoculation and is indicative of H_2S production. Absence of the precipitate is evidence of a negative reaction. The overall reactions for both pathways and their interpretation are illustrated in **Figure 24.1**.

Motility

SIM agar may also be used to detect motile organisms. Motility is recognized when culture growth (turbidity) of flagellated organisms is not restricted to the line of inoculation. Growth of nonmotile organisms is confined to the line of inoculation.

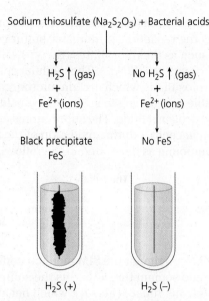

$$\begin{array}{l} CH_2-SH \\ | \\ CH-NH_2 \\ | \\ COOH \end{array} \xrightarrow[\text{desulfurase}]{\text{Cysteine}} \begin{array}{l} CH_3 \\ | \\ C=O \\ | \\ COOH \end{array} + NH_3 + H_2S\uparrow$$

Cysteine   Pyruvic   Ammonia   **H$_2$S gas**
             acid   (colorless)

Sodium thiosulfate (Na$_2$S$_2$O$_3$) + Bacterial acids

H$_2$S ↑ (gas)   No H$_2$S ↑ (gas)
+   +
Fe^{2+} (ions)   Fe^{2+} (ions)

Black precipitate   No FeS
FeS

H$_2$S (+)   H$_2$S (–)

Figure 24.1 Detectivhydrogen sulphide

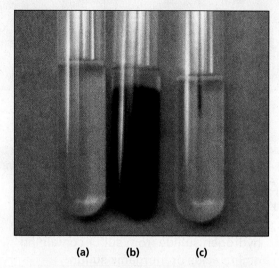

(a)   (b)   (c)

Figure 24.2 Hydrogen sulfide production test.
(a) Negative, **(b)** positive with motility, and
(c) positive with no motility

typhimurium `BSL-2` for the short version. 24- to
48-hour Trypticase soy broth cultures of the 13
organisms listed on page 148 for the long version.

Media

SIM agar deep tubes per designated student group:
5 for the short version, 14 for the long version.

Equipment

Microincinerator or Bunsen burner, inoculating
needle, test tube rack, and glassware marking pencil.

Procedure Lab One

1. Aseptically inoculate each experimental
   organism into its appropriately labeled tube
   by means of stab inoculation. The last tube
   will serve as a control.

2. Incubate all cultures for 24 to 48 hours at 37°C.

Procedure Lab Two

1. Examine all SIM cultures for the presence or
   absence of black coloration along the line of
   the stab inoculation. Refer to Figure 24.2, and
   record your results in the chart provided in
   the Lab Report.

2. Based on your observations, determine and
   record whether or not each organism was
   capable of producing hydrogen sulfide.

3. Observe all cultures for the presence (+) or
   absence (–) of motility. Record your results
   in the chart in the Lab Report.

CLINICAL APPLICATION

Identifying Intestinal Pathogens

While generally considered a self-limiting symptom,
diarrhea due to *Proteus* is initially difficult to differentiate from early stages of the more severe bloody
diarrhea (dysentery) associated with some *Shigella*
or *Salmonella* species. Bacteria belonging to the
genera *Salmonella* and *Proteus* enzymatically
metabolize inorganic sulfur compounds and sulfur-
containing amino acids, producing H$_2$S. The hydrogen sulfide test is one way to separate and identify
Shigella dysentariae, which does not produce H$_2$S,
from *Proteus* and *Salmonellla.*

AT THE BENCH

Materials

Cultures

24- to 48-hour Trypticase soy broth cultures of
Enterobacter aerogenes, Shigella dysenteriae
`BSL-2`, *Proteus vulgaris*, and *Salmonella*

Urease Test

LEARNING OBJECTIVE

Once you have completed this experiment, you should be able to

1. Determine the ability of microorganisms to degrade urea by means of the enzyme urease.

CLINICAL APPLICATION

Pathogens and the Urease Test

The urease test is primarily used to distinguish the small number of urease-positive enterics from other non–lactose-fermenting enteric bacteria. Many enterics can degrade urea but only a few are termed *rapid urease positive organisms*. While part of the normal flora, these commensals have been identified as opportunistic pathogens. Members of the gastrodeuodonal commensals are included among this group of organisms.

Principle

Urease, which is produced by some microorganisms, is an enzyme that is especially helpful in the identification of *Proteus vulgaris*. Although other organisms may produce urease, their action on the substrate urea tends to be slower than that seen with *Proteus* species. Therefore, this test serves to rapidly distinguish members of this genus from other non–lactose-fermenting enteric microorganisms.

Urease is a hydrolytic enzyme that attacks the nitrogen and carbon bond in amide compounds such as urea and forms the alkaline end product ammonia. This chemical reaction is illustrated in Figure 25.1.

The presence of urease is detectable when the organisms are grown in a urea broth medium containing the pH indicator phenol red. As the substrate urea is split into its products, the presence of ammonia creates an alkaline environment that causes the phenol red to turn to a deep pink. This is a positive reaction for the presence of urease. Failure of a deep pink color to develop is evidence of a negative reaction.

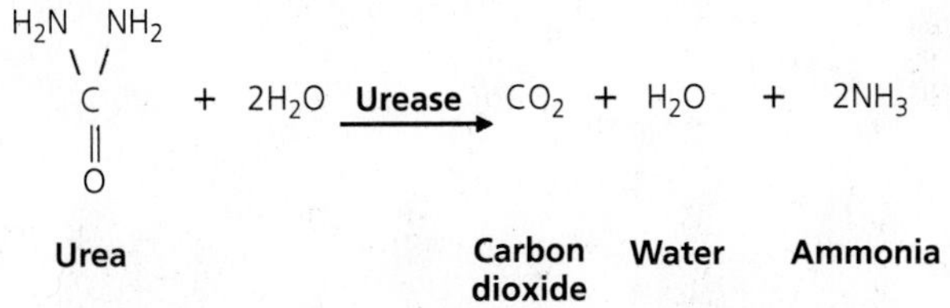

AT THE BENCH

Materials

Cultures

24- to 48-hour Trypticase soy broth cultures of *Escherichia coli*, *Proteus vulgaris*, *Klebsiella pneumoniae* BSL-2 , and *Salmonella typhimurium* BSL-2 for the short version. 24- to 48-hour Trypticase soy broth cultures of the 13 organisms listed on page 148 for the long version.

Media

Urea broth per designated student group: 5 for the short version, 14 for the long version.

Equipment

Microincinerator or Bunsen burner, inoculating loop, test tube rack, and glassware marking pencil.

Procedure Lab One

1. Using aseptic technique, inoculate each experimental organism into its appropriately labeled tube by means of loop inoculation. The last tube will serve as a control.

2. Incubate cultures 24 to 48 hours at 37°C.

$$\underset{\text{Urea}}{\underset{\|}{\overset{H_2N \quad NH_2}{\underset{O}{\overset{\backslash \;/}{C}}}}} + 2H_2O \xrightarrow{\text{Urease}} \underset{\substack{\text{Carbon} \\ \text{dioxide}}}{CO_2} + \underset{\text{Water}}{H_2O} + \underset{\text{Ammonia}}{2NH_3}$$

Figure 25.1 Enzymatic degradation of urea

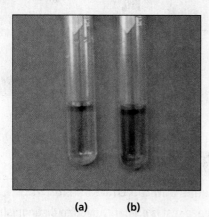

(a)   (b)

Figure 25.2 **Urease test. (a)** Negative and **(b)** positive

Procedure Lab Two

1. Examine all urea broth cultures for color (refer to Figure 25.2). Record your results in the chart in the Lab Report.

2. Based on your observations, determine and record whether or not each organism was capable of hydrolyzing the substrate urea.

LEARNING OBJECTIVE

Once you have completed this experiment, you should be able to

1. Differentiate among microorganisms that enzymatically transform different milk substrates into varied metabolic end products.

Principle

The major milk substrates capable of transformation are the milk sugar lactose and the milk proteins casein, lactalbumin, and lactoglobulin. To distinguish among the metabolic changes produced in milk, a pH indicator, the oxidation-reduction indicator litmus, is incorporated into the medium. Litmus milk now forms an excellent differential medium in which microorganisms can metabolize milk substrates depending on their enzymatic complement. A variety of different biochemical changes result, as follows:

Litmus milk
- Lactose fermentation
- Gas production
- Litmus reduction
- Curd formation
- Proteolysis
- Alkaline reaction

Lactose Fermentation

Organisms capable of using **lactose** as a carbon source for energy production utilize the inducible enzyme **β-galactosidase** and degrade lactose as follows:

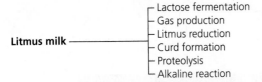

The presence of **lactic acid** is easily detected because litmus is purple at a neutral pH and turns pink when the medium is acidified to an approximate pH of 4.

Gas Formation

The end products of the microbial fermentation of lactose are likely to include the **gases $CO_2\uparrow$ + $H_2\uparrow$**. The presence of gas may be seen as separations of the curd or by the development of tracks or fissures within the curd as gas rises to the surface.

Litmus Reduction

Fermentation is an anaerobic process involving biooxidations that occur in the absence of molecular oxygen. These oxidations may be visualized as the removal of hydrogen (dehydrogenation) from a substrate. Since hydrogen ions cannot exist in the free state, there must be an immediate and concomitant electron acceptor available to bind these hydrogen ions, or else oxidation-reduction reactions are not possible and cells cannot manufacture energy. In the litmus milk test, **litmus** acts as such an acceptor. While in the oxidized state, the litmus is purple; when it accepts hydrogen from a substrate, it will become reduced and turn white or milk-colored. This oxidation of lactose, which produces lactic acid, butyric acid, $CO_2\uparrow$, and $H_2\uparrow$, is as follows:

Lactose $\longrightarrow$ Glucose $\longrightarrow$ Pyruvic acid $\longrightarrow$
- Lactic acid
- Butyric acid
- $CO_2 + H_2$

The excess hydrogen is now accepted by the hydrogen acceptor litmus, which turns white and is said to be reduced.

Curd Formation

The biochemical activities of different microorganisms grown in litmus milk may result in the production of two distinct types of curds (clots). Curds are designated as either acid or rennet, depending on the biochemical mechanism responsible for their formation.

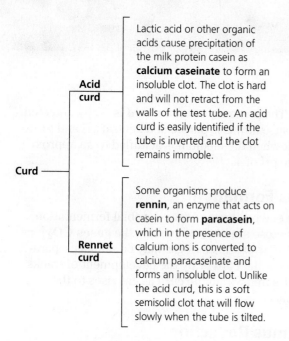

Curd —
- Acid curd — Lactic acid or other organic acids cause precipitation of the milk protein casein as **calcium caseinate** to form an insoluble clot. The clot is hard and will not retract from the walls of the test tube. An acid curd is easily identified if the tube is inverted and the clot remains immoble.
- Rennet curd — Some organisms produce **rennin**, an enzyme that acts on casein to form **paracasein**, which in the presence of calcium ions is converted to calcium paracaseinate and forms an insoluble clot. Unlike the acid curd, this is a soft semisolid clot that will flow slowly when the tube is tilted.

Proteolysis (Peptonization)

The inability of some microorganisms to obtain their energy by way of lactose fermentation means they must use other nutritional sources such as proteins for this purpose (see Figure 21.6). By means of proteolytic enzymes, these organisms hydrolyze the milk proteins, primarily casein, into their basic building blocks, namely amino acids. This digestion of proteins is accompanied by the evolution of large quantities of ammonia, resulting in an alkaline pH in the medium. The litmus turns deep purple in the upper portion of the tube, while the medium begins to lose body and produces a translucent, brown, whey-like appearance as the protein is hydrolyzed to amino acids.

Alkaline Reaction

An alkaline reaction is evident when the color of the medium remains unchanged or changes to a deeper blue. This reaction is indicative of the partial degradation of **casein** into **shorter polypeptide chains**, with the simultaneous release of alkaline end products that are responsible for the observable color change.

Figure 26.1 and 26.2 show the possible litmus milk reactions and their appearance following the appropriate incubation of the cultures.

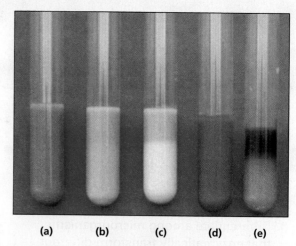

(a)   (b)   (c)   (d)   (e)

Figure 26.1 Litmus milk reactions.
(a) Uninoculated, **(b)** acid, **(c)** acid with reduction and curd, **(d)** alkaline, and **(e)** proteolysis.

CLINICAL APPLICATION

Differentiating *Enterobacteriaceae* and *Clostridium*.

The litmus milk test differentiates members of the Enterobacteriacaeae from other gram-negative bacilli based on the enterics' ability to reduce litmus. It is also used to differentiate members within the genus *Clostridium*. Watery diarrhea caused by *C. perfringes* (contaminated food) is generally considered self-limiting. But diarrhea caused by *C. difficile* may be associated with antibiotic use that has removed the normal flora of the colon.

AT THE BENCH

Materials

Cultures

24- to 48-hour Trypticase soy broth cultures of *Escherichia coli, Alcaligenes faecalis, Lactococcus lactis*, and *Pseudomonas aeruginosa* **BSL-2** for the short version. 24- to 48-hour Trypticase soy broth cultures of the 13 organisms listed on page 148 for the long version.

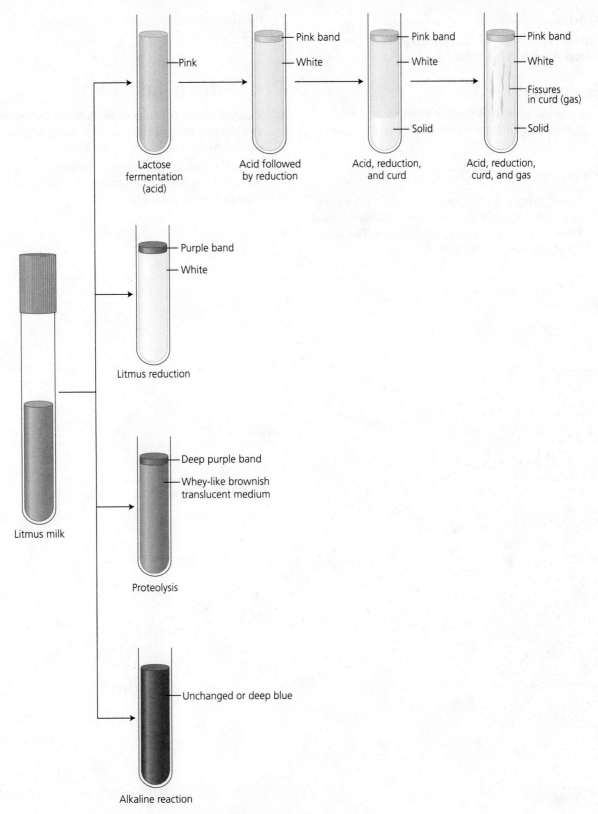

Figure 26.2 Summary of possible litmus milk reactions

Media

Litmus milk broth per designated student group: 5 for the short version, 14 for the long version.

Equipment

Microincinerator or Bunsen burner, inoculating loop, test tube rack, and glassware marking pencil.

Procedure Lab One

1. Using aseptic technique, inoculate each experimental organism into its appropriately labeled tube by means of a loop inoculation. The last tube will serve as a control.

2. Incubate all cultures for 24 to 48 hours at 37°C.

Procedure Lab Two

1. Examine all the litmus milk cultures for color and consistency of the medium. Record your results in the chart in the Lab Report.

2. Based on your observations, determine and record the type(s) of reaction(s) that have taken place in each culture.

Name: _____

Date: _____ Section: _____

Observations and Results

Bacterial Species	Appearance of Medium	Litmus Milk Reactions
E. coli		
E. aerogenes		
K. pneumoniae		
S. dysenteriae		
S. typhimurium		
P. vulgaris		
P. aeruginosa		
A. faecalis		
M. luteus		
L. lactis		
S. aureus		
B. cereus		
C. xerosis		
Alternate organism		
Control		

Review Questions

1. Distinguish between acid and rennet curds.

2. Describe the litmus milk reactions that may occur when proteins are metabolized as an energy source.

3. Explain how the litmus in the litmus milk acts as a redox indicator.

4. Can a litmus milk culture show a pink band at the top and a brownish translucent layer at the bottom? Explain.

5. Explain why litmus milk is considered a good differential medium.

Nitrate Reduction Test

LEARNING OBJECTIVE

Once you have completed this experiment, you should be able to

1. Determine the ability of some microorganisms to reduce nitrates (NO_3^-) to nitrites (NO_2^-) or beyond the nitrite stage.

Principle

The reduction of nitrates by some aerobic and facultative anaerobic microorganisms occurs in the absence of molecular oxygen, an anaerobic process. In these organisms, anaerobic respiration is an oxidative process whereby the cell uses inorganic substances such as nitrates (NO_3^-) or sulfates (SO_4^{2-}) to supply oxygen that is subsequently utilized as a final hydrogen acceptor during energy formation. The biochemical transformation may be visualized as follows:

Partial Reduction

$$NO_3^- + 2H^+ + 2e^- \xrightarrow{\text{Nitrate reductase}} NO_2^- + H_2O$$

Nitrate   Hydrogen   electrons   Nitrite   Water

Some organisms possess the enzymatic capacity to act further on nitrites to reduce them to ammonia (NO_3^+) or molecular nitrogen (N_2). These reactions may be described as follows:

Complete reduction

$$NO_2^- \longrightarrow NH_3^+$$

Nitrite   Ammonia

or

$$2NO_3^- + 12H^+ + 10e^- \longrightarrow N_2 + 6H_2O$$

Nitrate   Molecular nitrogen

Nitrate reduction can be determined by cultivating organisms in a nitrate broth medium. The medium is a nutrient broth supplemented with 0.1% potassium nitrate (KNO_3) as the nitrate substrate. In addition, the medium is made into a semisolid by the addition of 0.1% agar. The semi-solidity impedes the diffusion of oxygen into the medium, thereby favoring the anaerobic requirement necessary for nitrate reduction.

Following incubation of the cultures, an organism's ability to reduce nitrates to nitrites is determined by the addition of two reagents: Solution A, which is sulfanilic acid, followed by Solution B, which is α-naphthylamine. *Note: This should not be confused with Barritt's reagent.* Following reduction, the addition of Solutions A and B will produce an immediate cherry-red color.

$$NO_3^- \xrightarrow{\text{Nitrate reductase}} NO_2^- \text{ (Red color on addition of Solutions A and B)}$$

Cultures not producing a color change suggest one of two possibilities: (1) nitrates were not reduced by the organism, or (2) the organism possessed such potent **nitrate reductase** enzymes that nitrates were rapidly reduced beyond nitrites to ammonia or even molecular nitrogen. To determine whether or not nitrates were reduced past the nitrite stage, a small amount of zinc powder is added to the basically colorless cultures already containing Solutions A and B. Zinc reduces nitrates to nitrites. The development of red color therefore verifies that nitrates were not reduced to nitrites by the organism. If nitrates were not reduced, a negative nitrate reduction reaction has occurred. If the addition of zinc does not produce a color change, the nitrates in the medium were reduced beyond nitrites to ammonia or nitrogen gas. This is a positive reaction, as shown in Figure 27.1. Results of nitrate reduction tests are shown in Figure 27.2.

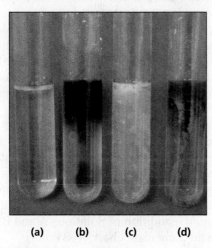

Sulfanilic acid (colorless) + α-naphthylamine (colorless) + Nitrous acid → Sulfobenzene azo-α-naphthylamine (red) + Water

Figure 27.1 Formation of colored complex indicative of NO₃⁻ reduction

(reaction structures: NH_2/SO_3H benzene, NH_2 naphthylamine + HNO_2 → N=N azo compound SO_2H/NH_2 + H_2O)

Figure 27.2 Nitrate reduction tests.
(a) Uninoculated, **(b)** positive with Solutions A + B,
(c) positive with Solutions A + B + zinc powder, and
(d) negative with Solutions A + B + zinc powder.

CLINICAL APPLICATION

Differentiating Mycobacterium Tuberculosis from Non-tubercle Mycobacterium

This test is used to identify intestinal bacteria that are able to reduce nitrates to nitrites. When presented with a patient that exhibits the symptoms of tuberculosis and is positive for tubercles on an x-ray, a sputum sample will be tested for *Mycobacterium*. To distinguish between *Mycobacterium tuberculosis* and other *Mycobacterium* species a nitrate reduction test will be used since *M. tuberculosis* is the only *Mycobacterium* species with this capacity.

AT THE BENCH

Materials

Cultures

24- to 48-hour Trypticase soy broth cultures of *Escherichia coli*, *Alcaligenes faecalis*, and *Pseudomonas aeruginosa* BSL-2 for the short version. 24- to 48-hour Trypticase soy broth cultures of the 13 organisms listed on page 148 for the long version.

Media

Trypticase nitrate broth per designated student group: 4 for the short version, 14 for the long version.

Reagents

Solution A (sulfanilic acid), Solution B (α-naphthylamine), and zinc powder.

Equipment

Microincinerator or Bunsen burner, inoculating loop, test tube rack, and glassware marking pencil.

Procedure Lab One

1. Using aseptic technique, inoculate each experimental organism into its appropriately labeled tube by means of a loop inoculation. The last tube will serve as a control.

2. Incubate all cultures for 24 to 48 hours at 37°C.

Procedure Lab Two

1. Add five drops of Solution A and then five drops of Solution B to all nitrate broth cultures. Observe and record in the Lab Report chart whether or not a red coloration develops in each of the cultures.

2. Add a minute quantity of zinc to the cultures in which no red color developed. Observe and record whether or not red coloration develops in each of the cultures.

3. On the basis of your observations, determine and record in the Lab Report chart whether or not each organism was capable of nitrate reduction. Identify the end product (NO_2^- or NH_3^+/N_2), if any, that is present.

Catalase Test

LEARNING OBJECTIVE

Once you have completed this experiment, you should be able to

1. Determine the ability of some microorganisms to degrade hydrogen peroxide by producing the enzyme catalase.

Principle

During aerobic respiration, microorganisms produce hydrogen peroxide and, in some cases, an extremely toxic superoxide. Accumulation of these substances will result in death of the organism unless they can be enzymatically degraded. These substances are produced when aerobes, facultative anaerobes, and microaerophiles use the aerobic respiratory pathway, in which oxygen is the final electron acceptor, during degradation of carbohydrates for energy production. Organisms capable of producing **catalase** rapidly degrade hydrogen peroxide as illustrated:

$$2H_2O_2 \xrightarrow{\text{Catalase}} 2H_2O + O_2\uparrow$$

Hydrogen peroxide   **Water**   **Free oxygen**

Aerobic organisms that lack catalase can degrade especially toxic superoxides using the enzyme **superoxide dismutase**; the end product of a superoxide dismutase is H_2O_2, but this is less toxic to the bacterial cells than are the superoxides.

The inability of strict anaerobes to synthesize catalase, peroxidase, or superoxide dismutase may explain why oxygen is poisonous to these microorganisms. In the absence of these enzymes, the toxic concentration of H_2O_2 cannot be degraded when these organisms are cultivated in the presence of oxygen.

Catalase production can be determined by adding the substrate H_2O_2 to an appropriately incubated Trypticase soy agar slant culture. If catalase is present, the chemical reaction mentioned is indicated by bubbles of free oxygen gas $O_2\uparrow$. This is a positive catalase test; the absence of bubble formation is a negative catalase test. Figure 28.1 shows the results of the catalase test using (a) the tube method, (b) the plate method, and (c) slide method.

CLINICAL APPLICATION

Differentiation of *Staphylococci*, *Streptococci*, and *Enterobacteriaceae*

The catalase test is used for the biochemical differentiation of catalase-positive *Staphylococci* and catalase-negative *Streptococci*, as well as members of the *Enterobacteriaceae*. With the increasing worry about methicillin-resistant strains of *Staphylococcus* in hospital settings, the catalase test is a quick and easy way to differentiate *S. aureus*, which may be methicillin-resistant *S. aureus* (MRSA), from other *Staphylococcus* species that have exhibited lower incidences of methicillin-resistance.

AT THE BENCH

Materials

Cultures

24- to 48-hour Trypticase soy broth cultures of *Staphylococcus aureus* BSL-2 , *Micrococcus luteus*, and *Lactococcus lactis* for the short version. 24- to 48-hour Trypticase soy broth cultures of the 13 organisms listed on page 148 for the long version.

Media

Trypticase soy agar slants per designated student group: 4 for the short version, 14 for the long version.

Reagent

3% hydrogen peroxide.

Equipment

Tube method: Microincinerator or Bunsen burner, inoculating loop, test tube rack, and glassware marking pencil.

(a) Tube method

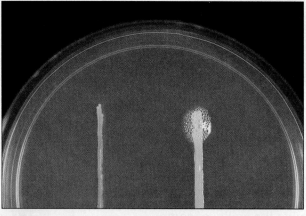

(b) Plate method

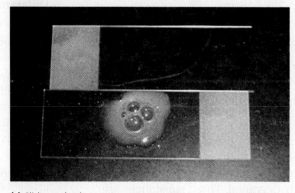

(c) Slide method

Figure 28.1 Catalase test. Negative results are shown on the left and positive results on the right in the **(a)** tube method and **(b)** plate method. Negative results are shown on the top and positive results on the bottom in the **(c)** slide method.

Slide method: Microincinerator or Bunsen burner, inoculating loop, glassware marking pencil, glass microscope slides (4 for the short version, 14 for the long version), Petri dish and cover.

Procedure Lab One

Tube Method

1. Using aseptic technique, inoculate each experimental organism into its appropriately labeled tube by means of a streak inoculation. The last tube will serve as a control.

2. Incubate all cultures for 24 to 48 hours at 37°C.

Procedure Lab Two

Tube method

1. Allow three or four drops of the 3% hydrogen peroxide to flow over the entire surface of each slant culture.

2. Examine each culture for the presence or absence of bubbling or foaming. Record your results in the chart in the Lab Report.

3. Based on your observations, determine and record whether or not each organism was capable of catalase activity.

Slide Method

1. Label slides with the names of the organisms.

2. Using a sterile loop, collect a small sample of the first organism from the culture tube and transfer it to the appropriately labeled slide.

3. Place the slide in the Petri dish.

4. Place one drop of 3% hydrogen peroxide on the sample. Do not mix. Place the cover on the Petri dish to contain any aerosols.

5. Observe for immediate presence of bubble formation. Record your results in the chart in the Lab Report.

6. Repeat Steps 2 through 5 for the remaining test organisms.

Oxidase Test

LEARNING OBJECTIVE

Once you have completed this experiment, you should be able to

1. Perform an experimental procedure that is designed to distinguish among groups of bacteria on the basis of cytochrome oxidase activity.

Principle

Oxidase enzymes play a vital role in the operation of the electron transport system during aerobic respiration. **Cytochrome oxidase** catalyzes the oxidation of a reduced cytochrome by molecular oxygen (O_2), resulting in the formation of H_2O or H_2O_2. Aerobic bacteria, as well as some facultative anaerobes and microaerophiles, exhibit oxidase activity. The oxidase test aids in differentiation among members of the genera *Neisseria* and *Pseudomonas*, which are oxidase-positive, and Enterobacteriaceae, which are oxidase-negative.

The ability of bacteria to produce cytochrome oxidase can be determined by the addition of the test reagent *p*-aminodimethylaniline oxalate to colonies grown on a plate medium. This light pink reagent serves as an artificial substrate, donating electrons and thereby becoming oxidized to a blackish compound in the presence of the oxidase and free oxygen. Following the addition of the test reagent, the development of pink, then maroon, and finally dark purple coloration on the surface of the colonies is indicative of cytochrome oxidase production and represents a positive test. No color change, or a light pink coloration on the colonies, is indicative of the absence of oxidase activity and is a negative test. The filter paper method may also be used and is described in this experiment.

CLINICAL APPLICATION

Test to Distinguish Family Enterobacteriaceae from Non-Enterobacteriaceae

Enterobacteriaceae are cytochrome oxidase–negative, while *Neisseria* and *Pseudomonas* are cytochrome oxidase–positive. The oxidase test is an important tool in the identification of *Neisseria meningitis*, the causative agent of bacterial meningitis, which has a significant morbidity and mortality rate. In addition, yeast of medical importance, such as *Candida,* can be separated from *Saccharomyces* and *Torulopsis* by this test.

AT THE BENCH

Materials

Cultures

24- to 48-hour Trypticase soy broth cultures of *Escherichia coli, Pseudomonas aeruginosa* **BSL-2** , and *Alcaligenes faecalis* for the short version. 24- to 48-hour Trypticase soy broth cultures of the 13 organisms listed on page 148 for the long version.

Media

Trypticase soy agar plates per designated student group: one for the short version, four for the long version.

Reagent

p-Aminodimethylaniline oxalate (Difco 0329-13-9).

Equipment

Plate method: Bunsen burner, inoculating loop, and glassware marking pencil.

Filter paper method: All of the above and filter paper (one for short version, four for long version).

Procedure Lab One

Plate Method

1. Prepare the Trypticase soy agar plate(s) for inoculation as follows:

   a. Short procedure: With a glassware marking pencil, divide the bottom of a Petri dish into three sections and label each section with the name of the test organism to be inoculated.

   b. Long procedure: Repeat Step 1a, dividing three plates into three sections and one plate into four sections to accommodate the 13 test organisms.

2. Using aseptic technique, make a single-line streak inoculation of each test organism on the agar surface of its appropriate section of the plate(s).

3. Incubate the plate(s) in an inverted position for 24 to 48 hours at 37°C.

Procedure Lab Two

Plate Method

1. Add two or three drops of the *p*-aminodimethylaniline oxalate to the surface of the growth of each test organism.

2. Observe the growth for the presence or absence of a color change from pink, to maroon, and finally to purple. Positive test (+), color change in 10–30 seconds; negative test (−), no color change, or light pink color. Refer to Figure 29.1. Record the results on the chart in the Lab Report.

3. Based on your observations, determine and record whether or not each organism was capable of producing cytochrome oxidase.

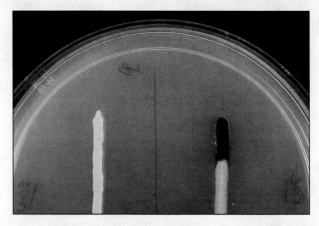

Figure 29.1 **Oxidase test.** Negative test, on left, results in no color change, and positive test, on right, results in a color change to purple.

Filter Paper Method

1. Prepare Petri dishes as described in Lab One Steps 1a and 1b.

2. Place filter paper in Petri dish.

3. With a sterile loop, obtain a heavy loopful of the first test organism and gently smear it on the filter paper.

4. Drop one or two drops of *p*-aminodimethylaniline oxalate reagent on the test organism.

5. Observe the organism for the appearance of a purple color within 30 seconds of contact with the oxidase reagent, indicating a positive test.

6. Repeat Steps 3 to 5 for the remaining test organisms.

7. Record your results in the chart in the Lab Report.

Utilization of Amino Acids

The study of the metabolism of amino acids began in the early part of the twentieth century. Some investigators found that the enteric microorganisms, such as *Proteus* and the so-called Providence species, were able to deaminate a variety of amino acids that provided a vehicle for distinguishing these microorganisms from other members of the large family of the Enterobacteriaceae. It was determined that 11 of the 22 amino acids were deaminated by amino acid oxidases, and it was phenylalanine deaminase that produced the most rapid enzymatic activity. Thus, phenylalanine deaminase became the most widely studied deaminase used to differentiate enteric organisms.

Likewise, some organisms were found to be capable of decarboxylating amino acids, providing a way to differentiate between the enteric genera and species. For instance, lysine decarboxylase is capable of differentiating between *Salmonella* and *Citrobacter*. Ornithine decarboxylase separates *Enterobacter* from *Klebsiella*. Decarboxylase enzymes are numerous, and each is specific for a particular substrate.

It is now evident that decarboxylases and deaminases play a vital role in the utilization of amino acids and the metabolism of nitrogen compounds.

PART A Decarboxylase Test

LEARNING OBJECTIVE

Once you have completed this experiment, you should be able to

1. Identify and differentiate organisms based on their ability to enzymatically degrade amino acid substrates.

Principle

Every biologically active protein is composed of the 20 essential amino acids. Structurally, amino acids are composed of an alpha carbon (—C—), an amino

group (—NH_2), a carboxyl group (—COOH), and a hydrogen atom (—H). Also attached to the alpha carbon is a side group or an atom designated by an (—R), which differs in each of the amino acids.

$$NH_2 - \overset{\overset{\displaystyle R}{|}}{\underset{\underset{\displaystyle H}{|}}{C}} - COOH$$

Decarboxylation is a process whereby some microorganisms that possess decarboxylase enzymes are capable of removing the carboxyl group to yield end products consisting of an **amine** or **diamine** plus **carbon dioxide**. Decarboxylated amino acids play an essential role in cellular metabolism since the amines produced may serve as end products for the synthesis of other molecules required by the cell. Decarboxylase enzymes are designated as adaptive (or induced) enzymes and are produced in the presence of specific amino acid substrates upon which they act. These amino acid substrates must possess at least one chemical group other than an amine (—NH_2) or a carboxyl group (—COOH). In the process of decarboxylation, organisms are cultivated in an acid environment and in the presence of a specific substrate. The decarboxylation end product (amines) results in a shift to a more alkaline pH.

In the clinical or diagnostic microbiology laboratory, three decarboxylase enzymes are used to differentiate members of the Enterobacteriaceae: lysine, ornithine, and arginine. Decarboxylase activity is determined by cultivating the organism in a nutrient medium containing glucose, the specific amino acid substrate, and bromthymol blue (the pH indicator). If decarboxylation occurs, the pH of the medium becomes alkaline despite the fermentation of glucose since the end products (amines or diamines) are alkaline. The function of the glucose in the medium is to ensure good microbial growth and thus more reliable results in the presence of the pH indicator. The presence of each decarboxylase enzyme can be tested for by supplementing decarboxylase broth with

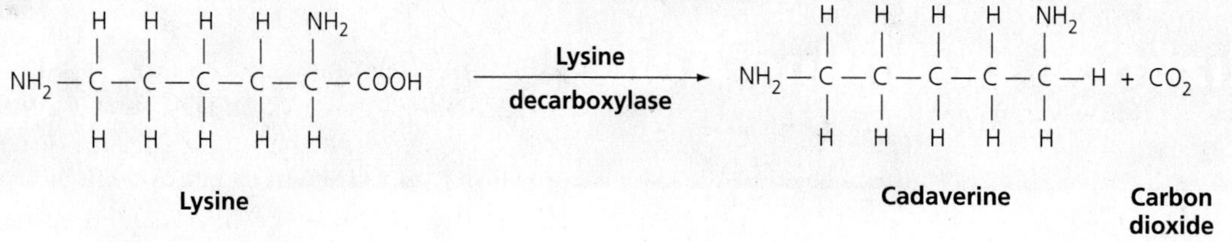

$$NH_2-\underset{\underset{H}{|}}{\overset{\overset{H}{|}}{C}}-\underset{\underset{H}{|}}{\overset{\overset{H}{|}}{C}}-\underset{\underset{H}{|}}{\overset{\overset{H}{|}}{C}}-\underset{\underset{H}{|}}{\overset{\overset{H}{|}}{C}}-\underset{\underset{H}{|}}{\overset{\overset{NH_2}{|}}{C}}-COOH \xrightarrow[\text{decarboxylase}]{\text{Lysine}} NH_2-\underset{\underset{H}{|}}{\overset{\overset{H}{|}}{C}}-\underset{\underset{H}{|}}{\overset{\overset{H}{|}}{C}}-\underset{\underset{H}{|}}{\overset{\overset{H}{|}}{C}}-\underset{\underset{H}{|}}{\overset{\overset{H}{|}}{C}}-\underset{\underset{H}{|}}{\overset{\overset{NH_2}{|}}{C}}-H + CO_2$$

Lysine   Cadaverine   Carbon dioxide

Figure 30.1 Degradation of lysine

the specific amino acid substrate, namely lysine, arginine, and ornithine. For example, **lysine decarboxylase** degrades l-lysine, forming the diamine end product **cadaverine** plus **carbon dioxide** as illustrated in Figure 30.1.

In the experiment that follows, the decarboxylation of L-lysine will be studied. It should be noted that decarboxylation reactions occur under anaerobic conditions that are satisfied by sealing the culture tubes with sterile mineral oil. In the sealed tubes, all of the unbound oxygen is utilized during the organisms' initial growth phase, and the pH of the medium becomes alkaline as carbon dioxide (CO_2) is produced in the culture tube. A pH indicator, such as bromcresol purple, is usually incorporated into the medium for the easy detection of pH changes. The production of acid end products will cause the bromcresol purple to change color from purple to yellow, indicating that acid has formed, the medium has been acidified, and the decarboxylase enzymes have been activated. The activated enzyme responds with the production of the alkalinizing diamine (cadaverine) and carbon dioxide, which will produce a final color change from yellow back to purple, thereby indicating that L-lysine has been decarboxylated. The development of a turbid purple color verifies a positive test for amino acid decarboxylation. The absence of a purple color indicates a negative result.

CLINICAL APPLICATION

Distinguishing between *Enterobacter* Species

The decarboxylase test identifies bacteria based on the production of ammonia from the amino acids lysine, ornithine, and arginine. The decarboxylase test can be used to differentiate the causative agent in many nosocomial infections of immuno-compromised patients. The bacterium *Enterobacter aerogenes* is lysine decarboxylase positive while other *Enterobacter* species are negative. The decarboxylase test is used primarily to identify bacteria within the *Enterobacteriaceae* family.

AT THE BENCH

Materials

Cultures

24-hour nutrient broth cultures of *Proteus vulgaris, Escherichia coli,* and *Citrobacter freundii* for the short version. 24- to 48-hour nutrient broth cultures of the 13 organisms listed on page 148 for the long version.

Media

Per designated student group: three tubes of Moeller's decarboxylase broth supplemented with L-lysine (10 g/l) (labeled LD+), three tubes of Moeller's decarboxylase broth without lysine (labeled LD−).

Equipment

Microincinerator or Bunsen burner, glassware marking pencil, inoculating loop and needle, sterile Pasteur pipettes, rubber bulbs, test tube rack, and sterile mineral oil.

Controls

A positive control organism for this test would be *Escherichia coli.*

Procedure Lab One

1. With a glassware marking pencil, label three tubes of the LD+ medium with the name of the organism to be inoculated. Similarly label three tubes of LD− medium. The use of (LD−) control tubes is essential since some bacterial strains are capable of turning substrate-free media positive. *Note: Control tubes should remain yellow after incubation, denoting that only glucose was fermented.* The presence of a positive control tube invalidates the test, and no interpretation is possible.

(a)   (b)   (c)

Figure 30.2 **Decarboxylase test (a)** Uninoculated, **(b)** negative, and **(c)** positive

2. Using aseptic technique, inoculate each experimental organism into its appropriately labeled tube using a loop inoculation.

3. Place a rubber bulb onto a sterile Pasteur pipette and overlay the surface of the inoculated culture tubes with 1 ml of sterile mineral oil. Hold the tubes in a slanted position while adding the mineral oil. *Note: Do not let the tip of the pipette touch the inoculated medium or the sides of the test tube walls.*

4. Repeat the above procedure for the remaining test cultures.

5. Incubate all tubes at 37°C for 24 to 48 hours.

Procedure Lab Two

1. Examine each culture tube for the presence of a color change. Refer to Figure 30.2.

2. Based on your observations, determine whether or not each organism was capable of performing decarboxylation of lysine.

3. Record your results in the chart in the Lab Report.

PART B Phenylalanine Deaminase Test

LEARNING OBJECTIVE

Once you have completed this experiment, you should be able to

1. Demonstrate the ability of some organisms to remove the amino group (–NH_2) from amino acids.

Principle

Microorganisms that contain deaminase enzymes are capable of removing the amino group (—NH_2) from amino acids and other NH_2-containing chemical compounds. During this process the amino acid, under the auspices of its specific deaminase, will produce keto acids and ammonia as end products. In the experiment to follow, the amino acid phenylalanine will be deaminated by **phenylalanine deaminase** and converted to the **keto acid phenylpyruvic acid** and **ammonia**. The organisms are cultured on a medium incorporating phenylalanine as the substrate. This chemical reaction is illustrated in Figure 30.3.

If the organism possesses phenylalanine deaminase, phenylpyruvic acid will be released into the medium and can be detected by the addition of a 10% to 12% ferric chloride solution to the surface of the medium. If a green color develops, the enzymatic deamination of the substrate has occurred and is indicative of a positive result. The absence of any color change indicates a negative result. The resultant green color produced upon the addition of ferric chloride ($FeCl_3$) is due to the formation of a keto acid (phenylpyruvic acid). It has been shown that α- and β-keto acids give a positive color reaction with either alcoholic or aqueous solutions of $FeCl_3$. Phenylpyruvic acid is an α-keto acid. The results should be read immediately following the addition of the reagent since the color produced fades quickly. When not in use, the ferric chloride reagent should be refrigerated and kept in a dark bottle to avoid exposure to light. The stability of this reagent varies and should be checked weekly with known positive cultures.

CLINICAL APPLICATION

Differentiating Intestinal Bacteria

The phenylalanine deaminase test uses the differential medium phenylalanine agar to detect bacteria containing the enzyme phenylalanine deaminase, and is used to differentiate the genera *Proteus, Morganella,* and *Providencia* from other gram-negative intestinal bacilli. These genera of enteric and environmental bacteria are known to cause UTIs and gastroenteritis. It is clinically important to distinguish them from other enteric bacteria due to their high level of antibiotic resistance.

Phenylalanine deaminase → Phenylpyruvic acid   Ammonia

Phenylalanine

Figure 30.3 Deamination of phenylalanine

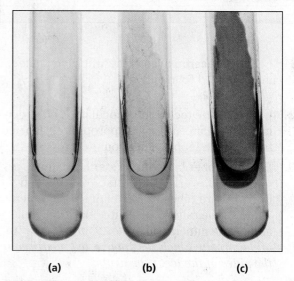

(a)   (b)   (c)

Figure 30.4 Phenylalanine deaminase test
(a) Uninoculated, **(b)** negative, and **(c)** positive

AT THE BENCH

Materials

Cultures

24-hour nutrient broth cultures of *Escherichia coli* and *Proteus vulgaris* for the short version. 24-hour nutrient broth cultures of the 13 organisms listed on page 148 for the long version.

Media

Two phenylalanine agar slants.

Reagents

10% to 12% ferric chloride solution.

Equipment

Microincinerator or Bunsen burner, glassware marking pencil, Pasteur pipettes, rubber bulbs, test tube racks, and inoculating loop.

Controls

A positive-control organism for this test would be *Escherichia coli.*

Procedure Lab One

1. Using aseptic technique, inoculate each experimental organism into its appropriately labeled tube using a streak inoculation.

2. Incubate cultures at 37°C for 24 to 48 hours.

Procedure Lab Two

1. Add 5 to 10 drops of the ferric chloride solution to each agar slant and mix gently. Ferric chloride is a chelating agent and binds to the phenylpyruvic acid to produce a green color on the slant (Figure 30.4).

2. Based on your observations, determine whether or not each organism was capable of amino acid deamination. *Note: Results should be read immediately following the addition of ferric chloride because the green color fades rapidly.*

3. Record your results in the Lab Report.

Name: _____

Date: _____ Section: _____

Observations and Results

PARTS A and B: Decarboxylase Test and Phenylalanine Deaminase Test

Bacterial Species	DECARBOXYLASE TEST				PHENYLALANINE DEAMINASE TEST	
	Color of Medium		Lysine Decarboxylase (+) or (−)		Color after FeCl₃	Deamination (+) or (−)
	LD+	LD−	LD+	LD−		
E. coli						
E. aerogense						
K. pneumoniae						
S. dysenteriae						
S. typhimurium						
P. vulgaris						
P. aeruginosa						
A. faecalis						
M. luteus						
L. lactis						
S. aureus						
B. cereus						
C. xerosis						
C. freundii						
Alternate organism						
Control						

Review Questions

1. A negative decarboxylase test is indicated by the production of a yellow color in the medium. Explain the reason for the development of this color.

2. Explain why deaminase activity must be determined immediately following the addition of ferric chloride.

3. What is the function of ferric chloride in the detection of deaminase activity?

4. Explain why the anaerobic environment is essential for decarboxylation of the substrate to occur.

5. Following a normal delivery, a nurse observes that the urine of the infant has a peculiar odor resembling that of burnt sugar or maple syrup. Subsequent examination by the pediatrician reveals that this child has maple syrup urine disease.

   a. What is this disease?

   b. How is it treated?

AT THE BENCH

Materials

Cultures

Number-coded 24- to 48-hour Trypticase soy agar slant cultures of the 13 bacterial species listed on page 148. You will be provided with one unknown pure culture.

Media

Two Trypticase soy agar slants, and one each of the following per student: phenol red sucrose broth, phenol red lactose broth, phenol red dextrose broth, SIM agar deep tube, MR-VP broth, tryptic nitrate broth, Simmons citrate agar slant, urea broth, litmus milk, Trypticase soy agar plate, nutrient gelatin deep tube, starch agar plate, and tributyrin agar plate.

Reagents

Crystal violet; Gram's iodine; 95% ethyl alcohol; safranin; methyl red; 3% hydrogen peroxide; Barritt's reagent, Solutions A and B; Kovac's reagent; zinc powder; and p-aminodimethylaniline oxalate.

Equipment

Microincinerator or Bunsen burner, inoculating loop and needle, staining tray, immersion oil, lens paper, bibulous paper, microscope, and glassware marking pencil.

Procedure Lab One

1. Perform a Gram stain of the unknown organism. Observe and record in the Lab Report chart the reaction and the morphology and arrangement of the cells.

2. Using aseptic inoculating technique, inoculate two Trypticase soy agar slants by means of a streak inoculation. Following incubation, you will use one slant culture to determine the cultural characteristics of the unknown microorganism. You will use the second as a stock

subculture should it be necessary to repeat any of the tests.

3. Exercising care in aseptic technique so as not to contaminate cultures and thereby obtain spurious results, inoculate the media for the following biochemical tests:

Medium	Test
a. Phenol red lactose broth b. Phenol red dextrose broth c. Phenol red sucrose broth	Carbohydrate fermentation
d. Litmus milk	Litmus milk reactions
e. SIM medium	Indole production H_2S production
f. Tryptic nitrate broth	Nitrate reduction
g. MR-VP broth	Methyl red test Voges-Proskauer test
h. Simmons citrate agar slant	Citrate utilization
i. Urea broth	Urease activity
j. Trypticase soy agar slant	Catalase activity
k. Starch agar plate	Starch hydrolysis
l. Tributyrin agar plate	Lipid hydrolysis
m. Nutrient gelatin deep tube	Gelatin liquefaction
n. Trypticase soy agar plate	Oxidase test

4. Incubate all cultures for 24 to 72 hours at 37°C.

Procedure Lab Two

1. Examine a Trypticase soy agar slant culture and determine the cultural characteristics of your unknown organism. Record your results in the Lab Report.

2. Perform biochemical tests on the remaining cultures, making reference to the specific laboratory exercise for each test. Record your observations and results.

3. Based on your results, identify the genus and species of the unknown organism. *Note: Results may vary depending on the strains of each species used and the length of time the organism has been maintained in stock culture. The observed results may not be identical to the expected results. Therefore choose the organism that best fits the results summarized in Table 31.1.*

Name: _____

Date: _____ Section: _____

Observations and Results

Description of Unknown's Characteristics		Student Culture no. Organism	_____ _____ _____
Experimental Procedure	**Observations**	**Results**	
Gram stain			
Acid-fast stain			
Shape and arrangement			
Cultural characteristics			
Litmus milk reactions			
Carbohydrate fermentations: Lactose			
Dextrose			
Sucrose			
H_2S production			
Nitrate reduction			
Indole production			
Methyl red test			
Voges-Proskauer test			
Citrate utilization			
Urease activity			
Catalase activity			
Starch hydrolysis			
Lipid hydrolysis			
Gelatin liquefaction			
Oxidase test			

PART 6

The Protozoa

Once you have completed the experiments in this section, you should be

1. Familiar with the distinguishing characteristics of protozoans.
2. Able to identify free-living and parasitic protozoans in microscopic views.

Introduction

The protozoa are a large and diverse group of unicellular, eukaryotic organisms. Most are free-living, but some are parasites. Their major distinguishing characteristics are

1. The absence of a cell wall; some, however, possess a flexible layer, a pellicle, or a rigid shell of inorganic materials outside of the cell membrane.
2. The ability during their entire life cycle or part of it to move by locomotor organelles or by a gliding mechanism.
3. Heterotrophic nutrition whereby the free-living forms ingest particulates such as bacteria, yeast, and algae, while the parasitic forms derive nutrients from the body fluids of their hosts.
4. Primarily asexual means of reproduction, although sexual modes occur in some groups.

Protozoan taxonomy is being continually updated as new technology enables classification based on molecular characteristics. To ease our discussion of protozoans, we follow a more traditional taxonomic scheme, dividing them into four groups based on means of locomotion.

1. **Sarcodina:** Motility results from the streaming of ectoplasm, producing protoplasmic projections called pseudopods (false feet). Prototypic amoebas include the free-living *Amoeba proteus* and the parasite *Entamoeba histolytica*.
2. **Mastigophora:** Locomotion is effected by one or more whiplike, thin structures called flagella. Free-living members include the genera *Cercomonas*, *Heteronema*, and *Euglena*, which are photosynthetic protists that may be classified as flagellated algae. The parasitic forms include *Trichomonas vaginalis*, *Giardia intestinalis* (formerly called *Giardia lamblia*), and the *Trypanosoma* species.
3. **Ciliophora:** Locomotion is carried out by means of short hairlike projections called cilia, whose synchronous beating propels the organisms. The characteristic example of free-living members of this group is *Paramecium caudatum*, and the parasitic example is *Balantidium coli*.
4. **Sporozoa:** Unlike other members of this phylum, sporozoa do not have locomotor organelles in their mature stage; however, immature forms exhibit some type of movement. All the members of this group are parasites. The most significant members belong to the genus *Plasmodium*, the malarial parasites of animals and humans.

Free-Living Protozoa

LEARNING OBJECTIVE

Once you have completed this experiment, you should be familiar with

1. The protozoa of pond water.

Principle

There are more than 20,000 known species of free-living protozoa. This manual does not present an in-depth study of this large and diverse population. Therefore, in this procedure, you will use Table 32.1 and Figure 32.1 and Figure 32.2 to become familiar with the general structural characteristics of representative protozoa, and you will identify these in a sample of pond water.

CLINICAL APPLICATION

Wet Mounts for Diagnosis

Wet mount slides, often utilizing stains, are routinely used in the examination of stool samples for infectious protozoans, such as *Entamoeba histolytica.* This organism causes amoebic dysentery and has been known to lead to severe liver damage. Although most infections are asymptomatic, carriers can still spread the disease. Diagnosis may require examination of several slide preparations.

AT THE BENCH

Materials

Cultures

Stagnant pond water and prepared slides of amoebas, paramecia, euglenas, and stentors.

Reagent

Methyl cellulose.

Equipment

Microscope, glass slides, coverslips, and Pasteur pipettes.

Procedure

1. Obtain a drop of pond water from the bottom of the culture and place it in the center of a clean slide.

2. Add a drop of methyl cellulose to the culture to slow down the movement of the protozoa.

3. Apply a coverslip in the following manner to prevent formation of air bubbles:

    a. Place one edge of the coverslip against the outer edge of the drop of culture.

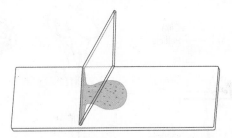

    b. After the drop of culture spreads along the inner aspect of the edge of the coverslip, gently lower the coverslip onto the slide.

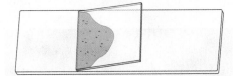

4. Examine your slide preparation under scanning, low-power, and high-power objectives with diminished light, and observe for the different protozoa present. Record your results in the Lab Report.

TABLE 32.1 Structural Characteristics of Free-Living Protozoa

Sarcodina

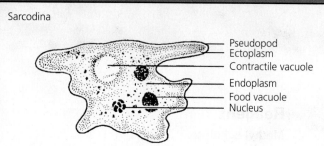

Pseudopod
Ectoplasm
Contractile vacuole
Endoplasm
Food vacuole
Nucleus

Amoeba

1. **Pseudopods:** Protoplasmic projections that function for locomotion
2. **Ectoplasm:** Outer layer of cytoplasm; clear in appearance
3. **Endoplasm:** Inner cytoplasmic region; granular in appearance
4. **Nucleus:** One present
5. **Food vacuoles:** Contain engulfed food undergoing digestion
6. **Contractile vacuole:** Large, clear circular structure that regulates internal water pressure

Mastigophora

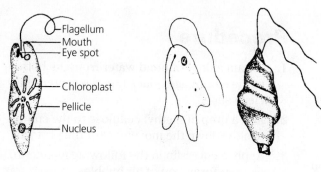

Flagellum
Mouth
Eye spot

Chloroplast

Pellicle

Nucleus

Euglena   *Cercomonas*   *Heteronema*

1. **Flagella:** One to several long whiplike structures that function for locomotion
2. **Pellicle:** Elastic layer outside of cell membrane
3. **Mouth:** Present but indistinct
4. **Chloroplast:** Organelles containing chlorophyll; present in photosynthetic forms only
5. **Eye spot:** Light-sensitive pigmented spot
6. **Nucleus:** One present

Ciliophora

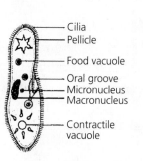

Cilia
Pellicle

Food vacuole

Oral groove
Micronucleus
Macronucleus

Contractile vacuole

Paramecium   *Stentor*   *Vorticella*

1. **Cilia:** Numerous, short, hairlike structures that function for locomotion
2. **Pellicle:** Outermost flexible layer
3. **Contractile vacuole** with radiating canals; regulates osmotic pressure
4. **Oral groove:** Indentation that leads to the mouth and gullet
5. **Food vacuoles:** Sites of digestion of ingested food
6. **Macronucleus:** A large nucleus that functions to control the cell's activities; one to several may be present
7. **Micronucleus:** A small nucleus that functions in conjugation, a mode of sexual reproduction

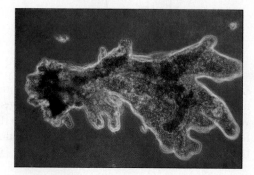

Figure 32.1 Amoeba

(a) *Euglena viridis*   **(b)** *Paramecium caudatum*

Figure 32.2 Euglena and Paramecium

Name: _____

Date: _____   Section: _____

Observations and Results

1. In the space provided, draw a representative sketch of several of the
    observed protozoa in stagnant pond water, indicate the magnifications used,
    and label their structural components. Identify each organism according to
    its class based on its mode of locomotion and its genus.

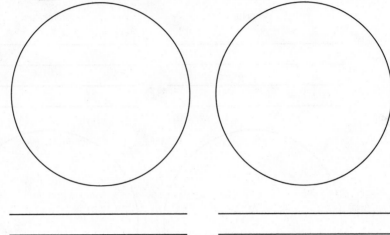

Magnification: _____   _____

Organelles of locomotion: _____   _____

Class: _____   _____

Genus: _____   _____

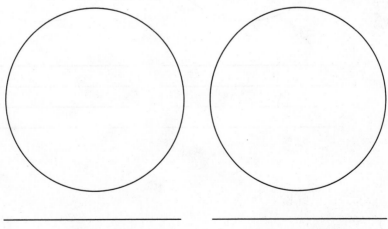

Magnification: _____   _____

Organelles of locomotion: _____   _____

Class: _____   _____

Genus: _____   _____

2. Draw representative sketches, indicate magnification, and label the structural components. Identify each organism according to its class based on locomotion and genus.

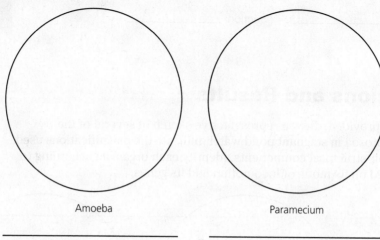

Amoeba   Paramecium

Magnification: _____   _____

Organelles of locomotion: _____   _____

Class: _____   _____

Genus: _____   _____

Euglena   Stentor

Magnification: _____   _____

Organelles of locomotion: _____   _____

Class: _____   _____

Genus: _____   _____

Review Questions

1. What are the distinguishing characteristics of the free-living members of Sarcodina, Mastigophora, and Ciliophora?

2. Identify and give the function of the following:
   a. Pseudopods:

   b. Contractile vacuole:

   c. Eye spot:

   d. Micronucleus:

   e. Pellicle:

   f. Oral groove:

3. People with AIDS are vulnerable to toxoplasmosis caused by the protozoan *Toxoplasma gondii*, resulting in infection of lungs, liver, heart, and brain, and often leading to death. About 25% of the world's population is infected, usually without developing symptoms. Why then are people with AIDS so susceptible to this disease?

Parasitic Protozoa

LEARNING OBJECTIVE

Once you have completed this experiment, you should be familiar with

1. Parasitic protozoan forms.

Principle

Unlike the life cycles of the free-living forms, the life cycles of parasitic protozoa vary greatly in complexity. Knowledge of the various developmental stages in these life cycles is essential in the diagnosis, clinical management, and chemotherapy of parasitic infections.

Parasites with the simplest or most direct life cycles not requiring an intermediate host are the following:

1. **Entamoeba histolytica:** A pseudopodian parasite of the class Sarcodina that causes amebic dysentery. Infective, resistant cysts are released from the lumen of the intestine through the feces and are deposited in water, in soil, or on vegetation. Upon ingestion, the mature quadrinucleated cyst wall disintegrates and the nuclei divide, producing eight active trophozoites (metabolically active cells) that move to the colon, where they establish infection.

2. **Balantidium coli:** The ciliated parasitic protozoan exhibits a life cycle similar to that of *E. histolytica* except that no multiplication occurs within the cyst. This organism resides primarily in the lumen and submucosa of the large intestine. It causes intestinal ulceration and alternating constipation and diarrhea.

3. **Giardia intestinalis:** The intestinal mastigophoric flagellate exhibits a life cycle

comparable to those of the above parasites. This organism is responsible for the induction of abdominal discomfort and severe diarrhea. Diagnosis is made by finding cysts in the formed stool and both cysts and trophozoites in the diarrhetic stool.

The mastigophoric hemoflagellate responsible for various forms of African sleeping sickness has a more complex life cycle. The *Trypanosoma* must have two hosts to complete its cyclic development: a vertebrate and an invertebrate, blood-sucking insect host. Humans are the definitive hosts harboring the sexually mature forms; the tsetse fly (*Glossina*) and the reduviid bug are the invertebrate hosts in which the developmental forms occur.

Table 33.1 illustrates the morphological characteristics of prototypic members of the parasitic protozoa except the Sporozoa.

Protozoa demonstrating the greatest degree of cyclic complexity are found in the class Sporozoa. They are composed of exclusively obligate parasitic forms, such as members of the genus *Plasmodium*, and are responsible for malaria in both humans and animals. The life cycle requires two hosts, a human being and the female *Anopheles* mosquito. It is significant to note that in this life cycle, the mosquito, and not the human, is the definitive host harboring the sexually mature parasite.

Malaria is initiated when a person is bitten by an infected mosquito, during which time infective sexually mature sporozoites are injected along with the insect's saliva. These parasites pass rapidly from the blood into the liver, where they infect the parenchymal cells. This is the **pre-erythrocytic stage**. The parasites develop asexually within the liver cells by a process called **schizogony**, producing **merozoites**. This cycle may be

TABLE 33.1 Structural Characteristics of Free-Living Protozoa

CLASS, ORGANISM, AND INFECTION	STRUCTURAL CHARACTERISTICS	LOCOMOTOR ORGANELLES	SITE OF INFECTION	ISOLATION OF PARASITIC FORM
SARCODINA *Entamoeba histolytica* Infection: Amebic dysentery Peripheral chromatin / Central karyosome / Uniform cytoplasm / Red blood cell Nuclei Chromatoid body Early　Late	**Trophozoite:** Shape: Variable Nucleus: Discrete nuclear membrane with central karyosome and peripheral chromatin granules Cytoplasm: Clear, red blood cells may be present	Pseudopods	Large intestine by ingestion of mature cysts	Diarrhetic stool
	Cyst: Shape: Round to oval with thick wall Nuclei: 1–4 present; mature cyst is quadrinucleated Chromatoid bodies: Sausage-shaped with rounded ends, present in young cysts only	None		Formed stool
MASTIGOPHORA *Trypanosoma gambiense* Infection: African sleeping sickness Flagellum Undulating membrane Nucleus Volutin granules Kinetoplast	**Trophozoite:** Shape: Crescent Nucleus: Large, central, and polymorphic Cytoplasm: Granular **Cyst:** None	Single flagellum along undulating membrane	Peripheral blood-stream by means of tsetse fly vector	Peripheral blood

repeated or the merozoites that are released from the ruptured liver cells may now infect red blood cells and initiate the **erythrocytic stage**. During this asexual development, the parasite undergoes a series of morphological changes that are of diagnostic value. These forms are designated as **signet rings**, **trophozoites**, **schizonts**, **segmenters**, **merozoites**, and **gametocytes**. The merozoites are capable of reinfecting other blood cells or liver cells. Ingestion of the **microgametocytes** (♂) and **macrogametocytes** (♀) by another mosquito during a blood meal initiates the sexual cycle

called **sporogamy**. Male and female gametes give rise to a zygote in the insect's gut. The zygote is then transformed into an **ookinete** that burrows through the gut wall to form an **oocyst** in which the sexually mature **sporozoites** develop, thereby completing the life cycle.

In this experiment, you will study the parasitic protozoa by using prepared slides and the diagnostic characteristics shown in Figure 33.1 on page 228 and Table 33.1. The purpose of the experiment is to help you understand life cycles of parasitic protozoa.

CLASS, ORGANISM, AND INFECTION	STRUCTURAL CHARACTERISTICS	LOCOMOTOR ORGANELLES	SITE OF INFECTION	ISOLATION OF PARASITIC FORM
MASTIGOPHORA *Giardia intestinalis* Infection: Dysentery — Nucleus — Karyosome — Median bodies — Axonemes — Nuclei — Median bodies — Retracted protoplasm — Axonemes	**Trophozoite:** Shape: Pear-shaped with concave sucking disc Nuclei: 2 bilaterally located with central karyosome and no peripheral chromatin Cytoplasm: Uniform and clear	4 pairs of flagella	Small intestine through ingestion of cysts	Diarrhetic stool
	Cyst: Shape: Oval to ellipsoidal Nuclei: 2–4 present and protoplasm retracted from cyst wall Axostyle Parabasal body	4 pairs of flagella within cyst		Formed stool
CILIOPHORA *Balantidium coli* Infection: Dysentery — Cytostome — Cilia — Micronucleus — Macronucleus — Cyst wall — Macronucleus	**Trophozoite:** Shape: Oval Nuclei: Kidney-shaped macronucleus and a micronucleus Cytoplasm: Vacuolated	Cilia	Large intestine by the ingestion of cysts	Diarrhetic stool
	Cyst: Shape: Round and thick-walled Nuclei: 1 macronucleus and a micronucleus that is not visible	Cilia within cyst		Formed stool

CLINICAL APPLICATION

Understanding Parasitic Protozoa

Parasitic protozoa can exist extracellularly or intracellularly and possess diverse morphologies. They rapidly reproduce, asexually or sexually, with short generation times. They are highly organ-, tissue-, or cell-specific organisms. Examples are *Plasmodium* species, which colonize red blood cells (malaria); *Trichomonas*, which colonize the urinary tract (vaginal infections); and *Entamoeba*, which colonizes the large intestine (severe diarrhea).

AT THE BENCH

Materials

Prepared Slides

Entamoeba histolytica trophozoite and cyst, *Giardia intestinalis* trophozoite and cyst (formerly *G. lamblia*), *Balantidium coli* trophozoite and cyst, *Trypanosoma gambiense*, and *Plasmodium vivax* in human blood smears.

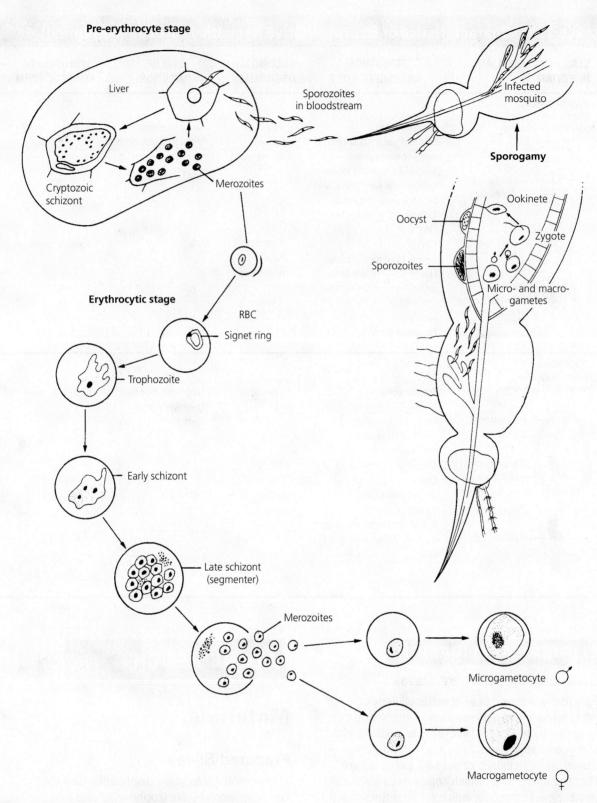

Pre-erythrocyte stage

Liver

Sporozoites in bloodstream

Infected mosquito

Cryptozoic schizont

Merozoites

Sporogamy

Ookinete

Oocyst

Zygote

Sporozoites

Micro- and macro-gametes

Erythrocytic stage

RBC

Signet ring

Trophozoite

Early schizont

Late schizont (segmenter)

Merozoites

Microgametocyte ♂

Macrogametocyte ♀

Figure 33.1 Life cycle of *Plasmodium vivax*

Equipment

Microscope, immersion oil, and lens paper.

Procedure

1. Examine all available slides under the oil-immersion objective. Use Table 33.1, Figure 33.1, and the photographs in Figure 33.2 through Figure 33.6 to identify the distinguishing microscopic characteristics of each parasite studied.

2. Record your observations in the Lab Report.

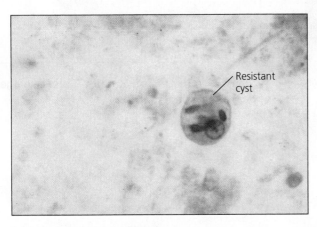

(a) *Entamoeba histolytica* (650×)

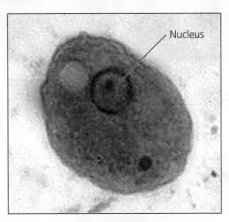

(b) A trophozoite of *Entamoeba histolytica* (2000×)

Figure 33.2 *Entamoeba histolytica*. **Causative agent of amebic dysentery.**

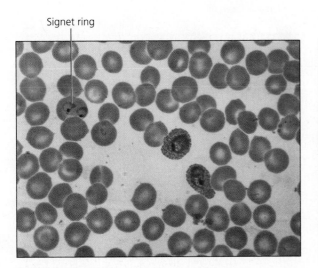

(a) Red blood cells infected by the ring stage (signet ring) (850×)

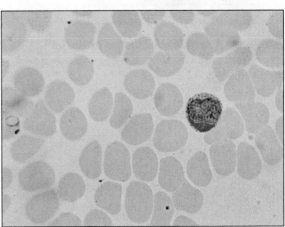

(b) Late schizont (segmenter) stage (1100×)

Figure 33.3 *Plasmodium vivax*. **Causative agent of malaria.**

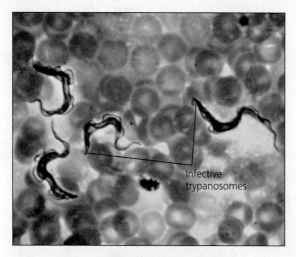

Figure 33.4 *Trypanosoma gambiense*. Causative agent of African sleeping sickness (1300×).

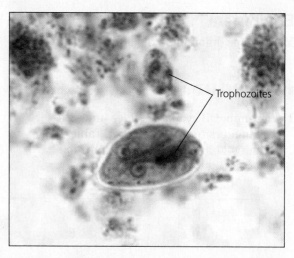

Figure 33.6 *Giardia intestinalis*. Causative agent of gastrointestinal diarrhea.

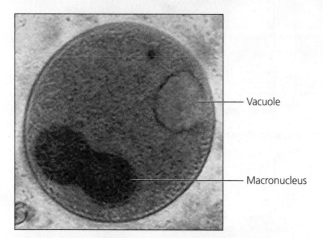

Figure 33.5 **A cyst of the ciliated protozoan** *Balantidium coli.* The cysts are spherical and lack surface cilia.

Name: _____

Date: _____   Section: _____

Observations and Results

Draw representative sketches of the parasitic organisms that you studied, and label the distinguishing structural characteristics you were able to observe.

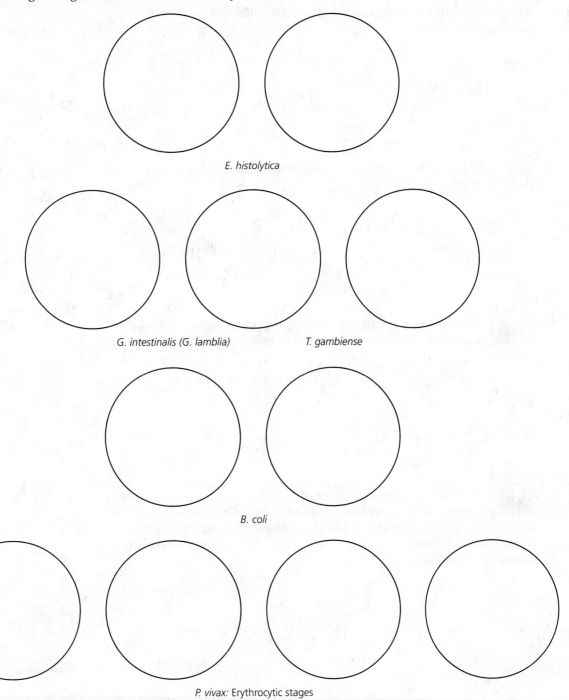

E. histolytica

G. intestinalis (G. lamblia)   T. gambiense

B. coli

P. vivax: Erythrocytic stages

Review Questions

1. Describe the developmental stages of the malarial parasite during sporogamy and schizogony.

2. What role does the invertebrate host play in the life cycle of the trypanosomes? Explain.

3. Distinguish between the pre-erythrocytic and erythrocytic stages in the life cycle of the malarial parasite.

4. In malarial infections, the sexually mature parasite is found in which host? Is this true for all other protozoan parasitic infections? Explain.

5. On returning from a trip overseas, an individual with persistent diarrhea is diagnosed as having an *E. histolytica* infection. Fecal examination reveals the presence of blood in the stool, suggesting damage to the intestinal mucosa. Explain why and how the mucosa was compromised by this parasite.

PART 7

The Fungi

When you have completed the experiments in this section, you should be familiar with

1. The macroscopic and microscopic structures of yeast and molds.
2. The basic mycological culturing and staining procedures.
3. The ability to identify selected common fungal organisms.

Introduction

The branch of microbiology that deals with the study of fungi (yeasts and molds) is called **mycology**. True fungi are separated into the following four groups on the basis of their sexual modes of reproduction:

1. **Zygomycetes:** Bread and terrestrial molds. Reproductive spores are external and uncovered. Sexual spores are zygospores, and asexual spores are sporangiospores.

2. **Ascomycetes:** Yeasts and molds. Sexual spores, called ascospores, are produced in a saclike structure called an ascus. Conidia are asexual spores produced on a conidiophore.

3. **Basidiomycetes:** Fleshy fungi, toadstools, mushrooms, puffballs, and bracket fungi. Reproductive spores, basidiospores, are separate from specialized stalks called basidia.

4. **Deuteromycetes:** Also called **Fungi Imperfecti** because no sexual reproductive phase has been observed.

The major characteristics of these four groups of fungi are shown in **Table P7.1**.

Nutritionally, fungi are heterotrophic, eukaryotic microorganisms that are enzymatically capable of metabolizing a wide variety of organic substrates. Fungi can have beneficial or detrimental effects on humans. Fungi that inhabit the soil play a vital role in decomposing dead plant and animal tissues, thereby maintaining a fertile soil environment. The fermentative fungi are of industrial importance in producing beer and wine, bakery products, cheeses, industrial enzymes, and antibiotics. The detrimental activities of some fungi include spoilage of foods by rots, mildews, and rusts found on fruit, vegetables, and grains. Some species are capable of producing toxins (for example, aflatoxin) and hallucinogens. A few fungal species are of medical significance because of their capacities to produce diseases in humans. Many of the pathogenic fungi are deuteromycetes and can be divided into two groups based on site of infection. The **superficial mycoses** cause infections of the skin, hair, and nails (for example, ringworm infections). The **systemic mycoses** cause infections of the subcutaneous and deeper tissues such as those of the lungs, genital areas, and nervous system.

Major Characteristics of the Four Groups of Fungi

CHARACTERISTICS	GROUP			
	ZYGOMYCETES	ASCOMYCETES	BASIDIOMYCETES	DEUTEROMYCETES
Mycelium	Nonseptate	Septate	Septate	Septate
Asexual spores	Found in sporangium; sporangiospores (nonmotile)	Formed on tip of conidiophore; conidia (nonmotile)	Same as the ascomycetes	Same as the ascomycetes
Sexual spores	Zygospores (motile), found in terrestrial forms; oospores, found in aquatic forms	Ascospores, contained in a saclike structure called the ascus	Basidiospores, carried on the outer surface of a club-shaped cell called the basidium	Fungi Imperfecti—no sexual reproductive phase observed; some members of the ascomycetes and basidiomycetes are Fungi Imperfecti
Common species	Bread molds, mildews, potato blight, *Rhizopus* species	Cup fungi, ergot, Dutch elm, yeast species	Smuts, rusts, puffballs, toadstools, mushrooms	*Aspergillus, Candida, Trichophyton, Cryptococcus, Blastomyces, Histoplasma, Microsporum,* and *Sporothrix*

Molds are the major fungal organisms that can be seen by the naked eye. We have all seen them growing on foods, such as bread or citrus fruit as a cottony, fuzzy, black, green, or orange growth, or as a mushroom with a visible cap attached to a stalk, depending on the mold. Examination with a simple hand lens shows that these organisms are composed of an intertwining branching mat called a **mycelium**. The filaments that make up this mycelial mat are called **hyphae**. Most of the mat grows on or in the surface of the nutrient medium so that it can extract nutrients; the mat is therefore called **vegetative mycelium**. Some of the mycelium mat rises upward from the mat and is referred to as **aerial mycelium**. Specialized hyphae are produced from the aerial mycelium and give rise to spores that are the reproductive elements of the mold. Figures 34.1, 34.2, 34.5, and 34.6 show the reproductive structures of some fungi.

The cultivation, growth, and observation of molds require techniques that differ from those used for bacteria. Mold cultivation requires the use of a selective medium such as Sabouraud agar or potato dextrose agar. These media favor mold growth because their low acidity (pH 4.5 to 5.6) discourages the growth of bacteria, which favor a neutral (pH 7.0) environment. The temperature requirements of molds are also different from those of bacteria, in that molds grow best at room temperature (25°C). In addition, molds grow at a much slower rate than bacteria do, requiring several days to weeks before visible colonies appear on a solid agar surface. Colony growth is shown in Figures 34.3 and 34.4.

PART A   Slide Culture Technique

LEARNING OBJECTIVES

When you have completed this experiment you should be

1. Acquainted with mold cultivation on glass slides.
2. Able to visualize and identify the structural components of molds.

Principle

Because the structural components of molds are very delicate, even simple handling with an inoculating loop may result in mechanical disruption of their components. The following slide culture technique is used to avoid such disruption. A deep concave slide containing a suitable nutrient

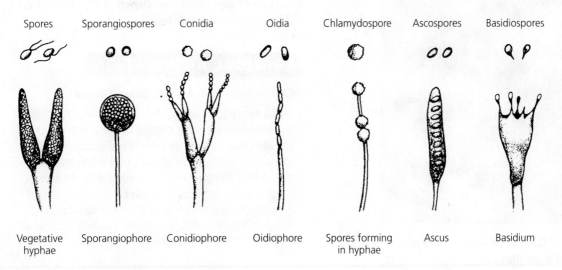

Figure 34.1 Spore and sporangia types

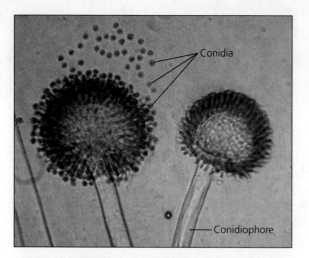

Figure 34.2 **Conidiophore and conidia of mold** *Aspergillus niger*

Figure 34.3 **Colony of *Penicillium chrysogenum***

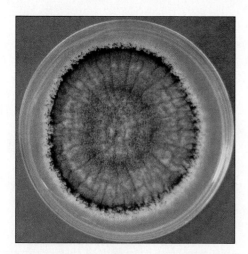

Figure 34.4 **Colony of *Aspergillus niger* on a Sabouraud agar plate**

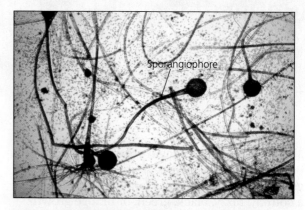

Figure 34.5 **Mucor mucedo**

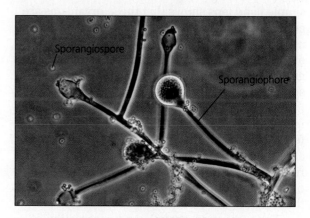

Figure 34.6 **Rhizopus stolonifer**

medium with an acidic pH, such as Sabouraud agar, is covered by a removable coverslip. Mold spores are deposited in the surface of the agar and incubated in a moist chamber at room temperature. Direct microscopic observation is then possible without fear of disruption or damage to anatomical components. Molds can be identified as to spore type and shape, type of sporangia, and type of mycelium, as shown in Figure 34.1 and Table 36.1 on pages 251–253.

CLINICAL APPLICATION

Cultivation of Fungi on Glass Slides

Since sporangia may be damaged during transfer to a glass slide, the slide culture technique prevents the disturbance and damage of the sporangia and other spore structures required for fungi identification. Intact samples can be used to distinguish a fungi like *Aspergillus niger,* which causes the most common fungal infection of the ear, from *Aspergillus flavus,* a fungal pathogen that may result in disseminating infection of the lungs.

Materials

Cultures

7- to 10-day Sabouraud agar cultures of *Penicillium chrysogenum* (formerly called *P. notatum*) and *Aspergillus niger*, *Mucor mucedo*, and *Rhizopus stolonifer*.

Media

Per student group: one Sabouraud agar deep tube.

Equipment

Microincinerator or Bunsen burner, waterbath, four concave glass slides, four coverslips, petroleum jelly, sterile Pasteur pipettes, toothpicks, four sterile Petri dishes, filter paper, forceps, inoculating loop and needle, four sterile U-shaped bent glass rods, thermometer, dissecting microscope, and beaker with 95% ethyl alcohol.

Procedure Lab One

1. Melt the deep tube of Sabouraud agar in a boiling water bath and cool to 45°C.

2. Place a piece of filter paper in the bottom of each Petri dish, lay a sterile bent glass rod in each dish, and replace the covers.

3. Using forceps, dip the concave slides and coverslips in a beaker of 95% ethyl alcohol, pass through Bunsen burner flame, remove from flame, and hold until all the alcohol has burned off the slides and coverslips.

4. Cool slides and coverslips. Place a slide, concave side up, with a coverslip to one side of the concavity, on the glass rod inside each Petri dish.

5. With a toothpick, add petroleum jelly to three sides surrounding the concavity of each slide. The fourth side will serve as a vent for air.

6. With a sterile Pasteur pipette, add one or two drops of cooled Sabouraud agar to the concavity of each slide.

7. Place a coverslip over the concave portion of each slide so that it is completely sealed.

8. With forceps, stand each slide upright inside its respective Petri dish until the agar solidifies, as illustrated below:

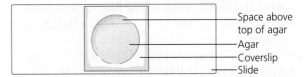
— Space above top of agar
— Agar
— Coverslip
— Slide

9. When agar is fully hardened, slide coverslips downward with forceps, and with a sterile needle inoculate each prepared slide with the spores from the test cultures.

10. Push the coverslips to their original positions, thereby sealing off the slide.

11. With a Pasteur pipette, moisten the filter paper with sterile water to provide a moist atmosphere. Remoisten filter paper when necessary during the incubation period.

12. Place the slide on the U-shaped bent rod, replace Petri dish cover, and label with the names of the organism and your initials.

13. Incubate the preparations for 7 days at 25°C.

Procedure Lab Two

1. Examine each mycological slide preparation under the low and high power of a dissecting microscope. Identify the mycelial mat, vegetative and reproductive hyphae, and spores. Use Table 36.1 on pages 251–253 to aid with your identification of mold structures.

2. Record your observations in the Lab Report.

PART B Mold Cultivation on Solid Surfaces

LEARNING OBJECTIVES

When you have completed this experiment, you should be

1. Acquainted with the technique of mold cultivation on agar plates.

2. Able to observe and identify colonial characteristics, such as growth rate, texture, pigmentation on the surface and reverse side, and folds or ridges on the surface.

Principle

Cultivating molds on solid surfaces allows you to observe the variations in gross colonial morphology among different genera of molds. These variations in colonial appearance play a major role in the identification of the filamentous fungi. Most microbiologists are familiar with the gross appearance of multicellular fungi, but even to the untrained, the macroscopic differences in colonial growths are obvious and recognizable. For example, most people have seen rotting citrus fruits (lemons and oranges) produce a blue-green velvety growth characteristic of *Penicillium* species. It is also common for stale cheese to show a grayish-white furry growth of *Mucor* species, and the black stalklike appearance of *Rhizopus* molds growing on bread is familiar to many.

In this part of the experiment, you will be able to visualize the gross appearance of the colonial growth of four different molds.

CLINICAL APPLICATION

Isolation of Fungi on Solid Media

Before a fungal species may be identified or studied it must first be isolated. Similar to using an agar plate for isolating a distinct bacterial species, agar plating may be used as a growth medium for the isolation of fungi spores. Once spores have been isolated from individual sporangia, subculturing on solid agar or slides will allow for characterization and genetic studies of the fungus.

AT THE BENCH

Materials

Cultures

7- to 10-day Sabouraud agar cultures of *Aspergillus niger, Penicillium chrysogenum* (formerly called *P. notatum*), *Mucor mucedo,* and *Rhizopus stolonifer.*

Media

Per designated student group: three Sabouraud agar plates and one potato dextrose agar plate.

Equipment

Microincinerator or Bunsen burner, four test tubes containing 2 ml of sterile saline, dissecting microscope, and an inoculating loop.

Procedure Lab One

1. Label the three Sabouraud agar plates as *Aspergillus niger, Penicillium chrysogenum,* and *Mucor mucedo,* and label the fourth plate containing potato dextrose agar as *Rhizopus stolonifer.*

2. Prepare a saline suspension of each mold culture. Label each of the four tubes of saline with the name of the organism. Using a sterile inoculating loop, scrape two loopfuls of mold culture into the corresponding tube of 2 ml of sterile saline and mix well by tapping the tube with your finger.

3. Using aseptic technique, inoculate each of the plates by placing a single loopful of mold suspension in the center of its respective agar plate. *Note: Do not spread the inoculum and do not shake or jostle the plates.*

4. Incubate all plates at room temperature, 25°C, for 7 to 10 days. *Note: Do not invert the plates.*

Procedure Lab Two

1. Examine each mold plate under the low and high power of a dissecting microscope. Refer to Table 36.1 for your identification of mold structures. *Note: Do not remove Petri dish covers.*

2. Record your observations in the Lab Report.

Name: _____

Date: _____ Section: _____

Observations and Results

Part A: Slide Culture Technique

Draw a representative microscopic field under low-power and high-power magnification and label the structural components of each test organism.

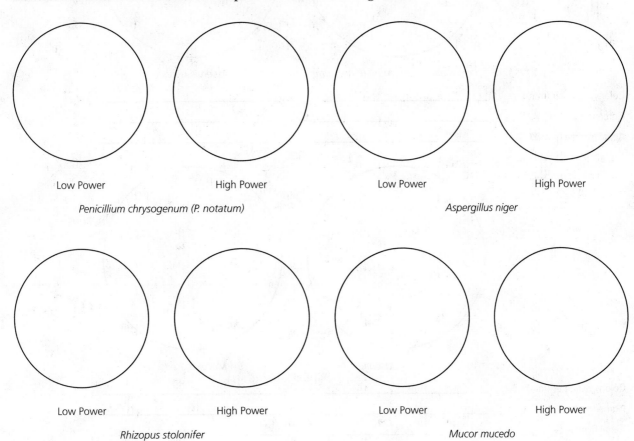

Low Power   High Power   Low Power   High Power

Penicillium chrysogenum (P. notatum)   *Aspergillus niger*

Low Power   High Power   Low Power   High Power

Rhizopus stolonifer   *Mucor mucedo*

Part B: Mold Cultivation on Solid Surfaces

Draw sketches of the mold colonies under low power, indicating the extent of growth (diameter in mm), pigmentation, and the presence or absence of aerial hyphae. Refer to Table 36.1 to aid with your identification of mold structures.

Penicillium chrysogenum (P. notatum)

Aspergillus niger

Colony diameter (mm): _____

Pigmentation: _____

Aerial hyphae (+ or –): _____

Rhizopus stolonifer

Mucor mucedo

Colony diameter (mm): _____

Pigmentation: _____

Aerial hyphae (+ or –): _____

Review Questions

1. Cite some beneficial and harmful aspects of molds.

2. What is the advantage of using Sabouraud agar?

3. In the slide culture technique, what is the purpose of the following?

   a. Moistened filter paper in the Petri dish:

b. A U-shaped glass rod in the Petri dish:

4. What is the advantage of the slide culture technique over that of a simple loop inoculation onto an agar plate (as in Part B)?

5. Why would it be advantageous to observe mold colonies on an agar plate?

6. Since dimorphism is a property of fungi, how do you account for the fact that molds grow preferentially *in vitro* rather than *in vivo?*

Yeast Morphology, Cultural Characteristics, and Reproduction

Once you have completed this experiment, you should

1. Know the morphology of different genera of yeast.

2. Understand the growth and fermentative properties of yeast cells.

3. Be familiar with the sexual and asexual modes of reproduction in yeast cells.

Principle

Yeasts are nonfilamentous unicellular fungi. Yeast cultures resemble bacteria when grown on the surface of artificial laboratory media; however, they are 5- to 10-times larger than bacteria. Yeast colonies are illustrated in Figure 35.1. Microscopically, yeast cells may be ellipsoidal, spherical, or in some cases, cylindrical (Figure 35.2). Unlike molds, yeast do not have aerial hyphae and supporting sporangia.

Yeast reproduce asexually by **budding** or by **fission**. In budding, an outgrowth from the parent cell (a **bud**) pinches off, producing a daughter cell (Figures 35.3a and 35.4). Fission occurs in certain species of yeast, such as those in the genus *Schizosaccharomyces*. During fission, the parent cell elongates, its nucleus divides, and it splits evenly into two daughter cells.

Some yeast may also undergo sexual reproduction when two sexual spores conjugate, giving rise to a zygote, or diploid cell. The nucleus of this cell divides by meiosis, producing four new haploid nuclei (sexual spores), called **ascospores**, contained within a structure called the **ascus** (Figure 35.3b). When the ascus ruptures, the ascospores are released and conjugate, starting the cycle again.

Yeasts are important for many reasons. *Saccharomyces cerevisiae* is referred to as baker's yeast and is used as the leavening agent in dough. Two major strains of yeast, *Saccharomyces carlsbergensis* and *Saccharomyces cerevisiae*, are used for brewing. The wine industry relies on wild yeast (present on the grape) for the fermentation of grape juice, which is supplemented with *Saccharomyces ellipsoideus* to begin the fermentation. Also, the high vitamin content of yeasts makes them particularly valuable as food supplements. As useful as some yeasts are, there are a few species that can create problems in the food industry or are harmful to humans. Undesired yeast must be excluded from the manufacture of

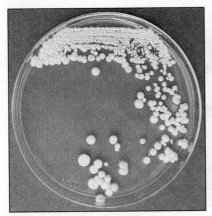 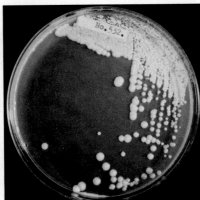 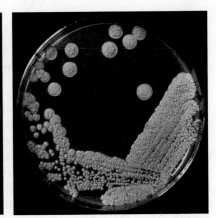

(a) *Saccharomyces cerevisiae*   **(b)** *Candida albicans*   **(c)** *Rhodotorula rubra*

Figure 35.1 Colonies of yeast cells

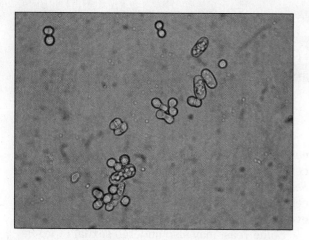

Figure 35.2 **Yeast Cells** *Schizosaccharomyces octosporus*

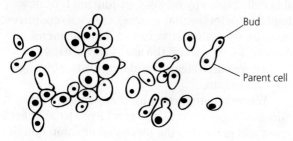

(a) Asexual reproductive yeast structures

Bud

Parent cell

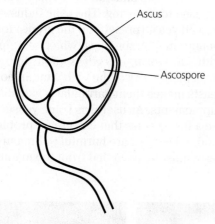

Ascus

Ascospore

(b) Sexual reproductive yeast structures

Figure 35.3 **Reproductive structures of yeast**

fruit juices, such as grape juice or apple cider, to prevent the fermentation of fruit sugars to alcohol. The contamination of soft cheese by some forms of yeast will destroy the product. Finally, some yeast such as *Candida albicans* are pathogenic and responsible for urinary tract and vaginal infections, known as **moniliasis**, and infections of the mouth called **thrush**.

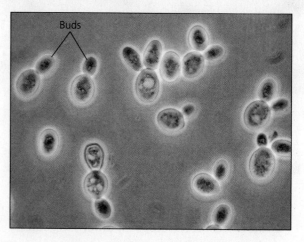

Buds

Figure 35.4 **Asexual yeast reproduction by budding** *Saccharomyces cerevisiae*

The cultural characteristics, the type of reproduction, and the fermentative activities used to identify the different genera of yeast will all be studied in this experiment.

CLINICAL APPLICATION

Opportunistic Yeast

Opportunistic yeast may cause serious or life-threatening infections in immunocompromised patients, such as people with cancer or AIDS. The yeast *Candida albicans* can cause relatively minor infections in healthy people, such as thrush or vaginal "yeast infections"; however, it can cause a dangerous bloodstream infection called *invasive candidiasis* in those with weakened immune systems. Similarly, the yeast *Cryptococcus neoformans* can cause a pulmonary infection that can lead to meningitis, most often in immunocompromised people.

AT THE BENCH

Materials

Cultures

7-day Sabouraud agar cultures of *Saccharomyces cerevisiae, Candida albicans, Rhodotorula rubra, Selenotila intestinalis,* and *Schizosaccharomyces octosporus.*

Media

Per designated student group: five tubes each of bromcresol purple glucose broth, bromcresol purple maltose broth, bromcresol purple lactose broth, and bromcresol purple sucrose broth, each containing a Durham tube; two glucose-acetate agar plates; and five test tubes (13 × 100mm) containing 2 ml of sterile saline.

Reagents

Water-iodine solution, lactophenol–cotton-blue solution.

Equipment

Microincinerator or Bunsen burner, inoculating loop and needle, 10 glass slides, 10 coverslips, 5 sterile Pasteur pipettes, glassware marking pencil, and microscope.

Procedure Lab One

Morphological Characteristics

Prepare a wet mount of each yeast culture in the following manner:

1. Suspend a loopful of yeast culture in a few drops of lactophenol–cotton-blue solution on a microscope slide and cover with a coverslip.
2. Examine all yeast wet-mount slide preparations under low and high power, noting the shape and the presence or absence of budding. Record your observations in the Lab Report.

Fermentation Studies

1. With a sterile loop, inoculate each experimental organism into appropriately labeled tubes of bromcresol purple glucose, maltose, lactose, and sucrose fermentation broths.
2. Incubate all cultures at 25°C for 4 to 5 days.

Sexual Reproduction

1. With a glassware marking pencil, divide the bottom of a glucose-acetate agar plate into three sections, and divide another glucose-acetate agar plate in half.

2. Label each section with the name of a test organism.
3. Label each tube of sterile saline with the name of a test organism.
4. With a sterile inoculating loop, suspend a heavy loopful of each test organism into its appropriately labeled tube of saline. Tap the tube with your finger to obtain a uniform cell suspension.
5. With a sterile Pasteur pipette, inoculate one drop of each test organism onto the surface of the appropriately labeled section on an agar plate. *Note: Allow the inoculum to diffuse into the agar for a few minutes. Do not swirl or rotate the plates.*
6. Incubate all plates at 25°C for 7 days. *Note: Visit the laboratory, if possible, during the incubation period and note when sporulation begins.*

Procedure Lab Two

Fermentation Studies

1. Examine all fermentation tubes for the presence of growth (turbidity), the presence or absence of acid (change in the color of medium), and the presence or absence of gas (bubble in Durham tube).
2. Record your results in the chart provided in the Lab Report.

Procedure Lab Three

Sexual Reproduction

1. Examine the glucose-acetate agar plates for the presence or absence of sporulation.
2. Prepare a water-iodine wet mount using a loopful of culture from each respective section on the glucose-acetate agar plate.
3. Observe the cells using the high-dry objective and record your observations in the Lab Report.

Name: _____

Date: _____ Section: _____

Observations and Results

Morphological Characteristics

Draw a representative field for each organism in the chart below. Note the shape and presence or absence of budding (+ or −).

Saccharomyces cerevisiae

Candida albicans

Shape: _____   _____

Budding (+ or −): _____   _____

Rhodotorula rubra

Selenotila intestinalis

Shape: _____   _____

Budding (+ or −): _____   _____

Schizosaccharomyces octosporus

Shape: _____

Budding (+ or −): _____

Fermentation Studies

Use a plus (+) or minus (−) in the chart below to record your results.

Organism	GLUCOSE			MALTOSE			LACTOSE			SUCROSE		
	T	A	G	T	A	G	T	A	G	T	A	G
Saccharomyces cerevisiae												
Candida albicans												
Rhodotorula rubra												
Selenotila intestinalis												
Schizosaccharomyces octosporus												

Note: T = turbidity, A = acid, and G = gas

Sexual Reproduction

In the circles below, draw representative reproductive structures and label the parts.

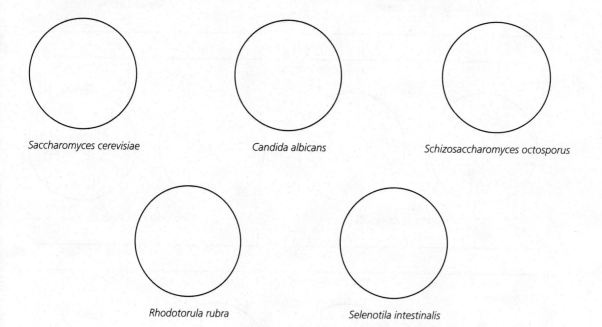

Saccharomyces cerevisiae

Candida albicans

Schizosaccharomyces octosporus

Rhodotorula rubra

Selenotila intestinalis

Review Questions

1. Indicate the significance of the following structures in the reproductive activities of yeast cells.

    a. Buds

    b. Ascus

    c. Ascospores

2. Why are yeast cells classified as fungi, and how do they differ from other fungi?

3. Why is yeast of industrial importance?

4. Why are yeasts significant from a medical perspective?

5. Why is it necessary to pasteurize fruit juices?

6. A female patient develops candidiasis (moniliasis) following prolonged antibiotic therapy for a bladder infection caused by Pseudomonas aeruginosa. How can you account for the development of this concurrent vaginal infection?

7. With regard to the fermentation of wine, what kind of wine would be produced if you washed the grapes prior to crushing them?

Identification of Unknown Fungi

LEARNING OBJECTIVE

When you have completed this experiment, you should be able to

1. Identify a fungal unknown based on colonial morphology and microscopic appearance.

CLINICAL APPLICATION

Identification of Fungal Infection

When presented with a patient with symptoms that suggest either an intestinal fungal infection or Crohn's Disease complications, isolation and identification of a fungal pathogen is required before the correct treatment may be prescribed. Using culturing techniques, an isolate of the fungi can be identified by mycelium morphology and genetic markers.

Principle

In this experiment, you will be provided with a number-coded pure culture of a representative fungal organism for cultivation and subsequent identification. Use **Table 36.1** to aid in identification of the unknown culture.

(Text continues on page 254.)

TABLE 36.1 Identification of Fungi

DIAGRAM	COLONIAL MORPHOLOGY	MICROSCOPIC APPEARANCE
Molds 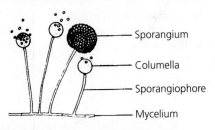 Sporangium, Columella, Collarette, Sporangiophore, Stolon, Mycelium, Rhizoid *Rhizopus:* Black bread mold; common laboratory contaminant	Rapidly growing white-colored fungus swarms over entire plate; aerial mycelium cottony and fuzzy	Spores are oval, colorless, or brown; nonseptate mycelium gives rise to straight sporangiophores that terminate with black sporangium containing a columella; rootlike hyphae (rhizoids) penetrate the medium
Sporangium, Columella, Sporangiophore, Mycelium *Mucor:* Food contaminant	Resembles the colonies of *Rhizopus* except that it lacks rhizoids and collarettes. Sporangiophore arises directly from mycelial mat. Note: Branching sporangiophores may occur with *Mucor.*	Spores are oval; nonseptate mycelium gives rise to single sporangiophores with globular sporangium containing a columella; there are no rhizoids

TABLE 36.1 | **Identification of Fungi (continued)**

DIAGRAM	COLONIAL MORPHOLOGY	MICROSCOPIC APPEARANCE
Molds (continued) 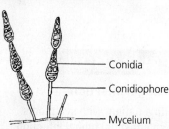 *Alternaria:* Normally found on plant material; also found in house dust	Grayish-green or black colonies with gray edges rapidly swarming over entire plate; aerial mycelium not very dense, appears grayish to white	Multicelled spores (conidia) are pear-shaped and attached to single conidiophores arising from a septate mycelium
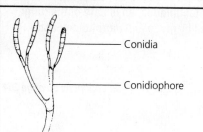 *Fusarium:* Found in soil; also likely in eye infections	Woolly, white, fuzzy colonies changing color to pink, purple, or yellow	Spores (conidia) are oval or crescent-shaped and attached to conidiophores arising from a septate mycelium; some spores are single cells, some are multicelled
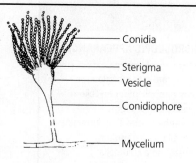 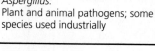 *Aspergillus:* Plant and animal pathogens; some species used industrially	White colonies become greenish-blue, black, or brown as culture matures	Single-celled spores (conidia) in chains developing at the end of the sterigma arising from the terminal bulb of the conidiophore, the vesicle; long conidiophores arise from a septate mycelium
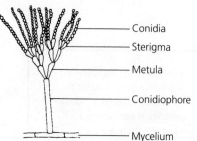 *Penicillium:* Antibiotic-producing citrus fruit contaminant; soil inhabitant	Mature cultures usually greenish or blue-green	Single-celled spores (conidia) in chains develop at the end of the sterigma arising from the metula of the conidiophore; branching conidiophores arise from a septate mycelium

TABLE 36.1   Identification of Fungi (continued)

DIAGRAM	COLONIAL MORPHOLOGY	MICROSCOPIC APPEARANCE
Molds (continued) 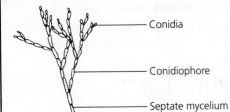 — Conidia — Conidiophore — Septate mycelium *Cladosporium:* Dead and decaying plants	Small, heaped colonies are greenish-black and powdery	Spores (conidia) develop at the end of complex conidiophores arising from a septate mycelium that is usually brownish
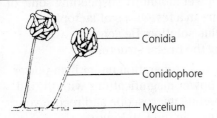 — Conidia — Conidiophore — Mycelium *Cephalosporium:* Antibiotic production	Rapidly growing compact and moist colonies becoming cottony with aerial hyphae that are gray or rose-colored	Single-celled conical or elliptical spores (conidia) held together in clusters at the tips of the conidiophores by a mucoid substance; erect, unbranched conidiophores arise from a septate mycelium
Yeast 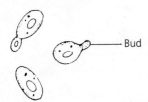 — Bud *Torula:* Cheese and food contaminant	Colonies are pink, moist, with unbroken, even edges	Cells are oval, colorless, and reproduce by budding
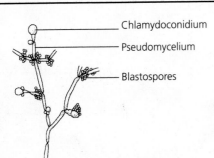 — Chlamydoconidium — Pseudomycelium — Blastospores *Candida:* Human pathogen	Colonies are small, round, moist, and colorless, with unbroken, even edges	Yeastlike fungus produces pseudomycelium

Materials

Cultures

Number-coded, 7-day Sabouraud broth spore suspensions of *Aspergillus, Mucor, Penicillium, Alternaria, Rhizopus, Cladosporium, Cephalosporium, Fusarium, Torula,* and *Candida.*

Media

One Sabouraud agar plate per student.

Reagent

Lactophenol–cotton-blue solution.

Equipment

Microincinerator or Bunsen burner, dissecting microscope, hand lens, sterile cotton swabs, glass slides, coverslips, inoculating loop, and glassware marking pencil.

Procedure Lab One

1. With a sterile inoculating loop, inoculate an appropriately labeled Sabouraud agar plate with one of the provided unknown cultures by placing one loopful in the center of the plate. *Note: Do not spread culture.*

2. Incubate the plates in a noninverted position for 1 week at 25°C in a moist incubator.

Procedure Lab Two

1. Observe mold cultures with a hand lens, noting and recording their colonial morphologies.

2. Prepare a wet mount by suspending some of the culture in a few drops of lactophenol–cotton-blue solution. Be gentle to avoid damaging the fungal structures.

3. Examine the preparation under high-power and low-power magnifications with the aid of a dissecting microscope and record your observations in the Lab Report.

Name: _____

Date: _____   Section: _____

Observations and Results

1. Draw representative microscopic fields of your culture below.
2. Using Table 36.1 and Figure 3.1 on page 26, identify your unknown fungal
    organism.

    a. Color pigmentation

    b. Diameter (mm)

    c. Texture (cottony, smooth, etc.)

    d. Margin (entire, undulating, lobular, etc.)

    e. Aerial hyphae (septate, nonseptate)

Diagram of microscopic appearance:

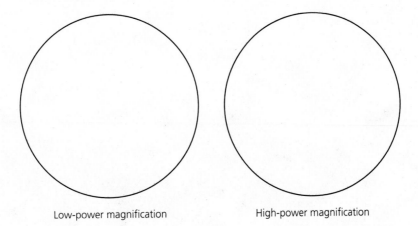

Low-power magnification   High-power magnification

Number assigned to
unknown culture: _____

Genus of fungal unkown: _____

PART 8

The Viruses

Once you have completed the experiments in this section, you should know

1. The chemical structures, morphologies, and replicative activities of bacterial viruses (bacteriophages).
2. How to perform a phage dilution procedure for the cultivation and enumeration of bacterial viruses.
3. How to isolate bacteriophages from sewage.

Introduction

Viruses are noncellular biological entities composed solely of a single type of nucleic acid surrounded by a protein coat called the capsid. Because of their limited and simplistic structures, viruses can be chemically defined as **nucleoproteins**. They are devoid of the sophisticated enzymatic and biosynthetic machinery essential for independent activities of cellular life. This lack of metabolic machinery mandates that they exist as parasites, and they cannot be cultivated outside of a susceptible living cell. Viruses are differentiated from cellular forms of life on the following bases:

1. They are ultramicroscopic and can only be visualized with the electron microscope.
2. They are filterable: They are able to pass through bacteria-retaining filters.
3. They do not increase in size.
4. They must replicate within a susceptible cell.
5. Replication occurs because the viral nucleic acid subverts the synthetic machinery of the host cell (namely, common host cell components and enzyme systems involved in decomposition, synthesis, and bioenergetics) for the purpose of producing new viral components.

6. Viruses are designated either RNA or DNA viruses because they contain one of the nucleic acids but never both.

Much of our knowledge of the mechanism of animal viral infection and replication has been based on our understanding of infection in bacteria by bacterial viruses, called **bacteriophages**, or **phages**. The bacteriophages were first described in 1915 almost simultaneously by Twort and d'Herelle. The name *bacteriophage*, which in Greek means "to eat bacteria," was coined by d'Herelle because of the destruction through lysis of the infected cell. Bacteriophages exhibit notable variability in their sizes, shapes, and complexities of structure. The T-even (T2, T4, and T6) phages illustrated in Figure P8.1 demonstrate the greatest morphological complexity.

Phage replication depends on the ability of the phage particle to infect a suitable bacterial host cell. Infection consists of the following sequential events:

1. **Adsorption:** Tail fibers of the phage particle bind to receptor sites on the host's cell wall.
2. **Penetration** (infection): Spiral protein sheath retracts, and an enzyme, early muramidase, perforates the bacterial cell wall, enabling the phage nucleic acid to pass

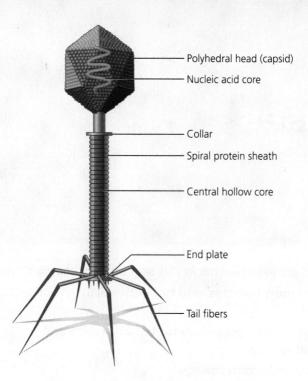

Polyhedral head (capsid)
Nucleic acid core
Collar
Spiral protein sheath
Central hollow core
End plate
Tail fibers

The functions of these structural components are as follows:

Component	Function
Capsid (protein coat)	Protection of nucleic acid from destruction by DNases
Nucleic acid core	Phage genome carrying genetic information necessary for replication of new phage particles
Spiral protein sheath	Retracts so that nucleic acid can pass from capsid into host cell's cytoplasm
End plate and **tail fibers**	Attachment of phage to specific receptor sites on a susceptible host's cell wall

Figure P8.1 Bacteriophage: Structural components and their functions

through the hollow core into the host cell's cytoplasm. The empty protein shell remains attached to the cell wall and is called the protein ghost.

3. **Replication:** The phage genome subverts the cell's synthetic machinery, which is then used for the production of new phage components.

4. **Maturation:** During this period, the new phage components are assembled and form complete, mature virulent phage particles.

5. **Release:** Late muramidase (lysozyme) lyses the cell wall, liberating infectious phage particles that are now capable of infecting new susceptible host cells, thereby starting the cycle over again.

Virulent phage particles that infect susceptible host cells always initiate the **lytic cycle** as described above. Other phage particles, called **temperate phages** or **lambda (λ) phages**, incorporate their nucleic acid into the host's chromosome. Lysis of the host cell does not occur until it is induced by exogenous physical agents such as ultraviolet or ionizing radiation or chemical mutagenic agents. Bacterial cells

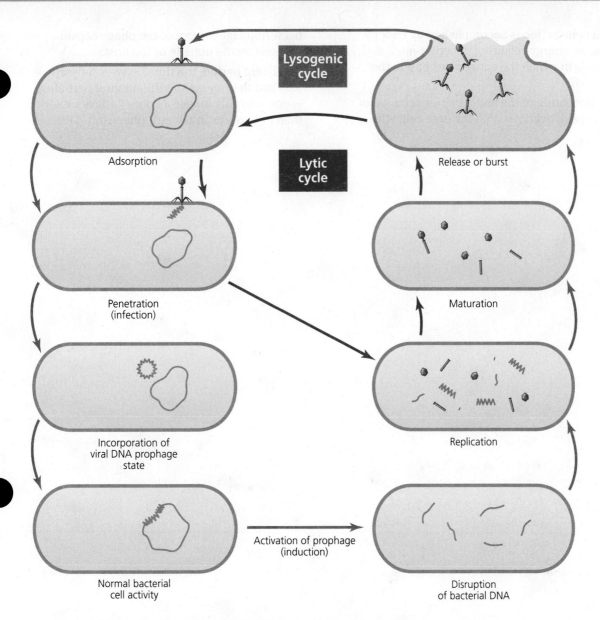

Lysogenic cycle

Adsorption

Penetration (infection)

Incorporation of viral DNA prophage state

Normal bacterial cell activity

Lytic cycle

Release or burst

Maturation

Replication

Activation of prophage (induction)

Disruption of bacterial DNA

Figure P8.2 **The lytic and lysogenic life cycles of a bacteriophage**

containing the incorporated phage nucleic acid, the **prophage**, are called **lysogenic cells**. Lysogenic cells appear and function as normal cells, and they reproduce by fission. When induced by physical or chemical agents, these cells will release a virulent prophage from the host's genome, which then initiates the lytic cycle. Figure P8.2 illustrates the lytic and lysogenic life cycles of a bacteriophage.

Animal viruses differ structurally from bacteriophages in that they lack the spiral protein sheath, end plate, and tail fibers. Their shapes may be helical or cuboidal (icosahedral, containing

20 triangular facets). Some animal viruses are designated as naked viruses because they are composed solely of nucleocapsids. In others, referred to as enveloped viruses, the nucleocapsid is surrounded by a lipid bilayer that may have glycoproteins associated with it.

The infectious process of the animal virus is very similar to bacteriophage infection. However, there are some notable differences:

1. Adsorption of the virus is to receptor sites that are located on the cell membrane of the host cell instead of the cell wall as in the bacterial host.

2. Viral penetration is accomplished by endocytosis, an energy-requiring, receptor-mediated process in which the entire virus enters the host cell.

3. The uncoating of the animal virus, removal of the capsid, occurs within the host cell; with bacteriophage infection, the phage capsid remains on the outside of the host.

4. The latent period, the time between adsorption and the release of virulent viral particles, is considerably longer—hours to days rather than minutes as in bacteriophage infection.

Cultivation and Enumeration of Bacteriophages

LEARNING OBJECTIVE

Once you have completed this experiment, you should be able to

1. Perform techniques for cultivation and enumeration of bacteriophages.

Principle

This exercise demonstrates the ability of viruses to replicate inside a susceptible host cell. For this purpose, you will be provided with a virulent phage and a susceptible host cell culture. This technique also enables you to enumerate phage particles on the basis of plaque formation in a solid agar medium. **Plaques** are clear areas in an agar medium previously seeded with a diluted phage sample and a host cell culture. Each plaque represents the lysis of a phage-infected bacterial cell.

The procedure requires the use of a double-layered culture technique in which the hard agar serves as a base layer, and a mixture of phage and host cells in a soft agar forms the upper overlay. Susceptible *Escherichia coli* cells multiply rapidly and produce a lawn of confluent growth on the medium. When one phage particle adsorbs to a susceptible cell, penetrates the cell, replicates, and goes on to lyse other host cells, the destroyed cells produce a single plaque in the bacterial lawn (see Figure 37.1). Each plaque can be designated as a **plaque-forming unit (PFU)** and used to quantitate the number of infective phage particles in the culture.

The number of phage particles contained in the original stock phage culture is determined by counting the number of plaques formed on the seeded agar plate and multiplying this by

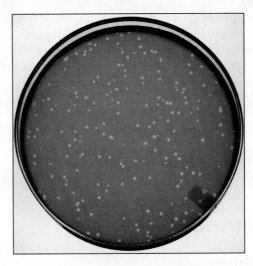

Figure 37.1 Plaque-forming units (PFUs).

the dilution factor. For a valid phage count, the number of plaques per plate should not exceed 300 nor be less than 30.

Example: 200 PFUs are counted in a 10^{-6} dilution.

$$200 \times 10^6 = 200 \times 10^6 \, or \, 2 \times 10^8$$

Plates showing greater than 300 PFUs are **too numerous to count (TNTC)**; plates showing fewer than 30 PFUs are **too few to count (TFTC)**.

The procedure covered in this experiment is based on protocols published by The American Society for Microbiology (www.asm.org) and is an example of the numerous procedures that can be found for the propagation and enumeration of bacteriophages. Refer to online sources such as ASM MicrobeLibrary (www.microbelibrary.org) or American Type Culture Collection (www.atcc.org) for alternate methods based on your needs or available laboratory equipment.

CLINICAL APPLICATION

Identification of Pathogenic Bacteria

Bacterial viruses (bacteriophages) are very common in all natural environments and are directly related to the number of bacteria present. They are most prevalent in soil, intestines of animals, sewage, and seawater. These viral particles have played an important role in the development of all types of viruses. Since many phages are specific about which bacteria they attack, a process called phage typing is used in clinical and diagnostic laboratories for the identification of pathogenic bacteria.

AT THE BENCH

Materials

Cultures

24-hour nutrient broth cultures of *Escherichia coli B* and T2 coliphage.

Media

Five each of the following per designated student group: 1.5% tryptone agar plates and 0.7% tryptone soft agar, 2 ml per tube; and nine tryptone broth tubes, 900 µl per tube.

Equipment

Microincinerator or Bunsen burner, waterbaths, thermometer, micropipette and tips, test tube rack, and glassware marking pencil.

Procedure Lab One

To perform the dilution procedure as illustrated in Figure 37.2, do the following:

1. Label all dilution tubes and media as follows:
    a. Five tryptone soft agar tubes: 10^{-5}, 10^{-6}, 10^{-7}, 10^{-8}, 10^{-9}.

b. Five tryptone hard agar plates: 10^{-5}, 10^{-6}, 10^{-7}, 10^{-8}, 10^{-9}.
    c. Nine tryptone broth tubes: 10^{-1} through 10^{-9}.

2. Place the five labeled soft tryptone agar tubes into a waterbath. Water should be of a depth just slightly above that of the agar in the tubes. Bring the waterbath to 100°C to melt the agar. Transfer the agar tubes to the second waterbath and maintain the melted agar at 45°C.

3. With micropipetter, aseptically perform a 10-fold serial dilution of the provided phage culture using the nine 900-µl tubes of tryptone broth.

4. To the tryptone soft agar tube labeled 10^{-5}, aseptically add 200 µl of the *E. coli B* culture and 100 µl of the 10^{-4} tryptone broth phage dilution. Rapidly mix by rotating the tube between the palms of your hands and pour the contents over the hard tryptone agar plate labeled 10^{-5}, thereby forming a double-layered plate culture preparation. Swirl the plate gently and allow to harden.

5. Using separate sterile micropipette tips, repeat Step 4 for the tryptone broth phage dilution tubes labeled 10^{-5} through 10^{-8} to effect the 10^{-6} through 10^{-9} tryptone soft agar overlays.

6. Following solidification of the soft agar overlay, incubate all plate cultures in an inverted position for 24 hours at 37°C.

Procedure Lab Two

1. Observe all plates for the presence of plaque-forming units that develop on the bacterial lawn.

2. Count the number of PFUs in the range of 30 to 300 on each plate.

3. Calculate the number of phage particles per ml of the stock phage culture based on your PFU count.

4. Record your results in the chart in the Lab Report.

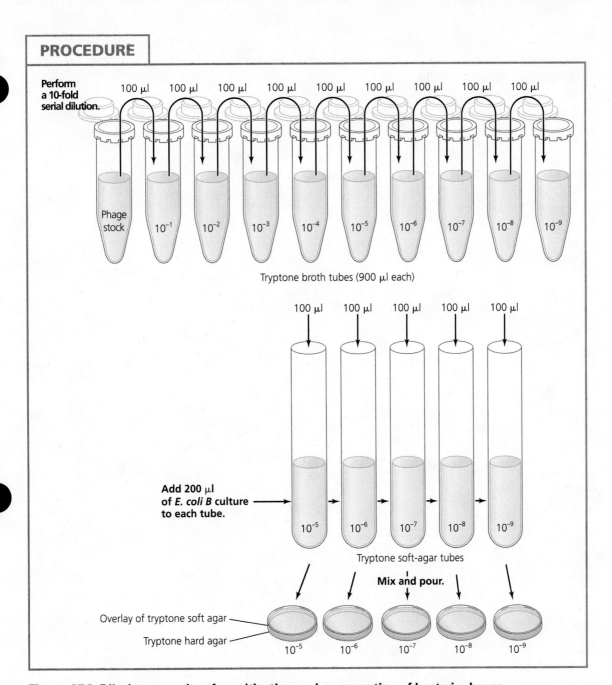

Perform a 10-fold serial dilution.

100 μl 100 μl 100 μl 100 μl 100 μl 100 μl 100 μl 100 μl 100 μl

Phage stock 10^{-1} 10^{-2} 10^{-3} 10^{-4} 10^{-5} 10^{-6} 10^{-7} 10^{-8} 10^{-9}

Tryptone broth tubes (900 μl each)

100 μl 100 μl 100 μl 100 μl 100 μl

Add 200 μl of *E. coli B* culture to each tube.

10^{-5} 10^{-6} 10^{-7} 10^{-8} 10^{-9}

Tryptone soft-agar tubes

Mix and pour.

Overlay of tryptone soft agar
Tryptone hard agar

10^{-5} 10^{-6} 10^{-7} 10^{-8} 10^{-9}

Figure 37.2 **Dilution procedure for cultivation and enumeration of bacteriophages**

Name: _____

Date: _____   Section: _____

Observations and Results

Phage Dilution	Number of PFUs	Calculation: PFUs × Dilution Factor	PFUs/ml of Stock Phage Culture
10^{-5}			
10^{-6}			
10^{-7}			
10^{-8}			
10^{-9}			

Review Questions

1. Discuss the effects of lytic and lysogenic infections on the life cycle of the host cell.

2. Discuss the factors responsible for the transformation of a lysogenic infection to one that is lytic.

3. Distinguish between the replicative and maturation stages of a lytic phage infection.

4. In this experimental procedure, why is it important to use a hard agar with a soft agar overlay technique to demonstrate plaque formation?

5. Explain what is meant by plaque-forming units.

6. Determine the number of PFUs per ml in a 10^{-9} dilution of a phage culture that shows 204 PFUs in the agar lawn.

7. The release of phage particles from the host bacterium always occurs by lysis of the cell and results in the death of the host. Animal viruses are released by either the lysis of the host cell or exocytosis, a reverse pinocytosis. Regardless of the mechanism of release, most infected cells die, while other viruses may escape the cell without damaging the host cell. Explain.

Isolation of Coliphages from Raw Sewage

LEARNING OBJECTIVE

Once you have completed this experiment, you should be able to

1. Isolate virulent coliphages from sewage.

Principle

Isolates of bacterial viruses (bacteriophages) can be obtained from a variety of natural sources, including soil, intestinal contents, raw sewage, and some insects, such as cockroaches and flies. Their isolation from these environments is not an easy task, because the phage particles are usually present in low concentrations. Therefore, isolation requires a series of steps:

1. Collection of the phage-containing sample at its source.

2. Addition of an enriched susceptible host cell culture to the sample to increase the number of phage particles for subsequent isolation.

3. Following incubation, centrifugation of the enriched sample for the removal of gross particles.

4. Filtration of the supernatant liquid through a bacteria-retaining membrane filter.

5. Inoculation of the bacteria-free filtrate onto a lawn of susceptible host cells grown on a soft agar plate medium.

6. Incubation and observation of the culture for the presence of phage particles, which is indicated by plaque formation in the bacterial lawn.

In the following experiment, you will use this procedure, as illustrated in **Figure 38.1**, for the isolation of *Escherichia coli* phage particles from raw sewage. Most bacteriophages that infect *E. coli* (coliphages) are designated by the letter T, indicating types. Seven types have been identified and are labeled T1 through T7. The T-even phages (T2, T4, and T6) differ from the T-odd phages in that the former vary in size, form, and chemical composition. All of the T phages are capable of infecting the susceptible *E. coli B* host cell.

CLINICAL APPLICATION

Phage Therapy

Phage therapy is the therapeutic use of bacteriophages to treat pathogenic bacterial infections. It is mainly used in Russia and the Republic of Georgia and is not universally approved elsewhere. In the West, no phage therapies are authorized for use on humans, although phages for killing food poisoning bacteria (*Listeria*) are now in use. They may also be used as a possible therapy against many strains of drug-resistant bacteria.

AT THE BENCH

Materials

Cultures

Lab One: 5-ml, 24-hour broth cultures of *E. coli B* and 45-ml samples of fresh sewage collected in screw-capped bottles. Lab Two: 10-ml, 24-hour broth cultures of *E. coli B*.

Media

Per designated student group: Lab One: One 5-ml tube of bacteriophage nutrient broth, 10 times normal concentration. Lab Two: Five tryptone agar plates and five 3-ml tubes of tryptone soft agar.

Equipment

Lab One: Sterile 250-ml Erlenmeyer flask and stopper. Lab Two: Sterile membrane filter apparatus, sterile 125-ml Erlenmeyer flask and stopper, 125-ml flask, 1000-ml beaker, centrifuge,

Pour in several
centrifuge tubes
and centrifuge

sewage sample
at 2500 rpm
for 20 minutes.

Remove tubes
and decant
supernatant
into a 125-ml beaker.

Filter supernatant
through membrane
filter into vacuum flask.

E. coli B enriched sewage
sample in 250-ml flask

Centrifuge

Flask and membrane
filter apparatus

Bacteria-free phage filtrate in vacuum flask

1 drop   2 drops   3 drops   4 drops   5 drops

Add 0.1 ml
to tubes 1–5

of molten
soft tryptone
agar.

1   2   3   4   5

E. coli B
culture

1   2   3   4   5

Plates of hard tryptone agar

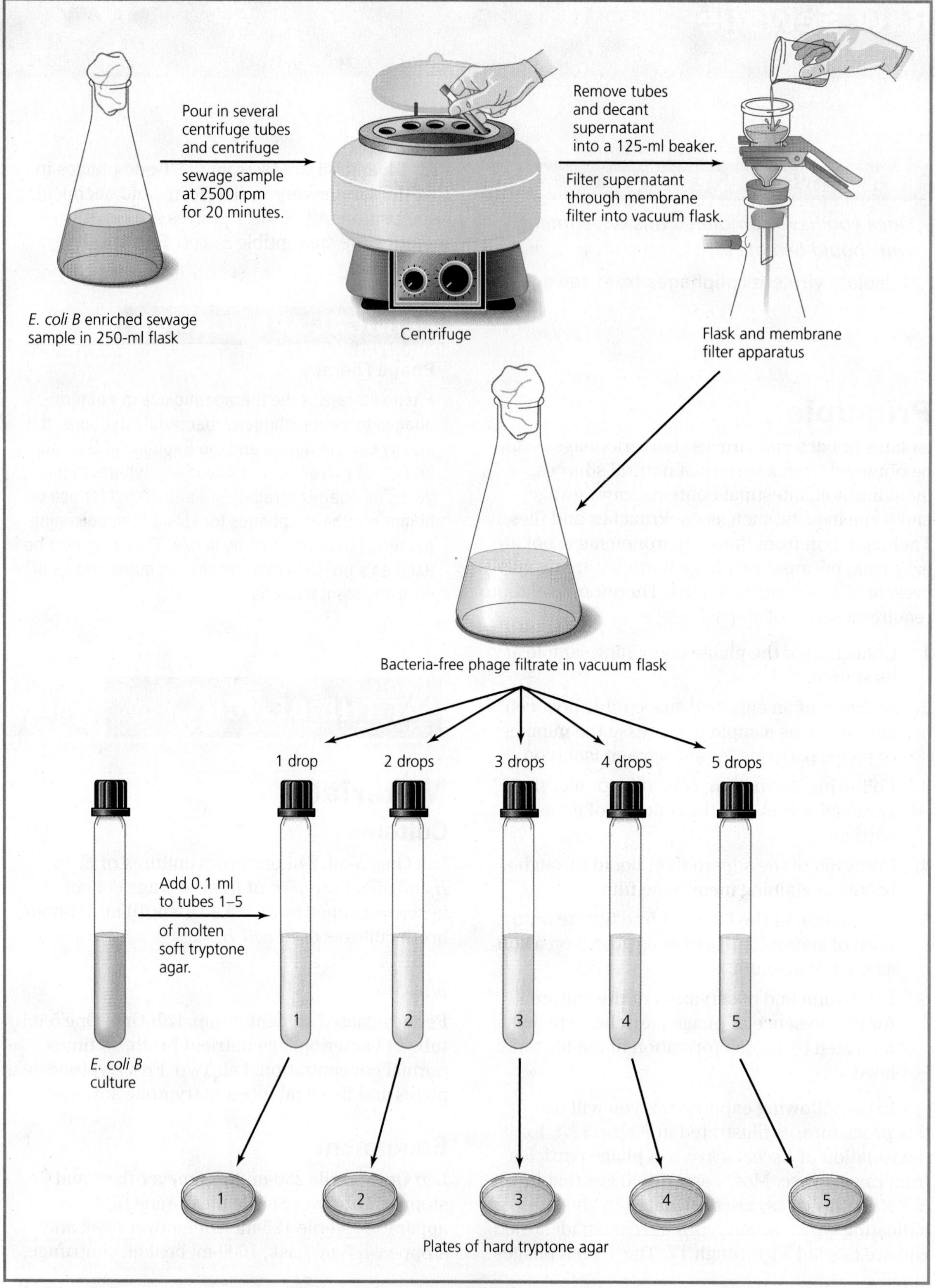

Figure 38.1 Procedure for isolation of coliphages from raw sewage

microincinerator or Bunsen burner, forceps, 1-ml sterile disposable pipettes, sterile Pasteur pipette, mechanical pipetting device, test tube rack, and glassware marking pencil.

Procedure Lab One

> ⚠ **Use disposable gloves. It is essential to handle raw sewage with extreme caution because it may serve as a vehicle for the transmission of human pathogens.**

Enrichment of Sewage Sample

1. Aseptically add 5 ml of bacteriophage nutrient broth, 5 ml of the *E. coli B* broth culture, and 45 ml of the raw sewage sample to an appropriately labeled sterile 250-ml Erlenmeyer flask.
2. Incubate the culture for 24 hours at 37°C.

Procedure Lab Two
Filtration and Seeding

1. Following incubation, pour the phage-infected culture into a 100-ml centrifuge bottle or several centrifuge tubes and centrifuge at 2500 rpm for 20 minutes.
2. Remove the centrifuge bottle or tubes, being careful not to stir up the sediment, and carefully decant the supernatant into a 125-ml beaker.

3. Pour the supernatant solution through a sterile membrane filter apparatus to collect the bacteria-free, phage-containing filtrate in the vacuum flask below. Refer to Experiment 48 for the procedure in assembling the filter membrane apparatus.
4. Melt the soft tryptone agar by placing the five tubes in a boiling waterbath and cool to 45°C.
5. Label the five tryptone agar plates and the five tryptone agar tubes 1, 2, 3, 4, and 5, respectively.
6. Using a sterile 1-ml pipette, aseptically add 0.1 ml of the *E. coli B* culture to all the molten soft-agar tubes.
7. Using a sterile Pasteur pipette, aseptically add 1, 2, 3, 4, and 5 drops of the filtrate to the respectively labeled molten soft-agar tubes. Mix and pour each tube of soft agar into its appropriately labeled agar plate.
8. Allow agar to harden.
9. Incubate all the plates in an inverted position for 24 hours at 37°C.

Procedure Lab Three

1. Examine all the culture plates for plaque formation, which is indicative of the presence of coliphages in the culture.
2. Indicate the presence (+) or absence (−) of plaques in each of the cultures in the chart in the Lab Report.

Name: _____

Date: _____ Section: _____

Observations and Results

Drops of Phage Filtrate	1	2	3	4	5
Plaque Formation (+) or (–)					

Based on your observations, what is the relationship between the number of plaques observed and the number of drops of filtrate in each culture?

Review Questions

1. Why is enrichment of the sewage sample necessary for the isolation of phage?

2. How is enrichment of the sewage sample accomplished?

3. How are bacteria-free phage particles obtained?

4. Why must you exercise caution when handling raw sewage samples?

Propagation of Isolated Bacteriophage Cultures

LEARNING OBJECTIVES

Once you have completed this experiment, you should be able to

1. Isolate bacteriophages from a plaque culture for later genetic studies or manipulations.

2. Enumerate the plaque-forming units isolated from an individual plaque.

CLINICAL APPLICATION

With the increase in the rates of antibiotic resistance in clinically relevant bacteria, pharmaceutical companies and researchers are looking for new therapeutic treatments in unlikely places. They are now looking at the possibility of treating a resistant bacterial infection with a virus. Current research is examining the clinical uses of bacteriophages as a means of treating bacterial infections in the absence of antibiotics.

Principle

This exercise will demonstrate the procedure for isolating and propagating a specific bacteriophage species from a single plaque picked from a lawn plate. Before a microbiologist or virologist may begin studying a new bacteriophage or begin genetic recombination studies an individual strain must be isolated. This is similar to what must be done before performing assays on bacterial species; a single colony must be chosen so that all the bacteria present will be genetic and metabolic clones of each other. These same practices will be followed when studying viruses.

What begins as a single virus infecting a single bacterium will eventually spread to neighboring cells. With the release of phage particles from an infected cell the phages will spread via diffusion to neighboring cells. Since the viruses have no mechanisms for propulsion, such as a flagella or fimbriae, the particles must rely on diffusion through the soft agar medium to spread from cell to cell. This exercise will use that occurrence to remove the phage particles from an isolated plaque.

AT THE BENCH

Materials

Cultures

Agar plates reserved from Experiments 37 or Experiment 38 that have individual plaques and a 24-hour nutrient broth culture of *Escherichia coli B*.

Media

Per designated student group: 10 ml of TRIS buffered saline (TBS); tryptone agar plates and tryptone soft agar, 2 ml per tube; and nine tryptone broth tubes, 0.9 ml per tube.

Equipment

Bunsen burner, waterbath, thermometer, 1.5-ml centrifuge tubes, 1-ml sterile pipettes, sterile glass Pasteur pipettes, rubber bulb, mechanical pipetting device, test tube rack, and glassware marking pencil.

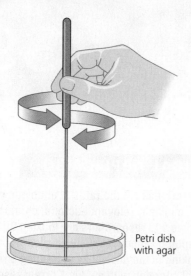

Figure 39.1 Use of a glass Pasteur pipette to remove an agar plug from a petri dish that contains a plaque of interest

Petri dish with agar

Procedure Lab One

Utilizing one or more plates reserved from Experiments 37 and 38, use the following procedure to isolate bacteriophages:

1. Place a rubber bulb on the end of a glass Pasteur pipette and use the end of the pipette to remove the plaque-containing agar from the selected plate as follows:

   a. Following the procedure illustrated in Figure 39.1, use the tapered end of the glass Pasteur pipette and plunge the pipette through the agar that surrounds the plaque.

   b. Give the pipette a few turns to reduce contact between the agar and the bottom of the petri dish.

   c. Lift up the pipette and the agar plug that contains the plaque.

2. Gently depress the bulb to dislodge the agar plug into a 1.5 ml centrifuge tube.

3. Add 1 ml of TRIS-buffered saline (TBS) to the tube and incubate at 4°C overnight or up to one week.

Procedure Lab Two

1. Label all dilution tubes and media as follows:

   a. Five tryptone soft agar tubes: 10^{-5}, 10^{-6}, 10^{-7}, 10^{-8}, 10^{-9}

   b. Five tryptone hard agar plates: 10^{-5}, 10^{-6}, 10^{-7}, 10^{-8}, 10^{-9}

   c. Nine tryptone broth tubes: 10^{-1} through 10^{-9}

2. Place the five labeled soft tryptone agar tubes into a waterbath. Water should be of a depth just slightly above that of the agar in the tubes. Bring the waterbath to 100°C to melt the agar. Transfer the agar tubes to the second waterbath and maintain the melted agar at 45°C.

3. With micropipetter, aseptically perform a 10-fold serial dilution of the provided phage culture using the nine 900-μl tubes of tryptone broth.

4. To the tryptone soft agar tube labeled 10^{-5}, aseptically add 200 μl of the *E. coli B* culture and 100 μl of the 10^{-4} tryptone broth phage dilution. Rapidly mix by rotating the tube between the palms of your hands and pour the contents over the hard tryptone agar plate labeled 10^{-5}, thereby forming a double-layered plate culture preparation. Swirl the plate gently and allow it to harden.

5. Using separate sterile micropipette tips, repeat Step 4 for the tryptone broth phage dilution tubes labeled 10^{-5} through 10^{-8} to effect the 10^{-6} through 10^{-9} tryptone soft agar overlays.

6. Following solidification of the soft agar overlay, incubate all plate cultures in an inverted position for 24 hours at 37°c.

Procedure Lab Three

1. Observe all plates for the presence of plaque-forming units that develop on the bacterial lawn.

2. Count the number of CFUs per plate in the range of 30 to 300 on each plate.

3. Calculate the number of phage particles and record your results.

Name: _____

Date: _____   Section: _____

Observations and Results

Phage Dilution	Number of PFUs	Calculation: PFUs × Dilution Factor	PFUs/mL of Recovered Phage Culture
10^{-5}			
10^{-6}			
10^{-7}			

Review Questions

1. How many bacteriophage particles were isolated from a single plaque? How many different strains of phage would be present?

2. You have tested a sewage sample for the presence of bacteriophages and have several plates with plaques present. Will all of these plaques be due to the same type or strain of virus? How would you go about answering this question?

PART 9

Physical and Chemical Agents for the Control of Microbial Growth

Once you have completed the experiments in this section, you should

1. Know the basic methods for inhibiting microbial growth and the modes of antimicrobial action.
2. Be able to demonstrate the effects of physical agents, moist heat, osmotic pressure, and ultraviolet radiation on selected microbial populations.
3. Be able to demonstrate the effects on selected microbial populations of chemical agents used as disinfectants, antiseptics, and antibiotics.

Introduction

Control of microorganisms is essential in the home, industry, and medical fields to prevent and treat diseases and to inhibit the spoilage of foods and other industrial products. Common methods of control involve chemical and physical agents that adversely affect microbial structures and functions, thereby producing a microbicidal or microbistatic effect. A **microbicidal effect** is one that kills the microbes immediately; a **microbistatic effect** inhibits the reproductive capacities of the cells and maintains the microbial population at a constant size.

Chemical Methods for Control of Microbial Growth

1. **Antiseptics:** Chemical substances used on living tissue that kill or inhibit the growth of vegetative microbial forms.
2. **Disinfectants:** Chemical substances that kill or inhibit the growth of vegetative microbial forms on nonliving materials.
3. **Chemotherapeutic agents:** Chemical substances that destroy or inhibit the growth of microorganisms in living tissues.

Physical Methods for Control of Microbial Growth

The modes of action of the different chemical and physical agents of control vary, although they all produce damaging effects to one or more essential cellular structures or molecules in order to cause cell death or inhibition of growth. Sites of damage that can result in malfunction are the cell wall, cell membrane, cytoplasm, enzymes, and nucleic acids. The adverse effects manifest themselves in the following ways.

1. **Cell-wall injury:** This can occur in one of two ways. First, lysis of the cell wall will leave the wall-less cell, called a protoplast, susceptible to osmotic damage, and a hypotonic environment may cause lysis of the vulnerable protoplast. Second, certain agents inhibit cell wall synthesis, which is essential during microbial cell reproduction. Failure to synthesize a missing segment of the cell wall results in an unprotected protoplast.

2. **Cell-membrane damage:** This may be the result of lysis of the membrane, which will cause immediate cell death. Also, the selective nature of the membrane may be affected without causing its complete disruption. As a

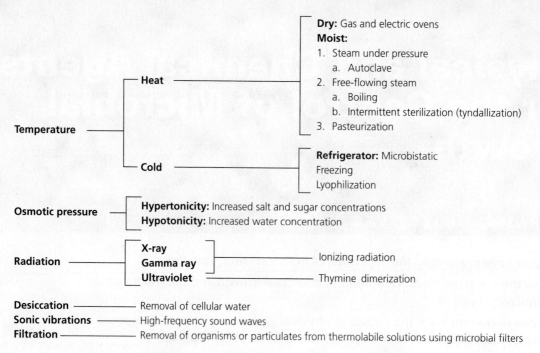

Figure P9.1 Physical methods used for the control of microbial growth

result, there may be a loss of essential cellular molecules or interference with the uptake of nutrients. In both cases, metabolic processes will be adversely affected.

3. **Alteration of the colloidal state of cytoplasm:** Certain agents cause denaturing of cytoplasmic proteins. Denaturing processes are responsible for enzyme inactivation and cellular death by irreversibly rupturing the molecular bonds of these proteins and rendering them biologically inactive.

4. **Inactivation of cellular enzymes:** Enzymes may be inactivated competitively or noncompetitively. Noncompetitive inhibition is irreversible and occurs following the application of some physical agent, such as mercuric chloride ($HgCl_2$), that results in the uncoiling of the protein molecule, rendering it biologically inactive. Competitive inhibition occurs when a natural substrate is forced to compete for the active site on an enzyme surface with a chemically similar molecular substrate, which can block the enzyme's ability to create end products. Competitive inhibitions are reversible.

5. **Interference with the structure and function of the DNA molecule:** The DNA molecule is the control center of the cell and may also represent a cellular target area for destruction or inhibition. Some agents have an affinity for DNA and cause breakage or

distortion of the molecule, thereby interfering with its replication and role in protein synthesis.

Figure P9.1 illustrates the acceptable physical methods used for the control of microbial growth.

Awareness of the mode of action of the physical and chemical agents is absolutely essential for their proper selection and application in microbial control. The exercises in this section are designed to acquaint you more fully with several commonly employed agents and their uses.

Governing Bodies for Laboratory Procedures

Numerous groups consisting of individuals involved in academics, microbiological research, industry, and government agencies have developed accepted procedures and practices for the research and development of new antibiotics and anti-microbial chemical agents. Groups that include Clinical and Laboratory Standards Institute (CLSI), the Association of Analytical Communities (AOAC), and the American Society for Microbiology (ASM) partner with government agencies, such as the Environmental Protection Agency (EPA) and the Centers for Disease Control and Prevention (CDC) to determine best practices for use in clinical and laboratory research that many of the procedures presented in Part 9 will be based on. Where possible, a formal agency guideline will be included for future reference.

Physical Agents of Control: Moist Heat

●

LEARNING OBJECTIVE

Once you have completed this experiment, you should understand

1. The susceptibility of microbial species to destruction by the application of moist heat.

Principle

Temperature has an effect on cellular enzyme systems and therefore a marked influence on the rate of chemical reactions and thus the life and death of microorganisms. Despite the diversity among microorganisms' temperature requirements for growth, extremes in temperature can be used in microbial growth control. Sufficiently low temperatures will inactivate enzymes and produce a static effect. High temperatures destroy cellular enzymes, which become irreversibly denatured.

The application of heat is a common means of destroying microorganisms. Both dry and moist heat are effective. However, moist heat, which (because of the hydrolyzing effect of water and its greater penetrating ability) causes coagulation of proteins, kills cells more rapidly and at lower temperatures than does dry heat. **Sterilization**, the destruction of all forms of life, is accomplished in 15 minutes at 121°C with moist heat (steam) under pressure; dry heat requires a temperature of 160°C to 180°C for 1½ to 3 hours.

Microbes exhibit differences in their resistance to moist heat. As a general rule, bacterial spores require temperatures above 100°C for destruction, whereas most bacterial vegetative cells are killed at temperatures of 60°C to 70°C in 10 minutes. Fungi can be killed at 50°C to 60°C, and fungal spores require 70°C to 80°C for 10 minutes for destruction. Because of this variability, moist heat can either sterilize or disinfect. Common applications include free-flowing steam

under pressure (autoclaving), free-flowing steam at 100°C (tyndallization), and the use of lower temperatures (pasteurization).

Free-flowing steam under pressure requires the use of an autoclave, a double-walled metal vessel that allows steam to be pressurized in the outer jacket (see **Figure 40.1**). At a designated pressure, the saturated steam is released into the inner chamber, from which all the air has been evacuated. The steam under pressure in the vacuumed inner chamber is now capable of achieving temperatures in excess of 100°C. The temperature is determined by the pounds of pressure applied per square inch:

Pressure (pounds/inch^2)	Temperature (°C)
0 (free-flowing steam)	100
10	115
15	121
20	126
25	130

A pressure of 15 pounds/inch^2 achieves a temperature of 121°C and sterilizes in 15 minutes. This is the usual procedure; however, depending on the heat sensitivity of the material to be sterilized, the operating pressure and time conditions can be adjusted.

Application of **free-flowing steam** requires exposure of the contaminated substance to a temperature of 100°C, which is achieved by boiling water. Exposures to boiling water for 30 minutes will result in disinfection only; all vegetative cells will be killed, but not necessarily the more heat-resistant spores.

Another procedure is **tyndallization**, also referred to as intermittent or fractional sterilization. This procedure requires exposure of the material to free-flowing steam at 100°C for 20 minutes on 3 consecutive days with intermittent incubation at 37°C. The steaming kills all vegetative cells. Any spores that may be present germinate during the period of incubation and are destroyed during

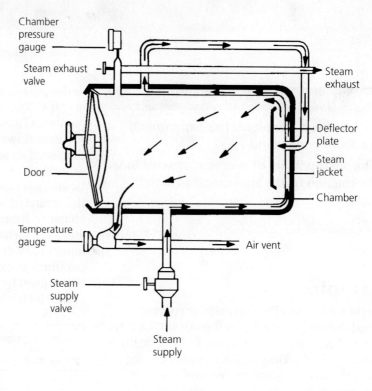

(a) An autoclave

(b) Schematic representation

Figure 40.1 The autoclave

subsequent exposure to a temperature of 100°C. Repeating this procedure for 3 days ensures germination of all spores and their destruction in the vegetative form. Because tyndallization requires so much time, it is used only for sterilization of materials that are composed of thermolabile chemicals and that might be subject to decomposition at higher temperatures.

Pasteurization exposes fairly thermolabile products, such as milk, wine, and beer, for a given period of time to a temperature that is high enough to destroy pathogens and some spoilage-causing microorganisms that may be present, without necessarily destroying all vegetative cells. There are three types of pasteurization: The high-temperature, short-time (HTST) procedure requires a temperature of 71°C for 15 seconds. The low-temperature, long-time (LTLT) method requires

63°C for 30 minutes, and the ultra high temperature (UHT) approach occurs at 138°C for 2 seconds.

CLINICAL APPLICATION

Autoclave Performance Testing

While the original "autoclave" was invented as a pressure cooker for food, modern autoclaves are precision instruments and require maintenance and periodic testing, especially if control of human pathogens is involved. Commonly, a sample of spores of the bacterium *Bacillus stearothermophilus* is sterilized in the chamber with a normal load, and then the sample is allowed to incubate—any growth indicates that the autoclave needs to be serviced.

Materials

Cultures

48- to 72-hour nutrient broth cultures (50 ml per 250-ml Erlenmeyer flask) of *Staphylococcus aureus* BSL-2 *and Bacillus cereus;* 72- to 96-hour Sabouraud broth cultures (50 ml per 250-ml Erlenmeyer flask) of *Aspergillus niger* and *Saccharomyces cerevisiae.*

Media

Per designated student group (pairs or groups of four): five nutrient agar plates, five Sabouraud agar plates, and one 10-ml tube of nutrient broth.

Equipment

Microincinerator or Bunsen burner, 800-ml beaker (waterbath), tripod and wire gauze screen with heat-resistant pad, thermometer, sterile test tubes, glassware marking pencil, and inoculating loop.

Procedure Lab One

1. Label the covers of each of the nutrient agar and Sabouraud agar plates, indicating the experimental heat temperatures to be used: 25°C (control), 40°C, 60°C, 80°C, and 100°C.

2. Score the underside of all plates with a glassware marking pencil into two sections. On the nutrient agar plates, label one section *S. aureus* BSL-2 and the other *B. cereus.* On the Sabouraud agar plates, label one section *A. niger* and the second *S. cerevisiae.*

3. Using aseptic technique, inoculate the nutrient agar and Sabouraud agar plates labeled 25°C by making a single-line loop inoculation of each test organism in its respective section of the plate.

4. Using a sterile pipette and mechanical pipetter, transfer 10 ml of each culture to four sterile test tubes labeled with the name of the organism and the temperature (40°C, 60°C, 80°C, and 100°C).

5. Set up the waterbath as illustrated in Figure 40.2, inserting the thermometer in an uncapped tube of nutrient broth.

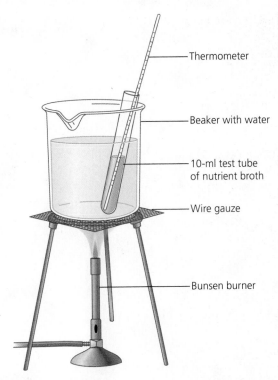

Figure 40.2 **Waterbath for moist heat experiment**

Thermometer

Beaker with water

10-ml test tube of nutrient broth

Wire gauze

Bunsen burner

6. Slowly heat the water to 40°C; check the thermometer frequently to ensure that it does not exceed the desired temperature. Place the four cultures of the experimental organisms into the beaker and maintain the temperature at 40°C for 10 minutes. Remove the cultures and aseptically inoculate each organism in its appropriate section on the two plates labeled 40°C.

7. Raise the waterbath temperature to 60°C and repeat Step 6 for the inoculation of the two plates labeled 60°C.

8. Raise the waterbath temperature to 80°C and repeat Step 6 for the inoculation of the two plates labeled 80°C.

9. Raise the waterbath temperature to 100°C and repeat Step 6 for the inoculation of the two plates labeled 100°C.

10. Incubate the nutrient agar plate cultures in an inverted position for 24 to 48 hours at 37°C and the Sabouraud agar plate cultures for 4 to 5 days at 25°C in a moist chamber.

Procedure Lab Two

1. Observe all plates for the amount of growth of the test organisms at each of the temperatures.

2. Record your results in the chart provided in the Lab Report.

Name: _____

Date: _____ Section: _____

Observations and Results

1. Record your results in the chart as 0 = none; 1+ = slight; 2+ =moderate;
   3+ = abundant.

Microbial Species	AMOUNT OF GROWTH				
	25°C	40°C	60°C	80°C	100°C
B. cereus					
S. aureus					
A. niger					
S. cerevisiae					

2. List the microbial organisms in order of their increasing heat resistance.

   _____

   _____

   _____

   _____

Review Questions

1. Account for the microbistatic effect produced by low temperatures as compared to the microbicidal effect produced by high temperatures.

2. Cite the advantages of each of the modes of sterilization: tyndallization and
   autoclaving.

3. Discuss the detrimental effects of control agents on the following: the cytoplasm, the cell wall, nucleic acids, and the cell membrane.

4. Explain why milk is subjected to pasteurization rather than sterilization.

5. *A. niger* and *B. cereus* cultures used in this experiment contained spores. Why is *B. cereus* more heat resistant?

6. Account for the fact that aerobic and anaerobic bacterial spore-formers are more heat resistant than the tubercle bacillus, which is also known to tolerate elevated temperatures.

Physical Agents of Control: Electromagnetic Radiations

Once you have completed this experiment, you should understand

1. The microbicidal effect of ultraviolet (UV) radiation on microorganisms.

Principle

Certain forms of electromagnetic radiation are capable of producing a lethal effect on cells and therefore can be used for microbial control. Electromagnetic radiations that possess sufficient energy to be microbicidal are the short-wavelength radiations, that is, 300 nanometer (nm) and below. These include UV, gamma rays, and x-rays. The high-wavelength radiations, those above 300 nm, have insufficient energy to destroy cells. The electromagnetic spectrum and its effects on molecules are illustrated in **Figure 41.1**.

Gamma radiation, originating from unstable atomic nuclei, and **x-radiation**, originating from outside of the atomic nucleus, are representative of **ionizing forms of radiation**. Both transfer their energy through quanta (photons) to the matter through which they pass, causing excitation and the loss of electrons from molecules in their paths. This injurious effect is nonspecific in that any molecule in the path of the radiation will undergo ionization. Essential cell molecules can be directly affected through loss of their chemical structures and activity brought about by the ionization. Also, water, the most abundant chemical constituent of cells, commonly undergoes radiation breakdown, with the ultimate production of highly reactive H^+, OH^-, and, in the presence of oxygen, HO_2 free radicals. These may combine with each other, frequently forming hydrogen peroxide (H_2O_2), which is highly toxic to cells lacking catalase or other peroxidases, or the highly reactive free radicals may combine with any cellular constituents, again resulting in cell damage.

Because of their high energy content and therefore ability to penetrate matter, x-ray and gamma radiations can be used as means of sterilization, particularly of thermolabile materials. They are not commonly used, however, because of the expense of the equipment and the special facilities necessary for their safe use.

Ultraviolet light, which has a lower energy content than ionizing radiations, is capable of producing a lethal effect in cells exposed to the low penetrating wavelengths in the range of 210 nm to 300 nm. Cellular components capable of absorbing ultraviolet light are the nucleic acids; DNA is the primary site of damage. As the pyrimidines especially absorb ultraviolet wavelengths, the major effect of this form of radiation is **thymine dimerization**, which is the covalent bonding of

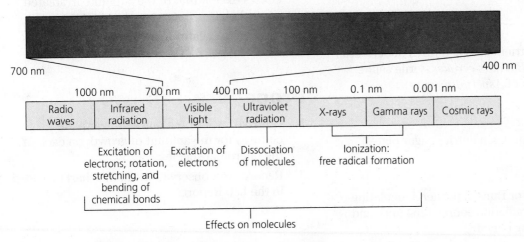

Figure 41.1 The electromagnetic spectrum and its effects on molecules

two adjacent thymine molecules on one nucleic acid strand in the DNA molecule. This dimer formation distorts the configuration of the DNA molecule, and the distortion interferes with DNA replication and transcription during protein synthesis.

Some cell types, including some microorganisms, possess enzyme systems for the repair of radiation-induced DNA damage. Two different systems may be operational: (1) The **excision repair system**, which functions in the absence of light; and (2) the **light repair system**, which is made operational by exposure of the irradiated cells to visible light in the wavelength range of 420 nm to 540 nm. The visible light serves to activate an enzyme that splits the dimers and reverses the damage.

Ultraviolet radiation, because of its low penetration ability, cannot be used as a means of sterilization, and its practical application is only for surface or air disinfection.

CLINICAL APPLICATION

Radiation Resistant Organisms

As lethal as radiation would seem to be, organisms vary tremendously in their ability to resist harm when exposed. Yeasts and bacterial spores are among the most resistant, while gram-positive bacteria are more resistant than gram-negative ones. Bacterial cocci also tend to be more resistant than bacilli, with viruses the most resistant of all. To effectively neutralize all contaminants, radiation dose and conditions must be carefully controlled.

AT THE BENCH

Materials

Cultures

24- to 48-hour nutrient broth cultures of *Serratia marcescens* and *Bacillus cereus*; sterile saline spore suspension of *Aspergillus niger*.

Media

Per student group: seven nutrient agar plates.

Equipment

Microincinerator or Bunsen burner, inoculating loop, ultraviolet radiation source (254 nm), and glassware marking pencil.

Procedure Lab One

> ⚠️ **Wear disposable gloves and do not expose your eyes to the ultraviolet light source.**

1. Divide all nutrient agar plates into three sections by marking the underside of each plate with a glassware marking pencil.
2. Label each of the sections on each plate with the name of the organism to be inoculated.
3. Using aseptic technique, inoculate all the plates by means of a streak inoculation *specifically* as shown in the following illustration:

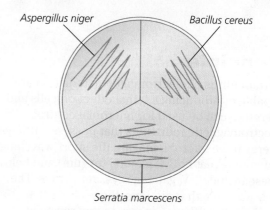

4. Label the cover of each inoculated plate with the exposure time to ultraviolet radiation as 0 (control), 15 seconds, 30 seconds, 1 minute, and 3 minutes. Label two plates as 5 minutes; one of these plates will serve as the irradiated, covered control.
5. Irradiate all inoculated plates for the designated period of time by placing them 12 inches below the ultraviolet light source. Make sure first to remove all Petri dish covers except that of the 5-minute irradiated control plate.
6. Incubate all plates in an inverted position for 4 to 5 days at 25°C.

Procedure Lab Two

1. Observe each of the nutrient agar plate cultures for the amount of growth on each of the microbial species.
2. Record your observations in the chart provided in the Lab Report.

Name: _____

Date: _____ Section: _____

Observations and Results

Record your observations in the chart as 0 = no growth; 1+ = slight groth;
2+ = moderate growth; 3+ = abundant growth.

	TIME OF IRRADIATION						
	Seconds			Minutes			
Microbial Species	0	15	30	1	3	5	5*
B. cereus							
S. marcescens							
A. niger							
*Irradiated, covered plate.							

Review Questions

1. Discuss the effects of ionizing radiation on cellular constituents.

2. Explain why x-rays can be used for sterilization, whereas ultraviolet rays
   can be used only for surface disinfection of materials.

3. Explain the mechanism of action of ultraviolet radiation on cells.

4. Account for the greater susceptibility of *S. marcescens* than that of *B. cereus* to the effects of ultraviolet radiation.

5. Why is it not essential to shield your hands from ultraviolet light, whereas you must exercise great care to shield your eyes from this type of radiation?

Chemical Agents of Control: Chemotherapeutic Agents

Chemotherapeutic agents are chemical substances used in the treatment of infectious diseases. Their mode of action is to interfere with microbial metabolism, thereby producing a bacteriostatic or bactericidal effect on the microorganisms, without producing a like effect in host cells. Chemotherapeutic agents act on a number of cellular targets. Their mechanisms of action include inhibition of cell-wall synthesis, inhibition of protein synthesis, inhibition of nucleic acid synthesis, disruption of the cell membrane, and inhibition of folic acid synthesis. These drugs can be separated into two categories:

1. **Antibiotics** are synthesized and secreted by some true bacteria, actinomycetes, and fungi that destroy or inhibit the growth of other microorganisms. Today, some antibiotics are laboratory synthesized or modified; however, their origins are living cells.

2. **Synthetic drugs** are synthesized in the laboratory.

To determine a therapeutic drug of choice, it is important to determine its mode of action,

possible adverse side effects in the host, and the scope of its antimicrobial activity. The specific mechanism of action varies among different drugs, and the short-term or long-term use of many drugs can produce systemic side effects in the host. These vary in severity from mild and temporary upsets to permanent tissue damage (Table 42.1).

Synthetic Agents

Sulfadiazine (a sulfonamide) produces a static effect on a wide range of microorganisms by a mechanism of action called **competitive inhibition**. The active component of the drug, sulfanilamide, acts as an **antimetabolite** that competes with the **essential metabolite**, p-aminobenzoic acid (PABA), during the synthesis of folic acid in the microbial cell. Folic acid is an essential cellular coenzyme involved in the synthesis of amino acids and purines. Many microorganisms possess enzymatic pathways for folic acid synthesis and can be adversely affected by sulfonamides. Human cells lack these enzymes, and the essential folic acid enters the cells in a

TABLE 42.1 Prototypic Antibiotics

ANTIBIOTIC	MODE OF ACTION	POSSIBLE SIDE EFFECTS
Penicillin	Prevents transpeptidation of the N-acetylmuramic acids, producing a weakened peptidoglycan structure	Penicillin resistance; sensitivity (allergic reaction)
Streptomycin	Has an affinity for bacterial ribosomes, causing misreading of codons on mRNA, thereby interfering with protein synthesis	May produce damage to auditory nerve, causing deafness
Chloramphenicol	Has an affinity for bacterial ribosomes, preventing peptide bond formation between amino acids during protein synthesis	May cause aplastic anemia, which is fatal because of destruction of RBC-forming and WBC-forming tissues
Tetracyclines	Have an affinity for bacterial ribosomes; prevent hydrogen bonding between the anticodon on the tRNA–amino acid complex and the codon on mRNA during protein synthesis	Permanent discoloration of teeth in young children
Bacitracin	Inhibits cell-wall synthesis	Nephrotoxic if taken internally; used for topical application only
Polymyxin	Destruction of cell membrane	Toxic if taken internally; used for topical application only
Rifampin	Inhibits RNA synthesis	Appearance of orange-red urine, feces, saliva, sweat, and tears
Quinolone	Inhibits DNA synthesis	Affects the development of cartilage

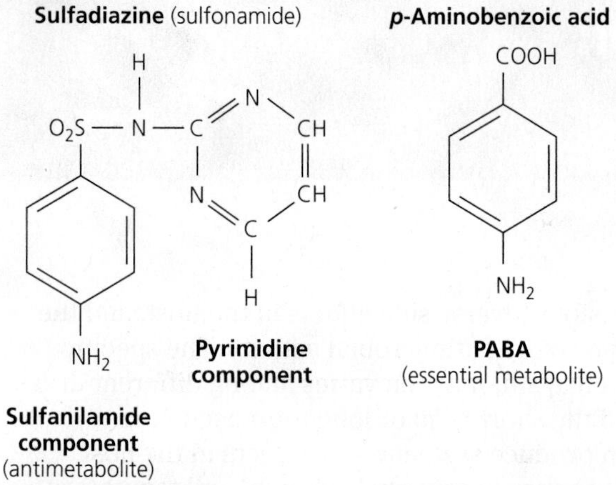

Sulfadiazine (sulfonamide)

p-Aminobenzoic acid

COOH

Pyrimidine component

PABA
(essential metabolite)

Sulfanilamide component
(antimetabolite)

Figure 42.1 **Chemical similarity of sulfanilamide and PABA**

preformed state. Therefore, these drugs have no competitive effect on human cells. The similarity between the chemical structure of the antimetabolite sulfanilamide and the structure of the essential metabolite PABA is illustrated in Figure 42.1.

PART A The Kirby-Bauer Antibiotic Sensitivity Test Procedure

LEARNING OBJECTIVE

Once you have completed this experiment, you should understand

1. The Kirby-Bauer procedure for the evaluation of the antimicrobial activity of chemotherapeutic agents.

Principle

The available chemotherapeutic agents vary in their scope of antimicrobial activity. Some have a limited spectrum of activity, being effective against only one group of microorganisms. Others exhibit broad-spectrum activity against a range of microorganisms. The drug susceptibilities of many pathogenic microorganisms are known, but it is sometimes necessary to test several agents to determine the drug of choice.

A standardized diffusion procedure with filter-paper discs on agar, known as the **Kirby-Bauer method**, is frequently used to determine the drug susceptibility of microorganisms isolated from infectious processes. This method allows the

rapid determination of the efficacy of a drug by measuring the diameter of the zone of inhibition that results from diffusion of the agent into the medium surrounding the disc. In this procedure, filter-paper discs of uniform size are impregnated with specified concentrations of different antibiotics and then placed on the surface of an agar plate that has been seeded with the organism to be tested. The medium of choice is Mueller-Hinton agar, with a pH of 7.2 to 7.4, which is poured into plates to a uniform depth of 5 mm and refrigerated after solidification. Prior to use, the plates are transferred to an incubator at 37°C for 10 to 20 minutes to dry off the moisture that develops on the agar surface. The plates are then heavily inoculated with a standardized inoculum by means of a cotton swab to ensure the confluent growth of the organism. The discs are aseptically applied to the surface of the agar plate at well-spaced intervals. Once applied, each disc is gently touched with a sterile applicator stick to ensure its firm contact with the agar surface.

Following incubation, the plates are examined for the presence of growth inhibition, which is indicated by a clear zone surrounding each disc (Figure 42.2). The susceptibility of an organism to a drug is assessed by the size of this zone, which is affected by other variables such as

1. The ability and rate of diffusion of the antibiotic into the medium and its interaction with the test organism.
2. The number of organisms inoculated.
3. The growth rate of the organism.

A measurement of the diameter of the zone of inhibition in millimeters is made, and its size

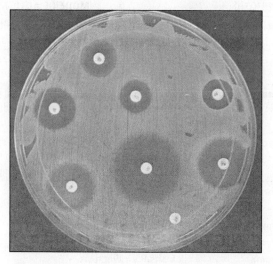

Figure 42.2 **Kirby-Bauer antibiotic sensitivity test**

TABLE 42.2 Zone Diameter Interpretive Standards for Organisms Other Than *Haemophilus* and *Neisseria gonorrhoeae*

ANTIMICROBIAL AGENT	DISC CONTENT	ZONE DIAMETER, NEAREST WHOLE MM		
		RESISTANT	INTERMEDIATE	SUSCEPTIBLE
Ampicillin				
when testing gram-negative bacteria	10 µg	≤13	14–16	≥17
when testing gram-positive bacteria	10 µg	≤28	—	≥29
Carbenicillin				
when testing *Pseudomonas*	100 µg	≤13	14–16	≥17
when testing other gram-negative organisms	100 µg	≤19	20–22	≥23
Cefoxitin	30 µg	≤14	15–17	≥18
Cephalothin	30 µg	≤14	16–17	≥18
Chloramphenicol	30 µg	≤12	13–17	≥18
Clindamycin	2 µg	≤14	15–20	≥21
Erythromycin	15 µg	≤13	14–22	≥23
Gentamicin	10 µg	≤12	13–14	≥15
Kanamycin	30 µg	≤13	14–17	≥18
Methicillin when testing staphylococci	5 µg	≤9	10–13	≥14
Novobiocin	30 µg	≤17	18–21	≥22
Penicillin G				
when testing staphylococci	10 units	≤28	—	≥29
when testing other bacteria	10 units	≤14	—	≥15
Rifampin	5 µg	≤16	17–19	≥20
Streptomycin	10 µg	≤11	12–14	≥15
Tetracycline	30 µg	≤14	15–18	≥19
Tobramycin	10 µg	≤12	13–14	≥15
Trimethoprim/sulfamethoxazole	1.25/23.75 µg	≤10	11–15	≥16
Vancomycin				
when testing enterococci	30 µg	≤14	15–16	≥17
when testing *Staphylococcus* spp.	30 µg	—	—	≥15
Sulfonamides	250 or 300 µg	≤12	—	≥17
Trimethoprim	5 µg	≤10	—	≥16

Source: Clinical and Laboratory Standards Institute. *Performance Standards for Antimicrobial Disk Susceptibility Tests*, Tenth Edition, 2008.

is compared to that contained in a standardized chart, which is shown in Table 42.2. Based on this comparison, the test organism is determined to be resistant, intermediate, or susceptible to the antibiotic.

The procedure given in this section is an approximation of the industry-accepted Performance Standards published by the Clinical and Laboratory Standards Institute (CLSI) in published standards documents M02-A12 and M07-A10, as well as the Manual of Antimicrobial Susceptibility Testing published by the American Society for Microbiology (ASM).

CLINICAL APPLICATION

Selection of Effective Antibiotics

Upon isolation of an infectious agent, a chemotherapeutic agent is selected and its effectiveness must be determined. This can be done using the Kirby-Bauer Antibiotic Sensitivity Test. This is the essential tool used in clinical laboratories to select the best agent to treat patients with bacterial infections.

Materials

Cultures

0.85% saline suspensions of *Escherichia coli*, *Staphylococcus aureus* BSL-2, *Pseudomonas aeruginosa* BSL-2, *Proteus vulgaris*, *Mycobacterium smegmatis*, *Bacillus cereus*, and *Enterococcus faecalis* BSL-2 adjusted to an absorbance of 0.1 at 600 nanometer (nm) or equilibrated to a 0.5 McFarland Standard. *Note: For enhanced growth of* M. smegmatis, *add Tween™ 80 (1 ml/liter of broth medium) and incubate for 3 to 5 days in a shaking waterbath, if available.*

Media

Per designated student group: seven Mueller-Hinton agar plates.

Antimicrobial-Sensitivity Discs

Penicillin G, 10 µg; streptomycin, 10 µg; tetracycline, 30 µg; chloramphenicol, 30 µg; gentamicin, 10 µg; vancomycin, 30 µg; and sulfanilamide, 300 µg.

Equipment

Sensi-Disc™ dispensers or forceps, micro-incinerator or Bunsen burner, sterile cotton swabs, glassware marking pencil, 70% ethyl alcohol, and millimeter ruler.

Procedure Lab One

1. Place agar plates right side up in an incubator heated to 37°C for 10 to 20 minutes with the covers adjusted so that the plates are slightly opened, allowing the plates to warm up and the surface to dry.

2. Label the bottom of each of the agar plates with the name of the test organism to be inoculated.

3. Using aseptic technique, inoculate all agar plates with their respective test organisms as follows:

   a. Dip a sterile cotton swab into a well-mixed saline test culture and remove excess inoculum by pressing the saturated swab against the inner wall of the culture tube.

   b. Using the swab, streak the entire agar surface horizontally, vertically, and around the outer edge of the plate to ensure a heavy growth over the entire surface.

4. Allow all culture plates to dry for about 5 minutes.

5. Using the Sensi-Disc dispenser, apply the antibiotic discs by placing the dispenser over the agar surface and pressing the plunger, depositing the discs simultaneously onto the agar surface (Figure 42.3, Step 1a). Or, if dispensers are not available, distribute the individual discs at equal distances with forceps dipped in alcohol and flamed (Figure 42.3, Step 1b).

6. Gently press each disc down with the wooden end of a cotton swab or with sterile forceps to ensure that the discs adhere to the surface of the agar (Figure 42.3, Step 2). *Note: Do not press the discs into the agar.*

7. Incubate all plate cultures in an inverted position for 24 to 48 hours at 37°C.

Procedure Lab Two

1. Examine all plate cultures for the presence or absence of a zone of inhibition surrounding each disc.

2. Using a ruler graduated in millimeters, carefully measure each zone of inhibition to the nearest millimeter (Figure 42.3, Step 3). Record your results in the chart provided in the Lab Report.

3. Compare your results with Table 42.2 and determine the susceptibility of each test organism to the chemotherapeutic agent. Record your results in the Lab Report.

PART B Synergistic Effect of Drug Combinations

LEARNING OBJECTIVE

Once you have completed this experiment, you should be able to

1. Perform the disc–agar diffusion technique for determination of synergistic combinations of chemotherapeutic agents.

PROCEDURE

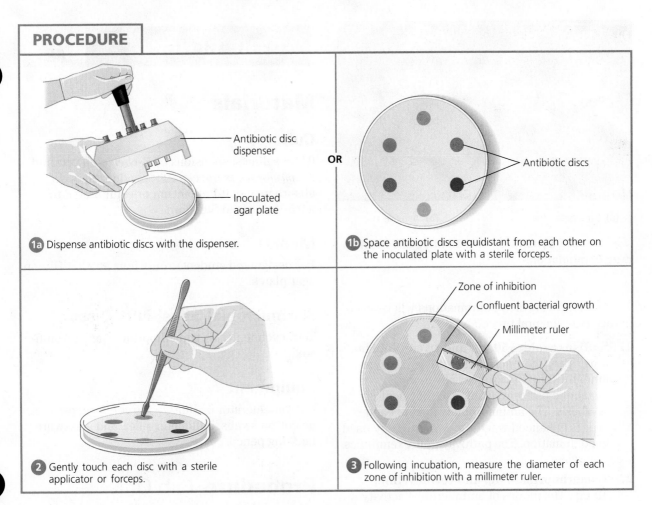

Figure 42.3 Kirby-Bauer antibiotic sensitivity procedure

Principle

Combination chemotherapy, the use of two or more antimicrobial or antineoplastic agents, is being employed in medical practice with ever-increasing frequency. The rationale for using drug combinations is the expectation that effective combinations might lower the incidence of bacterial resistance, reduce host toxicity of the antimicrobial agents (because of decreased dosage requirements), or enhance the agents' bactericidal activity. Enhanced bactericidal activity is known as **synergism**. Synergistic activity is evident when the sum of the effects of the chemotherapeutic agents used in combination is significantly greater than the sum of their effects when used individually. This result is readily differentiated from an **additive (indifferent) effect**, which is evident when the interaction of two drugs produces a combined effect that is no greater than the sum of their separately measured individual effects.

A variety of *in vitro* methods are available to demonstrate synergistic activity. In this experiment, a disc–agar diffusion technique will be performed to demonstrate this phenomenon. This technique uses the Kirby-Bauer antibiotic susceptibility test procedure, as described in Part A of this experiment, and requires both Mueller-Hinton agar plates previously seeded with the test organisms and commercially prepared, antimicrobial-impregnated discs. The two discs, representing the drug combination, are placed on the inoculated agar plate and separated by a distance (measured in mm) that is equal to or slightly greater than one-half the sum of their individual zones of inhibition when obtained separately. Following the incubation period, an additive effect is exhibited by the presence of two distinctly separate circles of inhibition. If the drug combination is synergistic, the two inhibitory zones merge to form a "bridge" at their juncture, as illustrated in Figure 42.4.

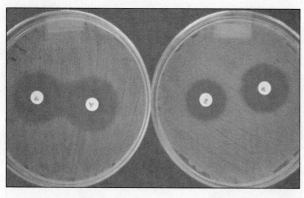

(a) Synergistic effect   (b) Additive effect

Figure 42.4 Synergistic and additive effects of drug combinations

The following drug combinations will be used in this experimental procedure:

1. **Sulfisoxazole, 150 µg, and trimethoprim, 5 µg.** Both antimicrobial agents are enzyme inhibitors that act sequentially in the metabolic pathway, leading to folic acid synthesis. The antimicrobial effect of each drug is enhanced when the two drugs are used in combination. The pathway thus exemplifies synergism.

2. **Trimethoprim, 5 µg, and tetracycline, 30 µg.** The modes of antimicrobial activity of these two chemotherapeutic agents differ; tetracycline acts to interfere with protein synthesis at the ribosomes. Thus, when used in combination, these drugs produce an additive effect.

CLINICAL APPLICATION

Multiple Drug Therapy

In antimicrobial therapy for drug-resistant bacteria, such as the opportunistic pathogen *Pseudomonas aeruginosa*, multiple drugs may be used to take advantage of synergistic effects. Research has shown that use of ampicillin to degrade gram-negative cell walls allows easier entry of kanamycin, which then inhibits protein synthesis. Combination therapies taking advantage of synergism also allow use of lower doses of each drug, which reduces overall toxic effects on the patient.

AT THE BENCH

Materials

Cultures

0.85% saline suspensions of *Escherichia coli* and *Staphylococcus aureus* BSL-2 adjusted to an absorbance of 0.1 at 600 nm or equilibrated to a 0.5 McFarland Standard.

Media

Per designated student group: four Mueller-Hinton agar plates.

Antimicrobial-Sensitivity Discs

Tetracycline, 30 µg; trimethoprim, 5 µg; and sulfisoxazole, 150 µg.

Equipment

Microincinerator or Bunsen burner, forceps, sterile cotton swabs, millimeter ruler, and glassware marking pencil.

Procedure Lab One

1. To inoculate the Mueller-Hinton agar plates, follow Steps 1 through 4 as described under the procedure in Part A of this experiment.

2. Using the millimeter ruler, determine the center of the underside of each plate and mark with a glassware marking pencil.

3. Using the glassware marking pencil, mark the underside of each agar plate culture at both sides from the center mark at the distances specified below:

   a. *E. coli*–inoculated plate for trimethoprim and sulfisoxazole combination sensitivity: 12.5 mm on each side of center mark.

   b. *S. aureus* BSL-2 –inoculated plate for trimethoprim and sulfisoxazole combination sensitivity: 14.5 mm on each side of center mark.

   c. *E. coli*–inoculated plate for trimethoprim and tetracycline combination sensitivity: 14.0 mm on each side of center mark.

d. *S. aureus* BSL-2 –inoculated plate for trimethoprim and tetracycline combination: 14.0 mm on each side of center mark.

4. Using sterile forceps, place the antimicrobial discs, in the combinations specified in Step 3, onto the surface of each agar plate culture at the previously marked positions. Gently press each disc down with the sterile forceps to ensure that it adheres to the agar surface.

5. Incubate all plate cultures in an inverted position for 24 to 48 hours at 37°C.

Procedure Lab Two

1. Examine all agar plate cultures to determine the zone of inhibition patterns exhibited. Distinctly separate zones of inhibition are indicative of an additive effect, whereas a merging of the inhibitory zones is indicative of synergism.

2. Record your observations and results in the chart provided in the Lab Report.

Determination of Penicillin Activity in the Presence and Absence of Penicillinase

43

LEARNING OBJECTIVES

Once you have completed this experiment, you should be able to

1. Employ a broth culture system for the determination of the minimal inhibitory concentration (MIC) of penicillin.

2. Demonstrate the reversal of penicillin inhibition against the test organism in the presence of penicillinase (β-lactamase).

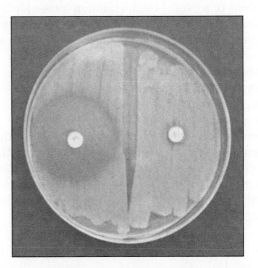

Figure 43.1 Molecular structure of benzylpenicillin (penicillin G)

Principle

In addition to the Kirby-Bauer paper disc–agar diffusion procedure, the broth tube dilution method may be used to determine the susceptibility of an organism to an antibiotic. The latter procedure, in which dilutions of the antibiotic are prepared in the broth medium, also permits the **minimal inhibitory concentration (MIC)** to be determined for the antibiotic under investigation. The **MIC** is the lowest concentration of an antimicrobial agent that inhibits the growth of the test microorganism. Quantitative data of this nature may be used by a clinician to establish effective antimicrobial regimens for the treatment of a bacterial infection in a host. These data are of particular significance when the toxicity of the antibiotic is known to produce major adverse effects in host tissues.

Penicillin is a potent antibiotic produced by the mold *Penicillium chrysogenum* (formerly called *P. notatum*). Sir Alexander Fleming's discovery of penicillin in 1928 provided the world with the first clinically useful antibiotic in the fight to control human infection. The activity of this antibiotic, as illustrated in Figure 43.1, is associated with the β-lactam ring within its molecular structure. Shortly after the clinical introduction of benzylpenicillin (penicillin G), pathogenic organisms, such as *Staphylococcus aureus*,

Figure 43.2 Pencillinase activity. Penicillin sensitivity is shown on the left; penicillin resistance is shown on the right

were found to be resistant to this "wonder drug." Research revealed that some organisms were genetically capable of producing β lactamase (penicillinase), an enzyme that breaks a bond in the β-lactam ring portion of the molecule. When the integrity of this ring is compromised, the inhibitory activity of the antibiotic is lost. Penicillinase activity is illustrated in Figure 43.2.

In this experiment, the MIC of penicillin will be determined against penicillin-sensitive and penicillinase-producing strains of *Staphylococcus aureus*. The procedure to be followed involves

Figure 43.3 **Minimal inhibitory concentration tube set-up**

CLINICAL APPLICATION

Wider Capability Seen in β-lactamases

Penicillinases are β-lactam ring breakers with specific activity against penicillin, while cephalosporins are generally not affected by them. New gene variants in gram-negative bacteria such as *Klebsiella pneumoniae* and *Neisseria gonorrhoeae* are now producing extended-spectrum β-lactamases (ESBLs), which not only hydrolyze penicillin, but also many cephalosporins and monobactams. These variants have been reported worldwide and now pose significant challenges in infection control.

specific concentrations of the penicillin prepared by means of a twofold serial dilution technique in an enriched broth medium. The tubes containing the antibiotic dilutions are then inoculated with a standardized concentration of the test organism (Figure 43.3) Table 43.1 illustrates the protocol for the antibiotic serial dilution–broth medium setup.

Following incubation, spectrophotometric absorbance readings will be used to determine the presence or absence of growth in the cultures. The culture that shows no growth in the presence of the lowest concentration of penicillin represents the minimal inhibitory concentration of this antibiotic against *S. aureus*.

AT THE BENCH

PART A MIC Determination Using a Spectrophotometer

Materials

Cultures

1:1000 brain heart infusion (BHI) broth dilutions of 24-hour BHI broth cultures of *Staphylococcus aureus* ATCC® 27661™ BSL-2 (penicillin-sensitive strain) and *Staphylococcus aureus* ATCC 27659 BSL-2 (penicillinase-producing strain).

TABLE 43.1 Antibiotic Serial Dilution–Broth Medium Setup

ADDITIONS (ML) TO:	TUBE NUMBER									
	1	2	3	4	5	6	7	8	9	10
Medium	0	2	2	2	2	2	2	2	2	2
Penicillin	2	2	Serial dilution (See protocol)							0
Test culture	2	2	2	2	2	2	2	2	2	2
Total volume	4	4	4	4	4	4	4	4	4*	4
Penicillin (µg/ml)	50	50	25	12.5	6.25	3.12	1.56	0.78	0.39	0
Control	(−)									(+)

*After 2 ml discarded

Media

Per designated student group: 40 ml of brain heart infusion broth in a 100-ml Erlenmeyer flask and 10 ml of sterile aqueous crystalline penicillin G solution (100 µg/ml).

Equipment

Sterile 13 × 100-mm test tubes, test tube racks, sterile 2-ml and 10-ml pipettes, mechanical pipetting device, microincinerator or Bunsen burner, spectrophotometer, glassware marking pencil, and disinfectant solution in a 500-ml beaker.

Procedure Lab One

1. Into each of two test tube racks, place a set of 10 sterile 13 × 100-mm test tubes labeled 1 through 10. Label one rack Set I—penicillin-sensitive and the other rack Set II—penicillin-resistant. *Refer to Table 43.1 for Steps 2 through 7.*

2. Using a sterile 10-ml pipette and mechanical pipetting device, add 2 ml of BHI broth to the tubes labeled 2 through 10 in Sets I and II. *Note: Discard the pipette into the beaker of disinfectant.*

3. With a 2-ml sterile pipette, add 2 ml of the penicillin solution to Tubes 1 and 2 in Sets I and II. Discard the pipette. *Note: Mix the contents of the tubes well.*

4. **Set I Serial Dilution:** Using a sterile 2-ml pipette, transfer 2 ml from Tube 2 to Tube 3. Mix well and transfer 2 ml from Tube 3 to Tube 4. Continue this procedure through Tube 9 into beaker. Discard 2 ml from Tube 9. Tube 10 receives no antibiotic and serves as a positive control. Discard the pipette. *Note: Remember to mix the contents of each tube well between transfers.*

5. **Set II Serial Dilution:** Using a sterile 2-ml pipette, repeat Step 4.

6. Using a sterile 2-ml pipette, add 2 ml of the 1:1000 dilution of the *S. aureus* ATCC 27661 **BSL-2** (penicillin-sensitive strain) to all tubes in Set I. Discard the pipette.

7. Repeat Step 6 to inoculate all the tubes in Set II with the 1:1000 dilution of *S. aureus* ATCC 27659 **BSL-2** (penicillinase-producing strain). Discard the pipette.

8. Incubate both sets of tubes for 12 to 18 hours at 37°C.

Procedure Lab Two

1. Follow the instructions for the use of the spectrophotometer as outlined in Experiment 13 to determine the absorbance readings for Tubes 2 through 10 in Sets I and II. Use the Number 1 tubes, the negative controls, as your blanks to adjust the spectrophotometer.

2. Record your absorbance readings in the chart in the Lab Report.

PART B | MIC Determination Using a Plate Reader

Materials

Cultures

1:1000 brain heart infusion (BHI) broth dilutions of 24-hour BHI broth cultures of *Staphylococcus aureus* ATCC® 27661™ **BSL-2** (penicillin-sensitive strain) and *Staphylococcus aureus* ATCC 27659 **BSL-2** (penicillinase-producing strain).

Media

Per designated student group: 40 ml of brain heart infusion and 10 ml of sterile aqueous crystalline penicillin G solution (100 µg/ml).

Equipment

Sterile 96-well plate with cover, micropipette with sterile tips, and a colorimetric plate reader.

Procedure Lab One

1. For each organism to be tested prepare a row of wells by adding 100 µl of BHI broth to wells 3 through 12 using a micropipette and sterile tips. *Refer to Table 43.2 on the following page for the remaining steps.*

TABLE 43.2 Antibiotic Serial Dilution-Plate Setup

Well	1	2	3	4	5	6	7	8	9	10	11	12
Medium	0	0	100	100	100	100	100	100	100	100	100	100
Penicillin (µl)	100	200				Serial dilution (see protocol)						0
Test Culture (µl)	100	100	100	100	100	100	100	100	100	100	100	100
Final volume	200	200	200	200	200	200	200	200	200	200	200*	200
Penicillin (µg/ml)	50	50	25	12.5	6.25	3.12	1.56	0.78	0.39	0.19	0.09	0
Contol	(−)											(+)

*After 200 µl discarded

2. Add 100 µl of penicillin G solution to well 1 and 200 µl to well 2 using a micropipette and sterile tips.

3. Perform a serial dilution of the penicillin G solution by transferring 100 µl of solution from well 2 into well 3 (which has 100 µl of BHI broth already added). Transfer 100 µl of the BHI/penicillin solution from well 3 into well 4, and repeat this procedure until well 11, when the 100 µl taken from well 11 will be discarded.

4. Using a micropipette and a sterile tip, add 100 µl of bacterial suspension to each well starting at well 12 and continuing to well 1. Discard the tip before the addition of new bacterial suspension to each row.

5. Cover plate and incubate at 37°C for 12 to 18 hours.

Procedure Lab Two

1. Follow the instructions for the use of the plate reader, as discussed in Experiment 12, to determine the absorbance readings for each well at 600 nanometer (nm). Wells number 1 and 12 should be used as the negative and positive controls respectively for this experiment to determine growth in each well.

2. Record your absorbance readings in the chart in the Lab Report.

Name: _____

Date: _____ Section: _____

Observations and Results

Part A: MIC Determination Using a Spectrophotometer

Absorbance Readings at 600 nm

Tube Number	2	3	4	5	6	7	8	9	10
Penicillin concentration (µg/ml)	50	25	12.5	6.25	3.12	1.56	0.78	0.39	0
Set I _____									
Set II _____									

Set I: Minimal inhibitory concentration: _____

Set II: Minimal inhibitory concentration: _____

Part B: MIC Determination Using a Plate Reader

Well Number	1	2	3	4	5	6	7	8	9	10	11	12
Penicillin concentration (µg/ml)		50	25	12.5	6.25	3.12	1.56	0.78	0.39	.19	.09	0
Organism 1:												
Organism II:												

Organism I: Minimal inhibitory concentration: _____

Organism II: Minimal inhibitory concentration: _____

Review Questions

1. 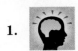 Was the ability of some microorganisms to produce β-lactamase present prior to their exposure to the antibiotic penicillin? Explain.

2. Can the results of an MIC test be used to determine whether an antibiotic is bactericidal or bacteriostatic? If not, set up an experimental procedure to determine whether the effect is bactericidal or bacteriostatic.

Chemical Agents of Control: Disinfectants and Antiseptics

Antiseptics and disinfectants are chemical substances used to prevent contamination and infection. Many are available commercially for disinfection and asepsis.

Table 44.1 shows the major groups of antimicrobial agents, their modes and ranges of action, and their practical uses.

TABLE 44.1 Chemical Agents—Disinfectants and Antiseptics

AGENT	MECHANISM OF ACTION	USE
Phenolic Compounds Phenol	1. Germicidal effect caused by alteration of protein structure resulting in protein denaturation. 2. Surface-active agent (surfactant) precipitates cellular proteins and disrupts cell membranes. (Phenol has been replaced by better disinfectants that are less irritating, less toxic to tissues, and better inhibitors of microorganisms.)	1. 5% solution: Disinfection. 2. 0.5% to 1% solutions: Antiseptic effect and relief of itching as it exerts a local anesthetic effect on sensory nerve endings.
Cresols	1. Similar to phenol. 2. Poisonous and must be used externally. 3. 50% solution of cresols in vegetable oil, known as Lysol®.	2% to 5% Lysol solutions used as disinfectants.
Hexachlorophene	Germicidal activity similar to phenol. (This agent is to be used with care, especially on infants, because after absorption it may cause neurotoxic effects.)	1. Reduction of pathogenic organisms on skin; added to detergents, soaps, lotions, and creams. 2. Effective against gram-positive organisms. 3. An antiseptic used topically.
Resorcinol	1. Germicidal activity similar to that of phenol. 2. Acts by precipitating cell protein.	1. Antiseptic. 2. Keratolytic agent for softening or dissolving keratin in epidermis.
Hexylresorcinol	Germicidal activity similar to that of phenol.	1. Treatment of worm infections. 2. Urinary antiseptic.
Thymol	1. Related to the cresols. 2. More effective than phenol.	1. Antifungal activity. 2. Treatment of hookworm infections. 3. Mouthwashes and gargle solutions.
Alcohols Ethyl: CH_3CH_2OH Isopropyl: $(CH_3)_2CHOH$	1. Lipid solvent. 2. Denaturation and coagulation of proteins. 3. Wetting agent used in tinctures to increase the wetting ability of other chemicals. 4. Germicidal activity increases with increasing molecular weight.	Skin antiseptics: Ethyl—50% to 70%. Isopropyl—60% to 70%.

TABLE 44.1 (continued)

AGENT	MECHANISM OF ACTION	USE
Halogens Chlorine compounds: Sodium hypochlorite (Dakin's fluid): NaOCl Chloramine: $CH_3C_6H_4SO_2NNaCl$	1. Germicidal effect resulting from rapid combination with proteins. 2. Chlorine reacts with water to form hypochlorous acid, which is bactericidal. 3. Oxidizing agent. 4. Noncompetitively inhibits enzymes, especially those dealing with glucose metabolism, by reacting with SH and NH_2 groups on the enzyme molecule.	1. Water purification. 2. Sanitation of utensils in dairy and restaurant industries. 3. Chloramine, 0.1% to 2% solutions, for wound irrigation and dressings. 4. Microbicidal.
Iodine compounds: Tincture of iodine Povidone-iodine solution (Betadine®)	1. Mechanism of action is not entirely known, but it is believed that it precipitates proteins. 2. Surface-active agent.	1. Tinctures of iodine are used for skin antisepsis. 2. Treatment of goiter. 3. Effective against spores, fungi, and viruses.
Heavy Metals Mercury compounds: Inorganic: Mercury bichloride Mercurial ointments	1. Mercuric ion brings about precipitation of cellular proteins. 2. Noncompetitive inhibition of specific enzymes caused by reaction with sulfhydryl group (SH) on enzymes of bacterial cells.	1. Inorganic mercurials are irritating to tissues, toxic systemically, adversely affected by organic matter, and incapable of acting on spores. 2. Mercury compounds are mainly used as disinfectants of laboratory materials.
Organic mercurials: Mercurochrome (merbromin) Merthiolate (thimerosal) Metaphen (nitromersol) Merbak (acetomeroctol)	1. Similar to those of inorganic mercurials, but in proper concentrations are useful antiseptics. 2. Much less irritating than inorganic mercurials.	1. Less toxic, less irritating; used mainly for skin asepsis. 2. Do not kill spores.
Silver compounds: Silver nitrate	1. Precipitate cellular proteins. 2. Interfere with metabolic activities of microbial cells. 3. Inorganic salts are germicidal.	Asepsis of mucous membrane of throat and eyes.
Surface-Active Agents Wetting agents: Emulsifiers, soaps, and detergents	1. Lower surface tension and aid in mechanical removal of bacteria and soil. 2. If active portion of the agent carries a negative electric charge, it is called an anionic surface-active agent. If active portion of the agent carries a positive electric charge, it is called a cationic surface-active agent. 3. Exert bactericidal activity by interfering with or by depressing metabolic activities of microorganisms. 4. Disrupt cell membranes. 5. Alter cell permeability.	Weak action against fungi, acid-fast microorganisms, spores, and viruses.

TABLE 44.1 (continued)

AGENT	MECHANISM OF ACTION	USE
Cationic agents: Quaternary ammonium compounds Benzalkonium chloride	1. Lower surface tension because of keratolytic, detergent, and emulsifying properties. 2. Their germicidal activities are reduced by soaps.	1. Bactericidal, fungicidal; inactive against spores and viruses. 2. Asepsis of intact skin. 3. Disinfectant for operating-room equipment. 4. Dairy and restaurant sanitization.
Anionic agents: Tincture of green soap Sodium tetradecyl sulfate	1. Neutral or alkaline salts of high-molecular-weight acids. Common soaps included in this group. 2. Exert their maximum activity in an acid medium and are most effective against gram-positive cells. 3. Same as all surface-active agents.	1. Cleansing agent. 2. Sclerosing agent in treatment of varicose veins and internal hemorrhoids.
Acids (H^+) Alkali (OH^-)	1. Destruction of cell wall and cell membrane. 2. Coagulation of proteins.	Disinfection; however, of little practical value.
Formaldehyde (liquid or gas)	Alkylating agent causes reduction of enzymes.	1. Room disinfection. 2. Alcoholic solution for instrument disinfection. 3. Specimen preservation.
Ethylene Oxide	Alkylating agent causes reduction of enzymes.	Sterilization of heat-labile material.
β-Propiolactone (liquid or gas)	Alkylating agent causes reduction of enzymes.	1. Sterilization of tissue for grafting. 2. Destruction of hepatitis virus. 3. Room disinfection.
Basic Dyes Crystal violet	Affinity for nucleic acids; interfere with reproduction in gram-positive organisms.	1. Skin antiseptic. 2. Laboratory isolation of gram-negative bacteria.

The efficiency of all disinfectants and antiseptics is influenced by a variety of factors, including the following:

1. **Concentration:** The concentration of a chemical substance markedly influences its effect on microorganisms, with higher concentrations producing a more rapid death. Concentration cannot be arbitrarily determined; the toxicity of the chemical to the tissues being treated and the damaging effect on nonliving materials must also be considered.

2. **Length of exposure:** All microbes are not destroyed within the same exposure time. Sensitive forms are destroyed more rapidly than resistant ones. The longer the exposure to the agent, the greater its antimicrobial activity. The toxicity of the chemical and environmental conditions must be considered in assessing the length of time necessary for disinfection or asepsis.

3. **Type of microbial population to be destroyed:** Microorganisms vary in their susceptibility to destruction by chemicals. Bacterial spores are the most resistant forms. Capsulated bacteria are more resistant than noncapsulated forms; acid-fast bacteria are more resistant than non–acid-fast; and older, metabolically less-active cells are more resistant than younger cells. Awareness of the types of microorganisms that may be present will influence the choice of agent.

4. **Environmental conditions:** Conditions under which a disinfectant or antiseptic affects the chemical agent are as follows:

a. **Temperature:** Cells are killed as the result of a chemical reaction between the agent and cellular component. As increasing temperatures increase the rate of chemical reactions, application of heat during disinfection markedly increases the rate at which the microbial population is destroyed.

b. **pH:** The pH conditions during disinfection may affect not only the microorganisms but also the compound. Extremes in pH are harmful to many microorganisms and may enhance the antimicrobial action of a chemical. Deviation from a neutral pH may cause ionization of the disinfectant; depending on the chemical agent, this may serve to increase or decrease the chemical's microbicidal action.

c. **Type of material on which the microorganisms exist:** The destructive power of the compound on cells is due to its combination with organic cellular molecules. If the material on which the microorganisms are found is primarily organic, such as blood, pus, or tissue fluids, the agent will combine with these extracellular organic molecules, and its antimicrobial activity will be reduced.

Numerous laboratory procedures are available for evaluating the antimicrobial efficiency of disinfectants or antiseptics. They provide a general rather than an absolute measure of the effectiveness of any agent because test conditions frequently differ considerably from those seen during practical use. The agar plate-sensitivity method, a commonly employed procedure, is presented.

PART A   Disc Diffusion Testing of Disinfectants and Antiseptics

LEARNING OBJECTIVE

Once you have completed this experiment, you should be able to

1. Evaluate the effectiveness of antiseptic agents against selected test organisms.

Principle

This procedure requires the heavy inoculation of an agar plate with the test organism. Sterile, color-coded filter-paper discs are impregnated with a different antiseptic and equally spaced on the inoculated agar plate. Following incubation, the agar plate is examined for zones of inhibition (areas of no microbial growth) surrounding the discs. A zone of inhibition is indicative of microbicidal activity against the organism. Absence of a zone of inhibition indicates that the chemical was ineffective against the test organism. *Note: The size of the zone of inhibition is not indicative of the degree of effectiveness of the chemical agent.* Antiseptic susceptibility is represented in Figure 44.1.

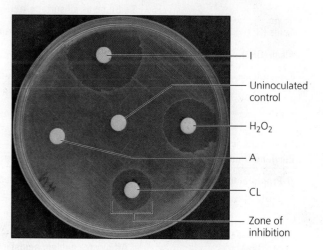

Figure 44.1 Antiseptic susceptibility test. Discs are saturated with chlorine bleach (CL), hydrogen peroxide (H_2O_2), isopropyl alchohol (A), and tincture of iodine (I).

CLINICAL APPLICATION

MRSA and Disinfection

Methicillin-resistant *Staphylococcus aureus* (MRSA) has achieved notoriety for causing infections that are difficult to treat with conventional antimicrobials, but these strains have also demonstrated resistance to disinfection. One study has shown that resistance to methicillin is directly related to lack of susceptibility to benzalkonium chloride and other disinfectants. It may be that adjusted contact times are necessary to adequately kill these troublesome strains.

Materials

Cultures

24- to 48-hour Trypticase soy broth cultures of *Escherichia coli*, *Bacillus cereus*, *Staphylococcus aureus* BSL-2 , and *Mycobacterium smegmatis*, and a 7-day-old Trypticase soy broth culture of *Bacillus cereus*.

Media

Per designated student group: five Trypticase soy agar plates.

Antiseptics/Disinfectants

Per designated student group: 10 ml of each of the following dispensed in 25-ml beakers: tincture of iodine, 3% hydrogen peroxide, 70% isopropyl alcohol, and 5% chlorine bleach.

Equipment

Four different-colored sterile Sensi-Discs, forceps, sterile cotton swabs, microincinerator or Bunsen burner, and glassware marking pencil.

Procedure Lab One

1. Aseptically inoculate the appropriately labeled agar plates with their respective test organisms by streaking each plate in horizontal and vertical directions and around the edge with a sterile swab.
2. Color-code the Sensi-Discs according to the chemical agents to be used (e.g., red = chlorine bleach).
3. Using forceps dipped in alcohol and flamed, expose five discs of the same color by placing them into the solution of one of the chemical agents. Drain the saturated discs on absorbent paper immediately prior to placing one on each of the inoculated agar plates. Place each disc approximately 2 cm in from the edge of the plate. Gently press the discs down with the forceps so that they adhere to the surface of the agar.

4. Impregnate the remaining discs as described in Step 3. Place one of each of the three remaining colored discs on the surface of each of the five inoculated agar plates equidistant from each other around the periphery of the plate.
5. Incubate all plate cultures in an inverted position for 24 to 48 hours at 37°C.

Procedure Lab Two

1. Observe all the plates for the presence of a zone of inhibition surrounding each of the impregnated discs.
2. Record your observations in the chart provided in the Lab Report.

PART B Modified Use Dilution Testing of Disinfectants and Antiseptics

LEARNING OBJECTIVE

Once you have completed this experiment, you should be able to

1. Evaluate the effectiveness of antiseptic agents against selected test organisms.

Principle

This procedure requires the adherence of dried bacterial cells to a treatable surface that can withstand exposure to the disinfectant being tested. The Environmental Protection Agency (EPA) accepted protocol recognizes guidelines published by the Association of Analytical Communities (AOAC) for *S. aureus* (Method 955.15) and *P. aeruginosa* (Method 964.02), which utilize a stainless steel carrier that will be dipped in the disinfectant or antiseptic to be tested. For this procedure, the carrier will be a glass slide that has the bacteria dried on its surface before being submerged in the test solution. The heated carrier is placed in a tube of broth and allowed to incubate up to 48 hours to determine if any cells have remained viable.

Materials

Cultures

24- to 48-hour Trypticase soy broth cultures of *Escherichia coli*, *Bacillus cereus*, *Staphylococcus aureus* BSL-2, and *Mycobacterium smegmatis*, and a 7-day-old Trypticase soy broth culture of *Bacillus cereus*.

Media

Per designated student group: twenty-five 50-ml tubes containing 20 ml of tryptic soy broth each.

Antiseptics/Disinfectants

Per designated student group: 10 ml of each of the following dispensed in 25-ml beakers: tincture of iodine, 3% hydrogen peroxide, 70% isopropyl alcohol, and 5% chlorine bleach.

Equipment

Sterile glass slides or cover slips, forceps, micro-incinerator or Bunsen burner, glassware marking pencil, and 70% ethyl alcohol.

Procedure Lab One

1. Aseptically add *E.coli* to five sterile glass slides or cover slips and allow to air dry for 10 minutes.
2. Separate the broth tubes into five sets and label each set for a different bacteria being tested. Also label each tube with the antiseptic or disinfectant treatment; reserving one tube per set as the untreated control.
3. Once the *E.coli* slides have dried, submerge each slide in one of the antiseptic/disinfectant solutions for 30 to 60 seconds.
4. Place slide on paper towel to dry before placing treated slide in appropriately labelled broth tube.
5. Repeat steps 1 through 4 for each bacterial culture to be tested.
6. Incubate all tubes, loosely capped, at 37°C for 24 to 48 hours.

Procedure Lab Two

1. Observe all tubes for the presence of bacterial growth signified by a cloudy appearance.
2. Record your observations in the chart provided in the Lab Report.

Name: _____

Date: _____ Section: _____

Observations and Results

Part A: Disc Diffusion Testing of Disinfectants and Antiseptics

1. Indicate the absence of a zone of inhibition as (0), and the presence of a zone of inhibition as (+).

Bacterial Species	ANTIMICROBIAL AGENT			
	Tincture of Iodine	3% Hydrogen Peroxide	70% Isopropyl Alcohol	5% Chlorine Bleach
E. coli gram-negative				
S. aureus gram-positive				
M. smegmatis acid-fast				
B. cereus spore-former gram-positive				
B. cereus spore-former gram-positive 7-day-old				

2. Indicate which of the antiseptics exhibited microbicidal activity against each of the following groups of microorganisms.

Bacterial Group	Tincture of Iodine	3% Hydrogen Peroxide	70% Isopropyl Alcohol	5% Chlorine Bleach
Gram-negative				
Gram-positive				
Acid-fast				
Spore-former				

3. Which of the experimental chemical compounds appears to have the broadest range of microbicidal activity? The narrowest range of microbicidal activity?

Part B: Modified Use Dilution Testing of Disinfectants and Antiseptics

1. Indicate the absence of a bacterial growth in each tube as (0) and the presence of growth as (+).

Bacterial Species	ANTIMICROBIAL AGENT			
	Tincture of Iodine	3% Hydrogen Peroxide	70% Isopropyl Alcohol	5% Chlorine Bleach
E. coli gram-negative				
S. aureus gram-positive				
M. smegmatis acid-fast				
B. cereus spore-former gram-positive				
B. cereus spore-former gram-positive 7-day-old				

2. Indicate which of the antiseptics exhibited microbicidal activity against each of the following groups of microorganisms.

Bacterial Group	Tincture of Iodine	3% Hydrogen Peroxide	70% Isopropyl Alcohol	5% Chlorine Bleach
Gram-negative				
Gram-positive				
Acid-fast				
Spore-former				

3. Which of the experimental chemical compounds appears to have the broadest range of microbicidal activity? The narrowest range of microbicidal activity?

Review Questions

1. Can the disinfection period (exposure time) be arbitrarily increased?
    Explain.

2.   A household cleanser is labeled germicidal. Explain what this
    means to you.

PART 10

Microbiology of Food

Once you have completed the experiments in this section, you should be familiar with

1. The endogenous and exogenous organisms that may be found in food products.
2. The analysis of food products as a means of determining their quality from the public health point of view.
3. The microbiological production of wine.

Introduction

Microbiologists have always been aware that foods, especially milk, have served as important inanimate vectors in the transmission of disease. Foods contain the organic nutrients that provide an excellent medium to support the growth and multiplication of microorganisms under suitable temperatures.

Food and dairy products may be contaminated in a variety of ways and from a variety of sources:

1. **Soil and water:** Food-borne organisms that may be found in soil and water and that may contaminate food are members of the genera *Alcaligenes, Bacillus, Citrobacter, Clostridium, Pseudomonas, Serratia, Proteus, Enterobacter,* and *Micrococcus*. The common soil and water molds include *Rhizopus, Penicillium, Botrytis, Fusarium,* and *Trichothecium*.

2. **Food utensils:** The type of microorganisms found on utensils depends on the type of food and the manner in which the utensils were handled.

3. **Enteric microorganisms of humans and animals:** The major members of this group are *Bacteroides, Lactobacillus, Clostridium, Escherichia, Salmonella, Proteus, Shigella, Staphylococcus,* and *Streptococcus*. These organisms find their way into the soil and water, from which they contaminate plants and are carried by wind currents onto utensils or prepared and exposed foods.

4. **Food handlers:** People who handle foods are especially likely to contaminate them because microorganisms on hands and clothing are easily transmitted. A major offending organism is *Staphylococcus*, which is generally found on hands and skin, and in the upper respiratory tract. Food handlers with poor personal hygiene and unsanitary habits are most likely to contaminate foods with enteric organisms.

5. **Animal hides and feeds:** Microorganisms found in water, soil, feed, dust, and fecal debris can be found on animal hides. Infected hides may serve as a source of infection for workers, or the microorganisms may migrate into the musculature of the animal and remain viable following its slaughter.

By enumerating microorganisms in milk and foods, the quality of a particular sample can be determined. Although the microorganisms cannot be identified, the presence of a high number suggests a good possibility that pathogens may be present. Even if a sample contains a low microbial count, it can still transmit infection.

In the laboratory procedures that follow, you will have an opportunity directly and indirectly to enumerate the number of microorganisms present in milk and other food products and to thereby determine the quality of the samples.

Microbiological Analysis of Food Products: Bacterial Count

LEARNING OBJECTIVES

Once you have completed this experiment, you should be able to determine

1. The total number of microorganisms present in food products.

2. The number of coliform bacteria in the selected food products.

Principle

Microorganisms in food may be harmful in some cases, while in other cases, they are beneficial. Certain microorganisms are necessary in preparation of foods, including cheese, pickles, yogurt, and sausage. However, other microorganisms are responsible for serious and sometimes fatal food poisoning and spoilage.

CLINICAL APPLICATION

Microorganisms in Food

Some of the pathogens that are tested for in food include: *Escherichia coli*, *Listeria monocytogenes*, *Salmonella* species, and *Aspergillus* fungi.

AT THE BENCH

Materials

Cultures

Samples of fresh vegetables, ground beef, and dried fruit.

Media

Per designated student group: nine brain–heart infusion agar deep tubes, three eosin–methylene blue (EMB) agar plates, three 99-ml sterile water blanks, and three 180-ml sterile water blanks.

Equipment

Bunsen burner, water bath, Quebec or electronic colony counter, balance, sterile glassine weighing paper, blender with three sterile jars, sterile Petri dishes, 1-ml pipettes, mechanical pipetting device, inoculation loop, and glassware marking pencil.

Procedure Lab One

Figure 45.1 illustrates the procedure.

1. Label three sets of three Petri dishes for each of the food samples to be tested and their dilutions (10^{-2}, 10^{-3}, 10^{-4}). Label the three EMB agar plates with the names of the food.

2. Melt the brain–heart infusion agar deep tubes in a water bath, cool, and maintain at 45°C.

3. Place 20 g of each food sample, weighed on sterile glassine paper, into its labeled blender jar. Add 180 ml of sterile water to each of the blender jars and blend each mixture for 5 minutes. You will have made a 1:10 (10^{-1}) dilution of each food sample.

4. Transfer 1 ml of the 10^{-1} ground beef suspension into its labeled 99-ml sterile water blank, thereby effecting a 10^{-3} dilution, and 0.1 ml to the appropriately labeled 10^{-2} Petri dish. Shake the 10^{-3} sample dilution, and using a different pipette, transfer 1 ml to the plate labeled 10^{-3} and 0.1 ml to the plate labeled 10^{-4}. Add a 15-ml aliquot of the molten and cooled agar to each of the three plates. Swirl the plates gently to obtain a uniform distribution, and allow the plates to solidify.

5. Repeat Step 4 for the remaining two 10^{-1} test food sample dilutions.

6. Aseptically prepare a four-way streak plate, as described in Experiment 3, and inoculate each 10^{-1} food sample dilution on its appropriately labeled EMB agar plate.

7. Incubate all plates in an inverted position for 24 to 48 hours at 37°C.

Procedure Lab Two

1. Following the instructions in the Lab Report, count and record the number of colonies on each plate.

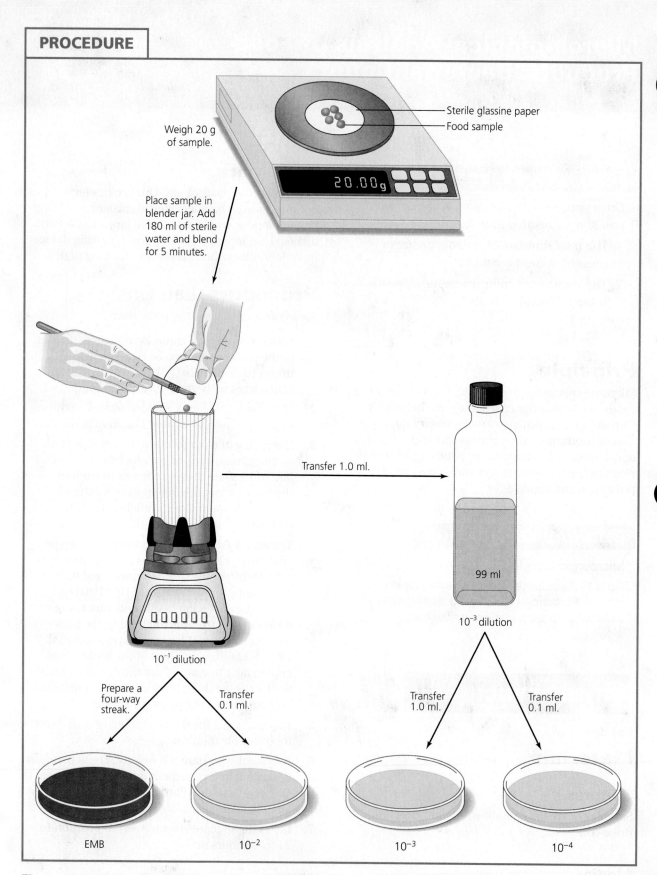

Weigh 20 g of sample.

Sterile glassine paper
Food sample

20.00g

Place sample in blender jar. Add 180 ml of sterile water and blend for 5 minutes.

Transfer 1.0 ml.

99 ml

10^{-3} dilution

10^{-1} dilution

Prepare a four-way streak.

Transfer 0.1 ml.

Transfer 1.0 ml.

Transfer 0.1 ml.

EMB

10^{-2}

10^{-3}

10^{-4}

Figure 45.1 **Preparation of a food sample for analysis**

Name: _____

Date: _____ Section: _____

Observations and Results

1. Using either the Quebec or electronic colony counter, count the number of colonies on each plate. Count only statistically valid plates that contain between 30 and 300 colonies. Designate plates with fewer than 30 colonies as **too few to count (TFTC)** and plates with more than 300 colonies as **too numerous to count (TNTC)**.

2. Determine the number of organisms per ml of each food sample on plates not designated as TFTC or TNTC by multiplying the number of colonies counted by the dilution factor.

3. Record in the chart below the number of colonies per plate and the number of organisms per milliliter of each food sample.

Type of Food	Dilution	Number of Colonies per Plate	Number of Organisms per ml
Ground beef	10^{-2}		
	10^{-3}		
	10^{-4}		
Fresh vegetables	10^{-2}		
	10^{-3}		
	10^{-4}		
Dried fruits	10^{-2}		
	10^{-3}		
	10^{-4}		

4. Examine the eosin–methylene blue agar plate cultures for colonies with a metallic green sheen on their surfaces, which is indicative of *E. coli.* Indicate in the chart below the presence or absence of *E. coli* growth and the possibility of fecal contamination of the food.

Sample	*E. coli* (+) or (−)	Fecal Contamination (+) or (−)
Ground beef		
Fresh vegetables		
Dried fruit		

Review Questions

1. Indicate some possible ways in which foods may become contaminated with enteric organisms.

2. Explain why it is not advisable to thaw and then refreeze food products without having cooked them.

3. Following a Fourth of July picnic lunch of ham, sour pickles, potato salad, and cream puffs, a group of students was admitted to the hospital with severe gastrointestinal distress. A diagnosis of staphylococcal food poisoning was made. Explain how the staphylococci can multiply in these foods and produce severe abdominal distress.

PART A   Alcohol Fermentation

LEARNING OBJECTIVE

Once you have completed this experiment, you should understand

1. Wine production by the fermentative activities of yeast cells.

Principle

Wine is a product of the natural fermentation of the juices of grapes and other fruits, including peaches, pears, plums, and apples, by the action of yeast cells. This biochemical conversion of juice to wine occurs when the yeast cells enzymatically degrade the fruit sugars, fructose and glucose, first to acetaldehyde and then to alcohol, as illustrated in Figure 46.1.

Grapes containing 20% to 30% sugar concentration will yield wines with an alcohol content of approximately 10% to 15%. Also present in grapes are acids and minerals whose concentrations are increased in the finished product and that are responsible for the characteristic tastes and bouquets of different wines. For red wine, the crushed grapes must be fermented with their skins to allow extraction of their color into the juice. White wine is produced from the juice of white grapes.

The commercial production of wine is a long and exacting process. First, the grapes are crushed or pressed to express the juice, which is called **must**. Potassium metabisulfite is added to the must to retard the growth of acetic acid bacteria, molds, and wild yeast that are endogenous to grapes in the vineyard. A wine-producing strain of yeast, *Saccharomyces cerevisiae* var. *ellipsoideus*, is used to inoculate the must, which is then incubated for 3 to 5 days under aerobic conditions at 21°C to 32°C. This is followed by an anaerobic incubation period. The wine is then aged for 1 year to 5 years in aging tanks or wooden barrels. During this time, the wine is clarified of any turbidity, thereby producing volatile esters that are responsible for characteristic flavors. The clarified product is then filtered, pasteurized at 60°C for 30 minutes, and bottled.

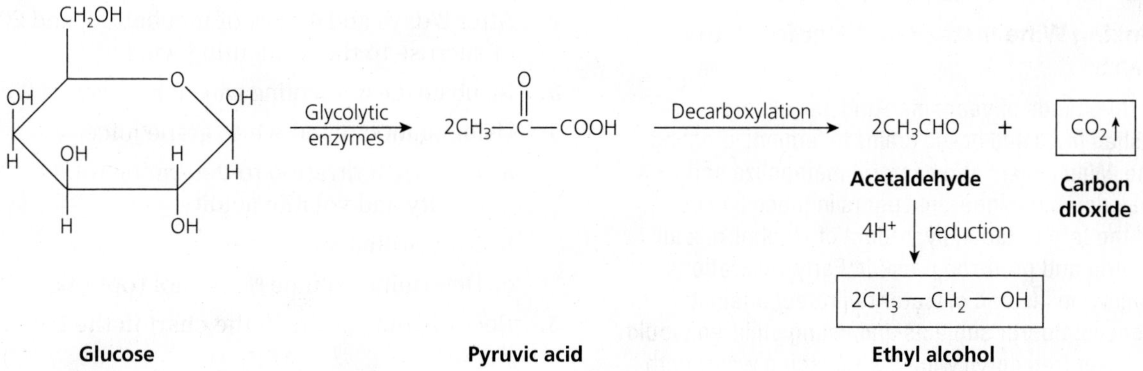

Figure 46.1 Biochemical pathway for alcohol production

This experiment is a modified method in which white wine is produced from white grape juice. You will examine the fermenting wine at 1-week intervals during the incubation period for:

1. Total acidity (expressed as % tartaric acid): To a 10-ml aliquot of the fermenting wine, add 10 ml of distilled water and 5 drops of 1% phenolphthalein solution. Mix and titrate to the first persistent pink color with 0.1N sodium hydroxide. Calculate total acidity using the following formula:

$$\% \text{ tartaric acid} = \frac{\text{ml alkali} \times \text{normality of alkali} \times 7.5}{\text{weight of sample in g*}}$$

*1 ml = 1 g

2. Volatile acidity (expressed as % acetic acid): Following titration, calculate volatile acidity using the following formula:

$$\% \text{ acetic acid} = \frac{\text{ml alkali} \times \text{normality of alkali} \times 6.0}{\text{weight of sample in g*}}$$

*1 ml = 1 g

3. Alcohol (expressed as volume %): Optional; can be determined by means of an ebulliometer.

4. Aroma: Fruity, yeast-like, sweet, none.

5. Clarity: Clear, turbid.

CLINICAL APPLICATION

Drinking Wine Instead of Water for Better Health

For thousands of years mankind has allowed crushed fruits and boiled grains to ferment, creating wine. Wild yeasts and bacteria metabolize and break down the inherent sugars in these liquids, and the fermentation byproduct of alcohol kills all bacteria and protozoa present. Early civilizations drank wine instead of water to protect against diseases. Poorer subjects and young children would drink watered down wine. By replacing water with wine in their daily diet, early civilizations were able to limit their exposure to pathogens.

AT THE BENCH

Materials

Cultures

50 ml of white grape juice broth culture of *Saccharomyces cerevisiae* var. *ellipsoideus* incubated for 48 hours at 25°C.

Media

Per designated student group: 500 ml of pasteurized Welch's® commercial white grape juice.

Reagents

1% phenolphthalein solution, 0.1N sodium hydroxide, and sucrose.

Equipment

1-liter Erlenmeyer flask, one-holed rubber stopper containing a 2-inch glass tube plugged with cotton, pan balance, spatula, glassine paper, 10-ml graduated cylinder, ebulliometer (optional), and burette or pipette for titration.

Procedure

1. Pour 500 ml of the white grape juice into the 1-liter Erlenmeyer flask. Add 20 g of sucrose and the 50 ml of *S. cerevisiae* grape juice broth culture (10% starter culture). Close the flask with the stopper containing a cotton plugged air vent.

2. After 2 days and 4 days of incubation, add 20 g of sucrose to the fermenting wine.

3. Incubate the fermenting wine for 21 days at 25°C.

4. Using uninoculated white grape juice:
   a. Perform a titration to determine total acidity and volatile acidity.
   b. Note aroma and clarity.
   c. Determine volume % alcohol (optional).

5. Record your results in the chart in the Lab Report.

6. At 7-day intervals, using samples of the fermenting wine, repeat Steps 4a though 4c and record your results in the Lab Report.

PART B | Lactic Acid Fermentation

LEARNING OBJECTIVE

Once you have completed this experiment, you should understand

1. Yogurt production by the fermentative activities of bacterial cells.

Principle

Yogurt is produced when bacterial species such as *Lactobacillus bulgarius* and *Streptococcus thermophiles* consume the sugars found in dairy products and produce lactic acid. The primary sugar utilized in this form of fermentation is the milk sugar lactose, a disaccharide sugar derived from galactose and glucose. Lactic acid fermentation occurs when sugars are broken down during glycolysis into lactate and release energy that powers the cell. Single sugar molecules are broken down during glycolysis into two molecules of pyruvate. As seen in Figure 46.2, in an anaerobic environment the enzyme lactate dehydrogenase converts pyruvate into lactic acid and allows for the oxidation of NADH back into NAD^+. This oxidation step frees up NAD^+ for the cell to continue glycolysis in the absence of oxygen. A decrease in pH due to the buildup of lactic acid causes the milk to clot, or form a soft gel that is characteristic of yogurt. The fermentation of lactose also produces the flavor compounds that are characteristic of yogurt.

Materials

Cultures

50 ml of bacterial cultures *Lactobacillus delbrueckii subsp. bulgaricus* and *Streptococcus thermophiles* that are each 24-hours old.

Media

Per designated student group: 400 ml of pasteurized heavy cream.

Equipment

Per designated student group: four 400 mL stoppered graduated Erlenmeyer flasks with stopper, sterile 10 mL serological pipettes, hot plates, varying ranges of pH paper, and glass markers.

Procedure

1. Aliquot 100 mL of heavy cream into four Erlenmeyer flasks labeled as
   a. Control (no bacteria)
   b. *Lactobacillus*
   c. *Streptococcus*
   d. Both

2. Record the pH of untreated cream and record.

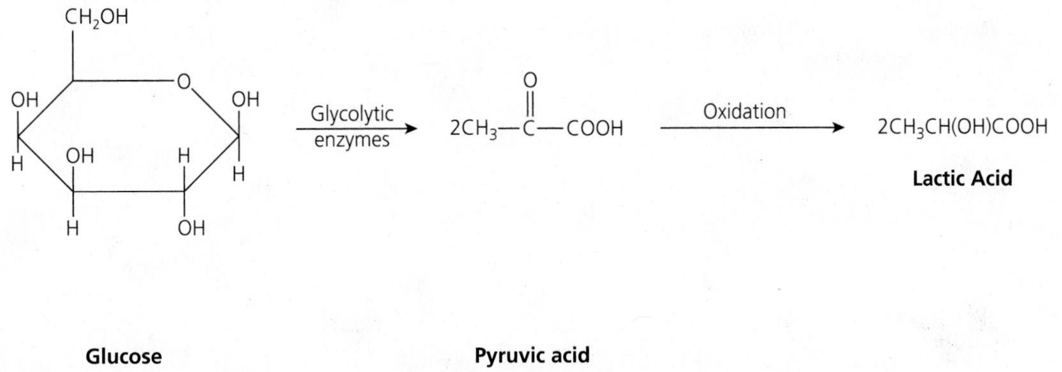

![Biochemical pathway diagram showing Glucose structure, converted by Glycolytic enzymes to $2CH_3-C(O)-COOH$ (Pyruvic acid), then by Oxidation to $2CH_3CH(OH)COOH$ (Lactic Acid)]

Glucose   **Pyruvic acid**

Figure 46.2 Biochemical pathway for lactic acid fermentation

3. Heat the cream on a hot plate to approximately 45°C, with stoppers placed loosely. *Caution: The high fat and high sugar content of the cream will easily burn, so monitor the temperature closely.*

4. After warming, remove flasks from the hot plate and add 10 mL of bacterial cultures to the appropriate flasks.

5. Incubate overnight at 42°C with loosened stoppers.

6. Store flasks at 4°C for 3 days while recording pH and cream consistency every 24 hours.

Name: _____

Date: _____ Section: _____

Observations and Results for Alcohol Fermentation

	Grape Juice	Fermenting Wine		
		7 days	14 days	21 days
% Tartaric acid				
% Acetic acid				
Volume % alcohol				
Aroma				
Clarity				

Observations and Results for Lactic Acid Fermentation

Day	Flask	pH	Consistency
0	Lactobacillus		
	Streptococcus		
	Both		
	Control		
1	Lactobacillus		
	Streptococcus		
	Both		
	Control		
2	Lactobacillus		
	Streptococcus		
	Both		
	Control		
3	Lactobacillus		
	Streptococcus		
	Both		
	Control		

Review Questions

1. What is the purpose of adding sulfite to the must?

2. Explain what occurs during the aging process in the commercial preparation of wine.

3. What are the chemical end products of fermentation?

4. Why is wine pasteurized? Would it be preferable to sterilize the wine? Explain.

5. What is the effect of lowered pH on the proteins found in dairy products during lactic acid fermentation?

PART 11

Microbiology of Water

LEARNING OBJECTIVES

Once you have completed the experiments in this section, you should be familiar with

1. The types of microorganisms present in water.
2. The methods to determine the potability of water using standard qualitative and quantitative procedures.

Introduction

The importance of potable (drinking) water supplies cannot be overemphasized. With increasing industrialization, water sources available for consumption and recreation have been adulterated with both industrial waste and animal and human wastes. As a result, water has become a formidable factor in disease transmission. Polluted waters contain vast amounts of organic matter that serve as excellent nutritional sources for the growth and multiplication of microorganisms. The presence of nonpathogenic organisms is not of major concern, but intestinal contaminants of fecal origin are important. These pathogens are responsible for intestinal infections such as **bacillary dysentery**, **typhoid fever**, **cholera**, and **paratyphoid fever**.

The World Health Organization (WHO) estimates that 1.7 million deaths per year result from unsafe water supplies. Most of these are from diarrheal diseases, and 90% of these deaths occur in children living in developing countries where sanitary facilities and potable water are at a minimum. The WHO indicates that about 3.4 million deaths annually are caused by dangerous waterborne enteric bacterial pathogens such as *Shigella dysenteriae*, *Campylobacter jejuni*, *Salmonella typhi*, and *Vibrio cholerae*.

In addition to bacterial infections, unsafe water supplies are responsible for numerous parasitological infections, including helminth diseases such as schistosomiasis and especially guinea worm *(Dracunculus medinensis)*, which infects about 200 million people worldwide each year. Intestinal, hepatic, and pulmonary flukes, including *Fasciolopsis buski*, *Clonorchis sinensis*, and *Paragonimus westermani*, are responsible for human infection and are all associated with unsafe water and sanitation. The parasitic protozoa *Entamoeba histolytica*, *Giardia intestinalis* (formerly called *G. lamblia*), and *Balantidium coli* are just a few of the protozoa responsible for major diarrheal disease in humans.

Although water-borne infections occur in the United States, their incidence in comparison to the rest of the world is much lower, and they occur sporadically. This can be attributed to the diligent attention given to our water supplies and sewage disposal systems.

Analysis of water samples on a routine basis would not be possible if each pathogen required detection. Therefore, water is examined to detect

Escherichia coli, the bacterium that indicates fecal pollution. Since *E. coli* is always present in the human intestine, its presence in water alerts public health officials to the possible presence of other human or animal intestinal pathogens. However, in the tropics and subtropics it is not considered a reliable indicator of fecal pollution because the soil in these regions naturally contains high levels of *E. coli*. Therefore, *E. coli* is present in the water anytime there is surface runoff. Both qualitative and quantitative methods are used to determine the sanitary condition of water.

Standard Qualitative Analysis of Water

EXPERIMENT

47

LEARNING OBJECTIVES

Once you have completed this experiment, you should be able to

1. Determine the presence of coliform bacteria in a water sample.
2. Obtain an index indicating the possible number of organisms present in the sample under analysis.
3. Confirm the presence of coliform bacteria in a water sample for which the presumptive test was positive.
4. Confirm the presence of coliform bacteria in a water sample, or if necessary, confirm a suspicious or doubtful result from the previous test.

Principle

The three basic tests to detect coliform bacteria in water are presumptive, confirmed, and completed (Figure 47.1). The tests are performed sequentially on each sample under analysis. They detect the presence of coliform bacteria (indicators of fecal contamination), the gram-negative, non–spore-forming bacilli that ferment lactose with the production of acid and gas that is detectable following a 24-hour incubation period at 37°C.

The Presumptive Test

The **presumptive test** is specific for detection of coliform bacteria. Measured aliquots of the water to be tested are added to a lactose fermentation broth containing an inverted gas vial. Because these bacteria are capable of using lactose as a carbon source (the other enteric organisms are not), their detection is facilitated by the use of this medium. In this experiment, you will use lactose fermentation broth containing an inverted Durham tube for gas collection.

Tubes of this lactose medium are inoculated with 10-ml, 1-ml, and 0.1-ml aliquots of the water sample. The series consists of at least three groups, each composed of five tubes of the specified medium. The tubes in each group are then inoculated with the designated volume of the water sample, as described under "Procedure: Lab One." The greater the number of tubes per group, the greater the sensitivity of the test. Development of gas in any of the tubes is *presumptive* evidence of the presence of coliform bacteria in the sample. The presumptive test also enables the microbiologist to obtain some idea of the number of coliform organisms present by means of the **most probable number (MPN) test**. The MPN is estimated by determining the number of tubes in each group that show gas following the incubation period (Table 47.1 on page 334).

The Confirmed Test

The presence of a positive or doubtful presumptive test immediately suggests that the water sample is nonpotable. Confirmation of these results is necessary because positive presumptive tests may be the result of organisms of noncoliform origin that are not recognized as indicators of fecal pollution.

The **confirmed test** requires that selective and differential media (e.g., eosin–methylene blue (EMB) or Endo agar) be streaked from a positive lactose broth tube obtained from the presumptive test. The nature of the differential and selective media was discussed in Experiment 15 but is reviewed briefly here. Eosin–methylene blue contains the dye methylene blue, which inhibits the growth of gram-positive organisms. In the presence of an acid environment, EMB forms a complex that precipitates out onto the coliform colonies, producing dark centers and a green metallic sheen. The reaction is characteristic for *Escherichia coli*, the major indicator of fecal pollution. Endo agar is a nutrient medium containing the dye fuchsin, which is present in the decolorized state. In the presence of acid produced by the coliform bacteria, fuchsin forms a dark pink complex that turns the *E. coli* colonies and the surrounding medium pink.

329

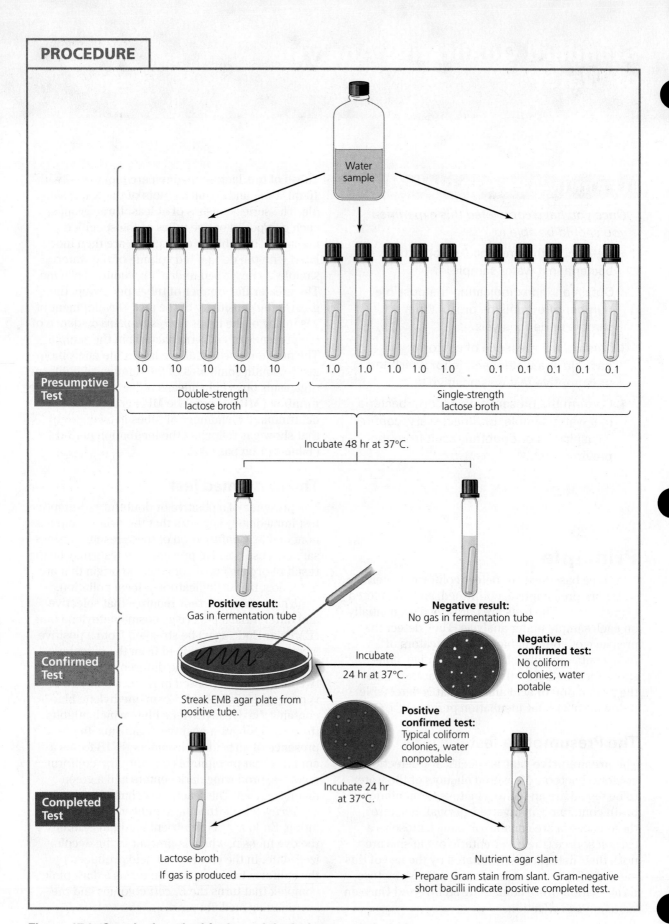

Figure 47.1 Standard method for bacteriological water analysis

The Completed Test

The **completed test** is the final analysis of the water sample. It is used to examine the coliform colonies that appeared on the EMB or Endo agar plates used in the confirmed test. An isolated colony is picked up from the confirmatory test plate and inoculated into a tube of lactose broth and streaked on a nutrient agar slant to perform a Gram stain. Following inoculation and incubation, tubes showing acid and gas in the lactose broth and presence of gram-negative bacilli on microscopic examination are further confirmation of the presence of *E. coli*, and they are indicative of a positive completed test.

Environmental Protection Agency

This step-wise method to determine the presence of coliform bacteria, indicating fecal contamination of sludge or treated water, is similar to one of the approved methods published by the Environmental Protection Agency (EPA). The EPA has the government mandate to protect the nation's waterways and terrestrial environments from human contamination or damage. The published "Method 1681: Fecal Coliforms in Sewage Sludge (Biosolids) by Multiple-Tube Fermentation using A-1 Medium" utilizes the process of presumptive tests followed by confirmed tests to determine the amount of fecal contamination in collected samples.

CLINICAL APPLICATION

Testing for Safe Water

Water used for human consumption and recreational use is routinely analyzed for safety. Water sources are regularly tested for the presence of *Escherichia coli* to determine the quality and safety of municipal water supplies. Several testing methods are available for this purpose including: most probable numbers (MPN), ATP testing, membrane filtration, and the use of pour plates.

AT THE BENCH

Materials

Cultures

Lab One: Water samples from sewage plant, pond, and tap. Lab Two: One 24–hour-old positive lactose broth culture from each of the three series from the presumptive test. Lab Three: One 24-hour, coliform-positive EMB or Endo agar culture from each of the three series of the confirmed test.

Media

Lab One (per designated student group): 15 double-strength lactose fermentation broths (LB2X) and 30 single-strength lactose fermentation broths (LB1X). Lab Two (three each per designated student group): eosin–methylene blue agar plates or Endo agar plates. Lab Three (three each per designated student group): nutrient agar slants and lactose fermentation broths.

Reagents

Lab Three: Crystal violet, Gram's iodine, 95% ethyl alcohol, and safranin.

Equipment

Lab One: Bunsen burner, 45 test tubes, test tube rack, sterile 10-ml pipettes, sterile 1-ml pipettes, sterile 0.1-ml pipettes, mechanical pipetting device, and glassware marking pencil. Lab Two: Bunsen burner, glassware marking pencil, and inoculating loop. Lab Three: Bunsen burner, staining tray, inoculating loop, lens paper, bibulous paper, microscope, and glassware marking pencil.

Procedure Lab One

Presumptive Test

> ⚠ **Exercise care in handling sewage waste water sample because enteric pathogens may be present.**

1. Set up three separate series consisting of three groups, a total of 15 tubes per series, in a test tube rack; for each tube, label the water source and volume of sample inoculated as illustrated.

Series 1: Sewage water	5 tubes of LB2X-10 ml
	5 tubes of LB1X-1 ml
	5 tubes of LB1X-0.1 ml
Series 2: Pond water	5 tubes of LB2X-10 ml
	5 tubes of LB1X-1 ml
	5 tubes of LB1X-0.1 ml
Series 3: Tap water	5 tubes of LB2X-10 ml
	5 tubes of LB1X-1 ml
	5 tubes of LB1X-0.1 ml

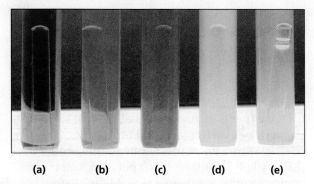

(a)   (b)   (c)   (d)   (e)

Figure 47.2 Possible MPN presumptive test results. (a) Uninoculated control tube, **(b, c)** inoculated tubes with no change, **(d)** inoculated tube with acid production only, and **(e)** inoculated tube with acid and gas production—the only positive result of the five tubes.

2. Mix sewage plant water sample by shaking thoroughly.

3. Flame bottle and then, using a 10-ml pipette, transfer 10-ml aliquots of water sample to the five tubes labeled LB2X-10 ml.

4. Flame bottle and then, using a 1-ml pipette, transfer 1-ml aliquots of water sample to the five tubes labeled LB1X-1 ml.

5. Flame bottle and then, using a 0.1-ml pipette, transfer 0.1-ml aliquots of water sample to the five tubes labeled LB1X-0.1 ml.

6. Repeat Steps 2 through 5 for the tap and pond water samples.

7. Incubate all tubes for 48 hours at 37°C.

Procedure Lab Two

Presumptive Test

1. Examine the tubes from your presumptive test after 24 and 48 hours of incubation. Your results are positive if the Durham tube fills 10% or more with gas in 24 hours, doubtful if gas develops in the tube after 48 hours, and negative if there is no gas in the tube after 48 hours. Refer to Figure 47.2 for a summary of possible MPN presumptive test results. Figure 47.3 shows actual results from an MPN presumptive test for a water sample. Record your results in the Lab Report.

2. Determine the MPN using Table 47.1, and record your results in the Lab Report.

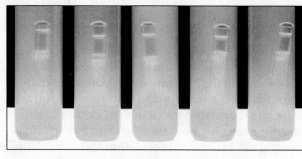

(a) Positive results (acid and gas) in five 10-ml tubes

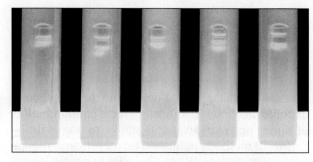

(b) Positive results (acid and gas) in five 1-ml tubes

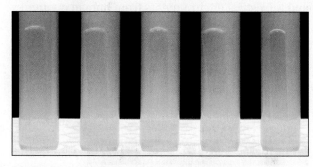

(c) Negative results (acid only) in five 0.1-ml tubes

Figure 47.3 MPN presumptive test results for a water sample. The results of this test (5 positive, 5 positive, and 5 negative) indicate 240 coliforms per 100 ml of water (see Table 47.1). This represents a positive presumptive test for the presence of coliforms in the tested water sample.

Confirmed Test

1. Label the covers of the three EMB plates or the three Endo agar plates with the source of the water sample (sewage, pond, and tap).

2. Using a positive 24-hour lactose broth culture from the sewage water series from the presumptive test, streak the surface of one EMB or one Endo agar plate, as described in Experiment 3, to obtain discrete colonies.

3. Repeat Step 2 using the positive lactose broth cultures from the pond and tap water series from the presumptive test to inoculate the remaining plates.

4. Incubate all plate cultures in an inverted position for 24 hours at 37°C.

Procedure Lab Three

Confirmed Test

1. Examine all the plates from your confirmed test for the presence or absence of *E. coli* colonies (refer to the description of the confirmed test in the experiment introduction, and see Figure 13.2 for an illustration of *E. coli* growth on EMB agar). Record your results in the Lab Report.

2. Based on your results, determine whether each of the samples is potable or nonpotable. The presence of *E. coli* is a positive confirmed test, indicating that the water is nonpotable. The absence of *E. coli* is a negative test, indicating that the water is not contaminated with fecal wastes and is therefore potable. Record your results in the Lab Report.

Completed Test

1. Label each tube of nutrient agar slants and lactose fermentation broths with the source of its water sample.

2. Inoculate one lactose broth and one nutrient agar slant with a positive isolated *E. coli* colony obtained from each of the experimental water samples during the confirmed test.

3. Incubate all tubes for 24 hours at 37°C.

Procedure Lab Four

Completed Test

1. Examine all lactose fermentation broth cultures for the presence or absence of acid and gas. Record your results in the Lab Report.

2. Prepare a Gram stain, using the nutrient agar slant cultures of the organisms that showed a positive result in the lactose fermentation broth (refer to Experiment 9 for the staining procedure).

3. Examine the slides microscopically for the presence of gram-negative short bacilli, which are indicative of *E. coli* and thus nonpotable water. In the Lab Report, record your results for Gram stain reaction and morphology of the cells.

TABLE 47.1 The MPN Index per 100 ml for Combinations of Positive and Negative Presumptive Test Results When Five 10-ml, Five 1-ml, and Five 0.1-ml Portions of Sample Are Used

NUMBER OF TUBES WITH POSITIVE RESULTS				95% CONFIDENCE LIMITS		NUMBER OF TUBES WITH POSITIVE RESULTS				95% CONFIDENCE LIMITS	
FIVE OF 10 ML EACH	FIVE OF 1 ML EACH	FIVE OF 0.1 ML EACH	MPN INDEX PER 100 ML	LOWER	UPPER	FIVE OF 10 ML EACH	FIVE OF 1 ML EACH	FIVE OF 0.1 ML EACH	MPN INDEX PER 100 ML	LOWER	UPPER
0	0	0	<2	0	6	4	2	1	26	7	67
0	0	1	2	<0.5	7	4	3	0	27	9	78
0	1	0	2	<0.5	7	4	3	1	33	9	78
0	2	0	4	<0.5	11	4	4	0	34	11	93
1	0	0	2	0.1	10	5	0	0	23	7	70
1	0	1	4	0.7	10	5	0	1	31	11	89
1	1	0	4	0.7	12	5	0	2	43	14	100
1	1	1	6	1.8	15	5	1	0	33	10	100
1	2	0	6	1.8	15	5	1	1	46	14	120
2	0	0	5	<0.5	13	5	1	2	63	22	150
2	0	1	7	1	17	5	2	0	49	15	150
2	1	0	7	1	17	5	2	1	70	22	170
2	1	1	9	2	21	5	2	2	94	34	230
2	2	0	9	2	21	5	3	0	79	22	220
2	3	0	12	3	28	5	3	1	110	34	250
3	0	0	8	2	22	5	3	2	140	52	400
3	0	1	11	4	23	5	3	3	180	70	400
3	1	0	11	5	35	5	4	0	130	36	400
3	1	1	14	6	36	5	4	1	170	58	400
3	2	0	14	6	36	5	4	2	220	70	440
3	2	1	17	7	40	5	4	3	280	100	710
3	3	0	17	7	40	5	4	4	350	100	710
4	0	0	13	4	35	5	5	0	240	70	710
4	0	1	17	6	36	5	5	1	350	100	1100
4	1	0	17	6	40	5	5	2	540	150	1700
4	1	1	21	7	42	5	5	3	920	220	2600
4	1	2	26	10	70	5	5	4	1600	400	4600
4	2	0	22	7	50	5	5	5	≥2400	700	---

Sources: pp 9–51, *Standard Methods for the Examination of Water and Wastewater,* 20th Edition (1998). M. J. Taras, A. E. Greenberg, R. D. Hoak, and M. C. Rand, eds. American Public Health Association, Washington, D.C. Copyright 1998, American Public Health Association, and *Bacteriological Analytical Manual (BAM),* 8th Edition, Food and Drug Administration, 1998.

Name: _____

Date: _____ Section: _____

Observations and Results

Presumptive Test

Using Table 47.1, determine and record the MPN.

Example: If gas appeared in all five tubes labeled LB2X–10, in two of the
tubes labeled LB1X–1, and in one labeled LB1X–0.1, the series would be read
as 5-2-1. From the MPN table, such a reading would indicate approximately
70 microorganisms per 100 ml of water, with a 95% probability that between
22 and 170 microorganisms are present.

	GAS															Reading	MPN	95% Probability Range
	LB2X–10					LB1X–1					LB1X–0.1							
	Tube					Tube					Tube							
Water Sample	1	2	3	4	5	1	2	3	4	5	1	2	3	4	5			
Sewage																		
Pond																		
Tap																		

Confirmed Test

	COLIFORMS		Potable	Nonpotable
Water Sample	EMB Plate	Endo Agar Plate		
Sewage				
Pond				
Tap				

Completed Test

		GRAM STAIN	POTABILITY	
Water Source	Lactose Broth A/G (+) or (–)	Reaction/ Morphology	Potable	Nonpotable
Sewage				
Pond				
Tap				

Review Questions

1. What is the rationale for selecting *E. coli* as the indicator of water potability?

2. Why is this procedure qualitative rather than quantitative?

3. Explain why it is of prime importance to analyze water supplies that serve industrialized communities.

4. Account for the presence of microorganisms in natural bodies of water and sewage systems. What is their function? Explain.

Quantitative Analysis of Water: Membrane Filter Method

LEARNING OBJECTIVE

Once you have completed this experiment, you should be able to

1. Determine the quality of water samples using the membrane filter method.

Principle

Bacteria-tight **membrane filters** capable of retaining microorganisms larger than 0.45 micrometer (μm) are frequently used for analysis of water. These filters offer several advantages over the conventional, multiple-tube method of water analysis: (1) results are available in a shorter period of time, (2) larger volumes of sample can be processed, and (3) because of the high accuracy of this method, the results are readily reproducible. A disadvantage involves the processing of turbid specimens that contain large quantities of suspended materials; particulate matter clogs the pores and inhibits passage of the specific volume of water.

A water sample is passed through a sterile membrane filter that is housed in a special filter apparatus contained in a suction flask. Following filtration, the filter disc that contains the trapped microorganisms is aseptically transferred to a sterile Petri dish containing an absorbent pad saturated with a selective, differential liquid medium. Following incubation, the colonies present on the filter are counted with the aid of a microscope.

This experiment is used to analyze a series of dilutions of water samples collected upstream and downstream from an outlet of a sewage treatment plant. EPA-approved guidelines for the determination of fecal contaminating organisms (EPA Method 1103.1), similar to what are seen in this experiment, are routinely utilized worldwide to examine water samples before treated water is released into a nation's waterways. A total count of coliform bacteria determines the potability of the water sources. Also, the types of fecal pollution, if any, are established by means of a fecal coliform count, indicative of human pollution, and a fecal streptococcal count, indicative of pollution from other animal origins. The ratio of the fecal coliforms to fecal streptococci per milliliter of sample is interpreted as follows: Between 2 and 4 indicates human and animal pollution; >4 indicates human pollution; and <0.7 indicates poultry and livestock pollution.

CLINICAL APPLICATION

Rapid Water Analysis

In the late 1950s, the membrane filter method was introduced as an alternative to the most probable number method (MPN). Microbiological analysis of water by the membrane filter procedure is a rapid method that isolates discrete bacteria that are able to be accurately counted, whereas the MPN method only allows for the approximate determination of the number of organisms and does not separate species without further testing.

AT THE BENCH

Materials

Cultures

Water samples collected upstream (labeled U) and downstream (labeled D) from an outlet of a sewage treatment plant.

Media

Per designated student group for analysis of one water sample: one 20-ml tube of m-Endo broth, one 20-ml tube of m-FC broth, one 20-ml tube of KF broth, four 90-ml sterile water blanks, and one 300-ml flask of sterile water.

Equipment

Sterile membrane filtration apparatus (i.e., Millipore®; Pall® Gelman; sterile, plastic, disposable membrane filters), 1-liter suction flask, 15 sterile membrane filters and absorbent pads, 15 sterile 50-mm Petri dishes, 12 10-ml pipettes, mechanical pipetting device, small beaker of 95% alcohol, membrane forceps, waterproof tape, watertight plastic bags, 44.5°C water bath, dissecting microscope, and glassware marking pencil.

Procedure Lab One

The following instructions are for analysis of one of the provided water samples using the Millipore system. Different samples may be assigned to individual groups.

> ⚠ **Use disposable gloves when handling the water samples in this experiment.**

1. Label the four 90-ml water blanks with the source of the water sample and dilution (10^{-1}, 10^{-2}, 10^{-3}, and 10^{-4}).

2. Using 10-ml pipettes, aseptically perform a 10-fold serial dilution of the assigned undiluted water sample, using the four 90-ml water blanks to effect the 10^{-1}, 10^{-2}, 10^{-3}, and 10^{-4} dilutions.

3. Arrange the 15 Petri dishes into three sets of five plates. Label each set as follows:

   a. For total coliform count (TCC) and dilutions (undiluted, 10^{-1}, 10^{-2}, 10^{-3}, and 10^{-4}).

   b. For fecal coliform count (FCC) and dilutions as in Step 3a.

   c. For fecal streptococcal count (FSC) and dilutions as in Step 3a.

Membrane Filter Technique

Refer to **Figure 48.1** as you read the instructions below.

1. Using sterile forceps dipped in 95% alcohol and flamed, add a sterile absorbent pad to all Petri dishes.

2. With sterile 10-ml pipettes, aseptically add the following:

   a. To each pad in the plates labeled TCC, 2 ml of m-Endo broth.

   b. To each pad in the plates labeled FCC, 2 ml of m-FC broth.

   c. To each pad in the plates labeled FSC, 2 ml of KF broth.

3. Aseptically assemble the sterile paper-wrapped membrane filter unit as follows:

   a. Unwrap and insert the sintered glass filter base into the neck of a 1-liter side-arm suction flask.

   b. With sterile forceps, place a sterile membrane filter disc, grid side up, on the sintered glass platform.

   c. Unwrap and carefully place the funnel section of the apparatus on top of the filter disc. Using the filter clamp, secure the funnel to the filter base.

   d. Attach a rubber hose from the side-arm on the vacuum flask to a vacuum source.

4. Using the highest sample dilution (10^{-4}) and a pipette, place 20 ml of the dilution into the funnel and start the vacuum.

   a. When the entire sample has been filtered, wash the inner surface of the funnel with 20 ml of sterile water.

5. Disconnect the vacuum, unclamp the filter assembly, and with sterile forceps, remove the membrane filter.

6. Place the filter on the medium-saturated pad in the Petri dish labeled TCC, 10^{-4}.

7. Aseptically place a new membrane on the platform, reassemble the filtration apparatus, and repeat Steps 4 through 6 twice, adding the filter discs to the 10^{-4} dilution plates labeled FCC and FSC.

8. Repeat Steps 4a through 7, using 20 ml of the 10^{-3}, 10^{-2}, and 10^{-1} dilutions and the undiluted samples.

9. Incubate the plates in an inverted position as follows:

   a. TCC and FSC plates for 24 hours at 37°C.

   b. FCC plates sealed with waterproof tape and placed in a weighted watertight plastic bag, which is then submerged in a 44.5°C water bath for 24 hours.

PROCEDURE

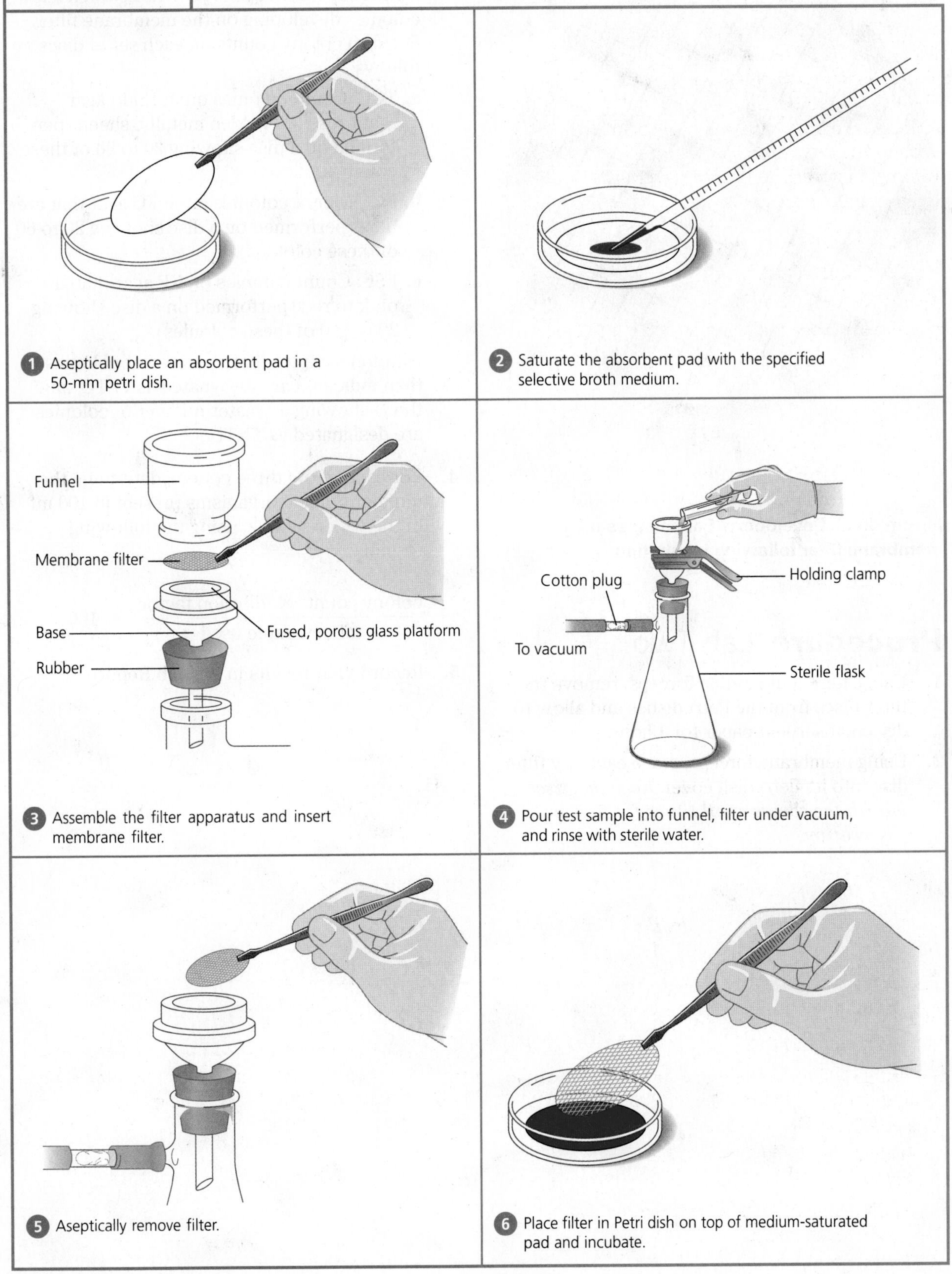

1. Aseptically place an absorbent pad in a 50-mm petri dish.

2. Saturate the absorbent pad with the specified selective broth medium.

Funnel

Membrane filter

Base

Rubber

Fused, porous glass platform

3. Assemble the filter apparatus and insert membrane filter.

Cotton plug

Holding clamp

To vacuum

Sterile flask

4. Pour test sample into funnel, filter under vacuum, and rinse with sterile water.

5. Aseptically remove filter.

6. Place filter in Petri dish on top of medium-saturated pad and incubate.

Figure 48.1 **Membrane filter technique**

Figure 48.2 Development of colonies on a membrane filter following incubation

Procedure Lab Two

1. Using sterile membrane forceps, remove the filter discs from the Petri dishes and allow to dry on absorbent paper for 1 hour.

2. Using membrane forceps, place each dry filter disc into its Petri dish cover. *Keep the discs within the covers at all times for further observation.*

3. Examine all filter discs under a dissecting microscope. Refer to Figure 48.2, which shows colonies developing on the membrane filter. Perform colony counts on each set of discs as follows:

   a. TCC: Count colonies on m-Endo agar that present a golden metallic sheen (performed on a disc showing 20 to 80 of these colonies).

   b. FCC: Count colonies on m-FC agar that are blue (performed on a disc showing 20 to 60 of these colonies).

   c. FSC: Count colonies on KF agar that are pink to red (performed on a disc showing 20 to 100 of these colonies).

   Dilution samples that show fewer colonies than indicated are designated as TFTC, and those showing a greater number of colonies are designated as TNTC.

4. For each of the three counts, determine the number of fecal organisms present in 100 ml of the water sample, using the following formula:

$$\frac{\text{colony count} \times \text{dilution factor}}{\text{ml of sample used}} \times 100$$

5. Record your results in the Lab Report.

Name: _____

Date: _____   Section: _____

Observations and Results

	UPSTREAM WATER					
	TCC		FCC		FSC	
Dilution	Count	Cells/100 ml	Count	Cells/100 ml	Count	Cells/100 ml
Undiluted						
10^{-1}						
10^{-2}						
10^{-3}						
10^{-4}						

	DOWNSTREAM WATER					
	TCC		FCC		FSC	
Dilution	Count	Cells/100 ml	Count	Cells/100 ml	Count	Cells/100 ml
Undiluted						
10^{-1}						
10^{-2}						
10^{-3}						
10^{-4}						

Determine the fecal coliform to fecal streptococcal (FC:FS) ratio. Record your results in the chart below.

	UPSTREAM WATER			DOWNSTREAM WATER		
	Cells/ml*			Cells/ml*		
Dilution	FCC	FSC	FC:FS Ratio	FCC	FSC	FC:FS Ratio
Undiluted						
1021						
1022						
1023						
1024						

$$*\text{Cells/ml} = \frac{\text{Cells/100ml}}{100}$$

Based on your FC:FS ratio, indicate the type of fecal pollution, if any, in the two samples:

    a. Upstream water sample:

    b. Downstream water sample:

Review Questions

1. What are the advantages of the membrane filter method in the analysis of water samples?

2. What are the disadvantages of the membrane filter method?

3. What is the purpose of determining the FC:FS ratio?

4. Cite some other microbiological applications of the membrane filter technique in environmental studies.

PART 12

Microbiology of Soil

LEARNING OBJECTIVES

Once you have completed the experiments in this section, you should be able to

1. Understand the characteristics and activities of soil microorganisms.
2. Enumerate soil microorganisms.
3. Demonstrate the ability of some soil microorganisms to produce antibiotics.
4. Demonstrate the use of enrichment cultures for the isolation of specific soil microorganisms.

Introduction

Soil is often thought of as an inert substance by the average layperson. However, contrary to this belief, it serves as a repository for many life forms, including a huge and diverse microbial population. The beneficial activities of these soil inhabitants far outweigh their detrimental effects.

Life on this planet could not be sustained in the absence of microorganisms that inhabit the soil. This flora is essential for degradation of organic matter deposited in the soil, such as dead plant and animal tissues and animal wastes. Hydrolysis of these macromolecules by microbial enzymes supplies and replenishes the soil with basic elemental nutrients. By means of enzymatic transformations, plants assimilate these nutrients into organic compounds essential for their growth and reproduction. In turn, these plants serve as a source of nutrition for animals and man. Thus, many soil microorganisms play a vital role in a number of elemental cycles, such as the nitrogen cycle, the carbon cycle, and the sulfur cycle.

Nitrogen Cycle

The nitrogen cycle is concerned with the enzymatic conversion of complex nitrogenous compounds in the soil and atmosphere into nitrogen compounds that plants are able to use for the synthesis of essential macromolecules, including nucleic acids, amino acids, and proteins. The four distinct phases in this cycle are as follows:

1. **Ammonification:** Soil microorganisms sequentially degrade nitrogenous organic compounds derived from dead plants and animals deposited in the soil. The degraded nitrogenous organic compounds are converted to inorganic nitrogen compounds and then to ammonia.

2. **Nitrification:** In this two-step process, (1) ammonia is oxidized to nitrite ions (NO_2^-) by an aerobic species of *Nitrosomonas*, and then (2) nitrites are converted to nitrate ions (NO_3^-) by another aerobic species, *Nitrobacter*. Nitrates are released into the soil and are assimilated as a nutritional source by plants.

3. **Denitrification:** Nitrates (NO_3^-) that are not used by plants are reduced to gaseous nitrogen ($N_2\uparrow$) and are liberated back into the atmosphere by certain groups of microorganisms.

4. **Nitrogen fixation:** This vital process involves the chemical combination of gaseous nitrogen ($N_2\uparrow$) with other elements to form fixed nitrogen (nitrogen-containing

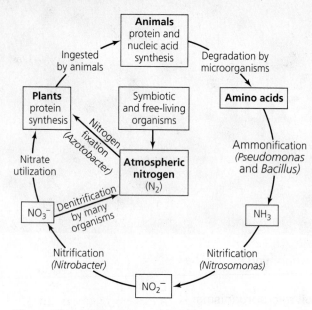

Figure P12.1 The nitrogen cycle

compounds), which are useful for plant growth. The two types of microorganisms involved in this process are free-living and symbiotic. Free-living microorganisms include *Azotobacter, Pseudomonas, Clostridium,* and *Bacillus,* as well as some species of yeast. Symbiotic microorganisms, such as *Rhizobium,* grow in tumor-like nodules in the roots of leguminous plants, and use nutrients in the plant sap to fix gaseous nitrogen as ammonia for its subsequent assimilation into plant proteins. Animals then consume the leguminous plants and convert plant protein to animal protein, completing the process. The nitrogen cycle is shown in Figure P12.1.

Carbon Cycle

Carbon dioxide is the major carbon source for the synthesis of organic compounds. The carbon cycle is basically represented by the following two steps:

1. Oxidation of organic compounds to carbon dioxide with the production of energy and heat by heterotrophs.
2. Fixation of carbon dioxide into organic compounds by green plants and some bacteria, the autotrophic soil flora.

Sulfur Cycle

Elemental sulfur and proteins cannot be utilized by plants for growth. They must first undergo enzymatic conversions into inorganic

sulfur-containing compounds. The basic steps in the sulfur cycle are

1. Degradation of proteins into hydrogen sulfide (H_2S) by many heterotrophic microorganisms.
2. Oxidation of H_2S to sulfur (S) by a number of bacterial genera, such as *Beggiatoa.*
3. Oxidation of sulfur to utilizable sulfate (SO_4^{2-}) by several chemoautotrophic genera, such as *Thiobacillus.*

Some soil microorganisms also play a role in the enzymatic transformation of other elements, such as phosphorus, iron, potassium, zinc, manganese, and selenium. These biochemical changes make the minerals available to plants in a soluble form.

Many members of the soil flora, because of their fermentative and synthetic capabilities, play an important role in the synthesis of a variety of industrial products:

1. **Food.** *Penicillium* spp. are used in the production of cheeses, including Camembert, Roquefort, and Brie.
2. **Beverages.** *Saccharomyces* spp. are utilized in the wine, beer, and ale industries.
3. **Vitamins.** *Eremothecium ashbyii* and *Pseudomonas denitrificans,* respectively, synthesize riboflavin (vitamin B_2) and cobalamin (vitamin B_{12}).
4. **Enzymes.** Amylases, pectinases, and proteases are produced by *Aspergillus* spp.
5. **Antibiotics.** *Penicillium* spp. (penicillin), *Streptomyces* spp. (kanamycins and tetracyclines), and *Bacillus* spp. (bacitracin).
6. **Steroids.** *Rhizopus, Streptomyces,* and *Curvularia* are microorganisms that are used to carry out specific reactions, bioconversions, to aid in the manufacture of these lipid compounds.
7. **Industrial chemicals.** *Clostridium acetobutylicum* is used in the production of acetone and butanol, and *Aspergillus niger* is used in the synthesis of citric acid.

The major adverse effect of soil organisms is the ability of some species to produce disease in plants and animals. Soil-borne human pathogens include members of the spore-forming bacterial genera *Clostridium* and *Bacillus,* and some fungal genera, such as *Cryptococcus* and *Coccidioides.*

Microbial Populations in Soil: Enumeration

LEARNING OBJECTIVES

Once you have completed this experiment, you should be

1. Familiar with the microbial soil flora.
2. Able to determine the number of bacteria and fungi present in a soil sample.

Principle

Soil contains myriads of microorganisms, including bacteria, fungi, protozoa, algae, and viruses. The most prevalent are bacteria, including the mold-like actinomycetes, and fungi:

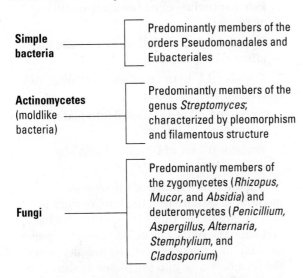

Simple bacteria — Predominantly members of the orders Pseudomonadales and Eubacteriales

Actinomycetes (moldlike bacteria) — Predominantly members of the genus *Streptomyces*; characterized by pleomorphism and filamentous structure

Fungi — Predominantly members of the zygomycetes (*Rhizopus*, *Mucor*, and *Absidia*) and deuteromycetes (*Penicillium*, *Aspergillus*, *Alternaria*, *Stemphylium*, and *Cladosporium*)

It is essential to bear in mind that the soil environment differs from one location to another and from one period of time to another. Therefore, factors, including moisture, pH, temperature, gaseous oxygen content, and organic and inorganic composition of soil are crucial in determining the specific microbial flora of a particular sample.

Just as the soil differs, microbiological methods used to analyze soil also vary. A single technique cannot be used to count all the different types of microorganisms present in a given soil sample because no one laboratory cultivation procedure can provide all the physical and nutritional requirements necessary for the growth of a greatly diverse microbial population. In this experiment, only the relative numbers of bacteria, actinomycetes, and fungi are determined. The method used is the serial dilution–agar plate procedure described in Experiment 18. Different media are employed to support the growth of these three types of microorganisms: glycerol yeast agar for the isolation of actinomycetes, Sabouraud agar for the isolation of fungi, and nutrient agar for the isolation of bacteria. The glycerol yeast agar and Sabouraud agar are supplemented with 10 μg of chlortetracycline (Aureomycin) per ml of medium to inhibit the growth of bacteria.

CLINICAL APPLICATION

Soil Testing

The enumeration of organisms in soil helps to establish the level of soil fertility as well as the types and kinds of pathogens it contains. From a clinical view, many bacterial pathogens originate from a soil environment. Current thought is that the ability of *Bacillus* species (for example, *B. subtilis* and *B. anthracis*) and *Mycobacterium* species (for example, *M. tuberculosis*) to survive in a soil environment—one that contains low nutrients, low moisture, and that necessitates sporulation or slow growth—aids the bacteria in infecting human tissues and surviving the immune response.

AT THE BENCH

Materials

Soil

1-g sample of finely pulverized, rich garden soil in a flask containing 99 ml of sterile water; flask labeled 1:100 dilution (10^{-2}).

Media

Per designated student group: four glycerol yeast agar deep tubes, four Sabouraud agar deep tubes, four nutrient agar deep tubes, and two 99-ml flasks of sterile water.

Equipment

Bunsen burner, 12 Petri dishes, Quebec colony counter, mechanical hand counter, sterile 1-ml pipettes, mechanical pipetting device, L-shaped bent glass rod, turntable (optional), 95% alcohol in a 500-ml beaker, and glassware marking pencil.

Procedure Lab One

Refer to Figure 49.1 as you read the following instructions.

1. Liquefy the glycerol yeast, Sabouraud, and nutrient agar deep tubes in an autoclave or by boiling. Cool the molten agar tubes and maintain in a waterbath at 45°C.

2. Divide the Petri dishes into three groups of four; using a glassware marking pencil, label the groups as nutrient agar, glycerol yeast extract agar, and Sabouraud agar. Then, label each set of Petri dishes as follows:

   - Nutrient agar: 10^{-4}, 10^{-5}, 10^{-6}, and 10^{-7} (to be used for enumeration of bacteria).
   - Glycerol yeast extract agar: 10^{-3}, 10^{-4}, 10^{-5}, and 10^{-6} (to be used for enumeration of actinomycetes).
   - Sabouraud agar: 10^{-2}, 10^{-3}, 10^{-4}, and 10^{-5} (to be used for enumeration of fungi).

3. With a glassware marking pencil, label the soil sample flask as Flask 1, and label the 99-ml sterile water Flasks 2 and 3.

4. Vigorously shake the provided soil sample dilution of 1:100 (10^{-2}) approximately 30 times, with your elbow resting on the table.

5. With a sterile 1-ml pipette, transfer 1 ml of the provided soil sample dilution to Flask 2 and shake vigorously as before. The final dilution is 1:10,000 (10^{-4}).

6. Using another sterile 1-ml pipette, transfer 1 ml of Dilution 2 to Flask 3 and shake vigorously as before. The final dilution is 1:1,000,000 (10^{-6}).

7. Using sterile 1-ml pipettes and aseptic technique, add the proper amount of each dilution into each Petri dish as indicated in a–c and shown in Figure 49.1.

   a. **For actinomycetes**—in plates labeled glycerol yeast extract agar:

      Transfer 0.1 ml of Dilution 1 into plate to effect a 10^{-3} dilution.

      Transfer 1 ml of Dilution 2 into plate to effect a 10^{-4} dilution.

      Transfer 0.1 ml of Dilution 2 into plate to effect a 10^{-5} dilution.

      Transfer 1 ml of Dilution 3 into plate to effect a 10^{-6} dilution.

   b. **For molds**—in plates labeled Sabouraud agar:

      Transfer 1 ml of Dilution 1 into plate to effect a 10^{-2} dilution.

      Transfer 0.1 ml of Dilution 1 into plate to effect a 10^{-3} dilution.

      Transfer 1 ml of Dilution 2 into plate to effect a 10^{-4} dilution.

      Transfer 0.1 ml of Dilution 2 into plate to effect a 10^{-5} dilution.

   c. **For bacteria**—in plates labeled nutrient agar:

      Transfer 1 ml of Dilution 2 into plate to effect a 10^{-4} dilution.

      Transfer 0.1 ml of Dilution 2 into plate to effect a 10^{-5} dilution.

      Transfer 1 ml of Dilution 3 into plate to effect a 10^{-6} dilution.

      Transfer 0.1 ml of Dilution 3 into plate to effect a 10^{-7} dilution.

8. Check the temperature of the molten agar medium to be sure that the temperature is 45°C. Remove the tubes from the water bath and wipe the outside surface dry with a paper towel. Using the pour-plate technique, pour the liquefied agar into the plates as shown in Figure 18.2 on page 138 and rotate gently to ensure uniform distribution of the cells in the medium.

9. Incubate the plates in an inverted position at 25°C. Perform colony counts on nutrient agar plate cultures in 2 to 3 days and on the remaining agar plate cultures in 4 to 7 days.

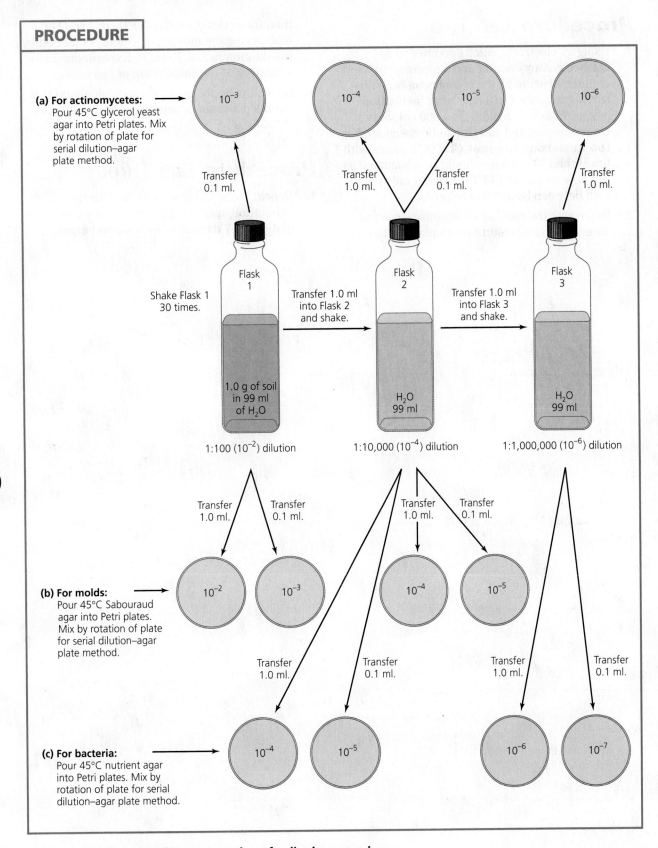

(a) **For actinomycetes:**
Pour 45°C glycerol yeast agar into Petri plates. Mix by rotation of plate for serial dilution–agar plate method.

(b) **For molds:**
Pour 45°C Sabouraud agar into Petri plates. Mix by rotation of plate for serial dilution–agar plate method.

(c) **For bacteria:**
Pour 45°C nutrient agar into Petri plates. Mix by rotation of plate for serial dilution–agar plate method.

Figure 49.1 Procedure for enumeration of soil microorganisms

Procedure Lab Two

1. Using an electronic colony counter or a Quebec colony counter and a mechanical hand counter, observe all the colonies on each nutrient agar plate 2 to 3 days after incubation begins. Plates with more than 300 colonies cannot be counted and should be designated as **too numerous to count (TNTC)**; plates with fewer than 30 colonies should be designated as **too few to count (TFTC)**. Count only plates with between 30 and 300 colonies.

2. Determine the number of organisms per milliliter of original culture on all plates other than those designated as TFTC or TNTC by multiplying the number of colonies counted by the dilution factor. Refer to Experiment 20 for examples of the calculation of cell counts.

3. Record your observations and calculated cell counts per gram of sample in the Lab Report chart.

Procedure Lab Three

1. Repeat Steps 1–3 from Lab Two for the Sabouraud agar and glycerol yeast extract agar plates 4 to 7 days after incubation begins.

Name: _____

Date: _____ Section: _____

Observations and Results

Organism	Dilution	Number of Colonies	Organisms per Gram of Soil
Bacteria	10^{-4}		
	10^{-5}		
	10^{-6}		
	10^{-7}		
Actinomycetes	10^{-3}		
	10^{-4}		
	10^{-5}		
	10^{-6}		
Molds	10^{-2}		
	10^{-3}		
	10^{-4}		
	10^{-5}		

Based on your results, which of the three types of soil organisms was most abundant in your sample? Least abundant?

Review Questions

1. Would you expect to be able to duplicate your results if a soil sample were taken from the same location at a different time of the year? Explain.

2. In the experiment performed, why wasn't the same medium used for enumeration of all three types of soil organisms?

3. Would you expect to be able to isolate an anaerobic organism from any of your cultures? Explain.

4. Explain why most microorganisms are present in the upper layers of the soil.

5. Following the nuclear disaster at Chernobyl, the regional microbial flora was destroyed. What impact did this have on higher forms of plant and animal life in this area?

Isolation of Antibiotic-Producing Microorganisms and Determination of Antimicrobial Spectrum of Isolates

LEARNING OBJECTIVES

Once you have completed this experiment, you should be able to

1. Isolate antibiotic-producing microorganisms.
2. Determine the spectrum of antimicrobial activity of the isolated antibiotic.

Principle

Soil is the major repository of microorganisms that produce **antibiotics** capable of inhibiting the growth of other microorganisms. Clinically useful antibiotics have been isolated from five groups of soil microorganisms—*Streptomyces*, *Amycolatopsis* (including some species formerly classified as *Streptomyces*), *Bacillus*, *Penicillium*, and *Acremonium*—that represent three microbial types, namely, actinomycetes, true bacteria, and molds.

Although soils from all parts of the world are continually screened in industrial laboratories for the isolation of new antibiotic-producing microorganisms, industrial microbiology is directing its energies toward chemical modification of existing antibiotic substances. This is accomplished by adding or replacing chemical side chains, reorganizing intramolecular bonding, or producing mutant microbial strains capable of excreting a more potent form of the antibiotic. The establishment of chemical congeners has been responsible for the circumvention of antibiotic resistance, minimizing adverse side effects in the host and increasing the effective spectrum of a given antibiotic.

In Part A of this experiment, you will use the **crowded-plate** technique for isolation of antibiotic-producing microorganisms from two soil samples, one of which is seeded with *Streptomyces griseus* to serve as a positive control. **Figure 50.1** illustrates the procedure to be followed. In Part B, isolates exhibiting antibiotic activity will be screened against several different microorganisms to establish their effectiveness.

CLINICAL APPLICATION

Testing New Antibiotics

Soil is the major reservoir housing microorganisms that produce antibiotics, which are used offensively to reduce competition for available nutrients. The most prolific antibiotic producers are within the phylum *Actinobacteria*. The genus *Streptomyces* are the major producers of currently used antibiotics (for example, neomycin, streptomycin, and tetracyclin) along with the genus *Actinomycetes* (for example, erythromycin). Interestingly, hundreds of new antibiotics are isolated annually using the crowded plate technique and other techniques, but most have a limited spectrum and only a few are found to be clinically acceptable.

AT THE BENCH

Materials

Cultures

For Part B: 24-hour Trypticase soy broth cultures of *Escherichia coli*, *Staphylococcus aureus* BSL-2 , *Mycobacterium smegmatis*, and *Pseudomonas aeruginosa* BSL-2 .

Soil Suspensions

For Part A: 1:500 dilution of soil sample suspension (0.1 g of soil per 50 ml of tap water) to serve as an unknown; 1:500 dilution of soil sample seeded with *S. griseus* (0.1 g of soil per 50 ml of tap water) to serve as a positive control.

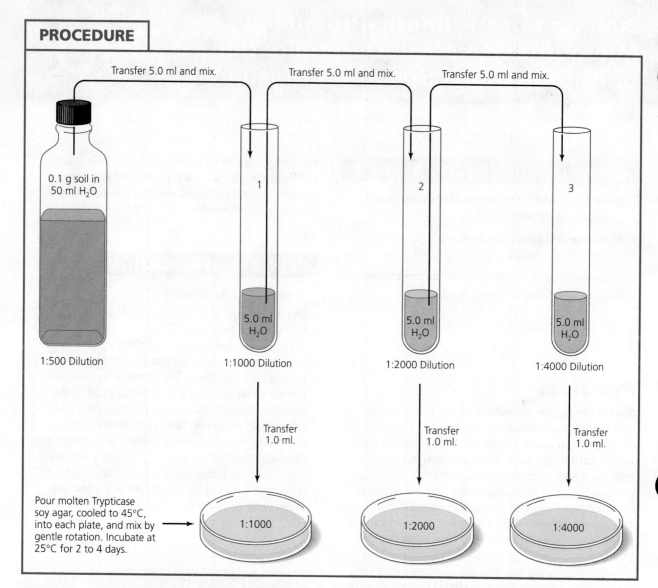

Figure 50.1 Crowded-plate technique for isolation of antibiotic-producing microorganisms

Media

Per designated student group: Part A: Six 15-ml Trypticase soy agar deep tubes, and two Trypticase soy agar slants. Part B: Two Trypticase soy agar plates.

Equipment

Part A: 500-ml beaker, test tubes, test tube rack, sterile Petri dishes, inoculating needle, hot plate, thermometer, 1-ml and 5-ml pipettes, mechanical pipetting device, and magnifying hand lens. Part B: Bunsen burner, inoculating loop, and glassware marking pencil.

PART A Isolation of Antibiotic-Producing Microorganisms

Procedure Lab One

1. Label two sets of three sterile Petri dishes with the types of soil samples being used and dilutions (1:1000, 1:2000, and 1:4000).

2. Place six Trypticase soy agar deep tubes into a beaker of water and bring to 100°C on a hot plate. Once agar is liquefied, add cool water to the water bath. Cool to 45°C, checking the temperature with a thermometer.

3. Prepare a serial dilution of the unknown and positive control 1:500 soil samples as follows (refer to Figure 50.1):

    a. Label three test tubes 1, 2, and 3. With a pipette, add 5 ml of tap water to each tube.

    b. Shake the provided 1:500 soil sample thoroughly for 5 minutes to effect a uniform soil-water suspension.

    c. Using a 5-ml pipette, transfer 5 ml from the 1:500 dilution to Tube 1 and mix. The final dilution is 1:1000.

    d. Using another pipette, transfer 5 ml from Tube 1 to Tube 2 and mix. The final dilution is 1:2000.

    e. Using another pipette, transfer 5 ml from Tube 2 to Tube 3 and mix. The final dilution is 1:4000.

    f. Using separate 1-ml pipettes, transfer 1 ml of the 1:1000, 1:2000, and 1:4000 dilutions to their appropriately labeled Petri dishes.

    g. Pour one tube of molten Trypticase soy agar, cooled to 45°C, into each plate and mix by gentle rotation.

    h. Allow all plates to solidify.

4. Incubate all plates in an inverted position for 2 to 4 days at 25°C.

Procedure Lab Two

1. Examine all crowded-plate dilutions for colonies exhibiting zones of growth inhibition. Use a hand magnifying lens if necessary. Record in the Lab Report the number of colonies showing zones of inhibition.

2. Aseptically isolate one colony showing a zone of growth inhibition from each soil culture with an inoculating needle and streak onto Trypticase soy agar slants labeled with the soil sample from which the isolate was obtained.

3. Incubate the slants for 2 to 4 days at 25°C. These will serve as stock cultures of antibiotic-producing isolates to be used in Part B.

PART B Determination of Antimicrobial Spectrum of Isolates

Procedure Lab One

1. Label the Trypticase soy agar plates with the soil sample source of the isolate.

2. Using the aseptic technique, make a single-line streak inoculation of each isolate on the surface of an agar plate so as to divide the plate in half as shown:

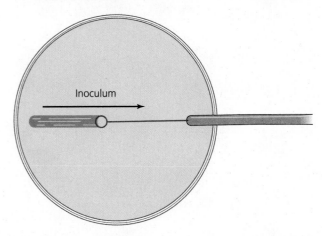

Inoculum

3. Incubate the plates in an inverted position for 3 to 5 days at 25°C.

Procedure Lab Two

1. Following incubation, on the bottom of each plate draw four lines perpendicular to the growth of the antibiotic-producing isolate as shown:

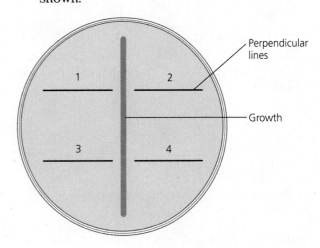

Perpendicular lines

Growth

2. Aseptically make a single-line streak inoculation of each of the four test cultures following the inoculation template on each plate. Start close to, but not touching, the growth of the antibiotic-producing isolate and streak toward the edge of the plate.

3. Incubate the plates in an inverted position for 24 hours at 37°C.

Procedure Lab Three

1. Examine all plates for inhibition of test organisms, and record your observations in the Lab Report.

Name: _____

Date: _____ Section: _____

Observations and Results

Part A: Isolation of Antibiotic-Producing Microorganisms

	NUMBER OF COLONIES SHOWING INHIBITION ZONE		
	Dilutions		
Soil Sample	**1:1000**	**1:2000**	**1:4000**
Unknown			
Positive control			

Part B: Determination of Antimicrobial Spectrum of Isolates

1. Draw a representation of the observed antibiotic activity against the test organisms.

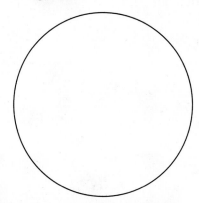

Antibiotic-Producing Isolate 1

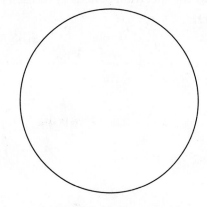

Antibiotic-Producing Isolate 2

2. Based on your observations, record in the chart the presence (+) or absence (−) of antibiotic activity against each of the test organisms and the spectrum of antimicrobial activity (broad or narrow).

	TEST ORGANISMS				
Soil Sample	**E. coli** **Gram-negative**	**S. aureus** **Gram-positive**	**P. aeruginosa** **Gram-negative**	**M. smegmatis** **Acid-fast**	**Spectrum**
Unknown					
Positive control					

Review Questions

1. Why is it frequently advantageous to modify antibiotics in industrial laboratories?

2. Is the ability to produce antibiotics limited only to bacterial species? Explain.

3. Do you feel that sufficient test organisms were used in Part B to determine fully the spectrum of activity of each isolated antibiotic? Explain.

Isolation of *Pseudomonas* Species by Means of the Enrichment Culture Technique

LEARNING OBJECTIVE

Once you have completed this experiment, you should

1. Understand the enrichment culture technique for the isolation of a specific microbial cell type.

Principle

The enrichment culture technique is used for the isolation of a specific type of microorganism from an environment that is replete with different types of microbes. In such an environment, the desired organism may be present only in very small numbers because of the competitive activities of this diverse microbial population. Under these circumstances, the use of conventional enriched media is not suitable for the selection of a specific cell type. These special-purpose media are supplemented with a variety of enriching nutrients capable of supporting the growth of many organisms rather than a single cell type in the test sample. Enrichment broths, on the other hand, are designed to contain a limited number of specific substrates that will preferentially promote the growth of the desired microorganisms.

The enrichment culture technique employs a specifically designed enrichment broth for the initial inoculation of the test sample. Once growth occurs in the primary culture, it is sequentially transferred into a fresh medium of the same composition until the desired microorganisms are predominant in the culture. These organisms are capable of exponential growth because of their ability to adapt to the medium and to enzymatically use the incorporated substrate(s)

as an energy source. Most of the competitors, however, are incapable of utilizing the substrate(s) and therefore remain in the lag phase of the growth curve. In some instances, the organisms to be isolated do not grow more rapidly than their competitors. Instead, they produce a growth inhibitor that greatly suppresses the growth of the competing population. After the serial transfer through the broth medium, the culture is streaked on an agar plate of the same composition as the enrichment broth for the isolation and subsequent identification of the discrete colonies.

The use of the enrichment culture technique has a wide range of applications in clinical, industrial, and environmental microbiology. Enrichment methods may be used to isolate and cultivate specific soil microorganisms for the production of industrial products such as steroids, enzymes, and vitamins. Likewise, a beneficial environmental application may involve the isolation by enrichment of petroleum-utilizing microorganisms, such as *Pseudomonas*, that would be capable of degrading environmentally destructive oil spills in waterways.

In this experimental procedure, a compost or a rich garden soil sample will be used to isolate *Pseudomonas* species by means of the enrichment culture procedure. Members of the genus *Pseudomonas* can utilize mandelic acid aerobically as their sole carbon and energy source. Therefore, this compound is the most important factor in the enrichment broth, which also contains a number of inorganic salts. The pseudomonads are gram-negative, motile organisms that generally produce a diffusible yellow-green pigment. In addition, they commonly reduce nitrates (NO_3^-) and produce an alkaline or proteolytic reaction in litmus milk. The schema for the experimental procedure to be followed is illustrated in **Figure 51.1** on the following page.

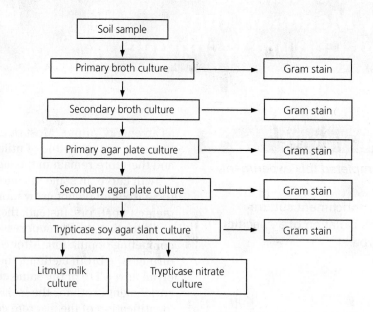

Figure 51.1 **Enrichment culture procedure schema**

CLINICAL APPLICATION

Medical Use for the Enrichment Culture Technique

Medically, the enrichment culture technique is routinely used for the isolation of intestinal pathogens from fecal samples when these organisms may be present only in low concentrations during the infectious process. With hundreds of different bacterial species composing our intestinal flora, identifying a new bacterial pathogen, such as *Salmonella* or a new strain of *E. coli,* within that population through normal plating techniques may not be possible. By increasing the number of bacteria present in a media that is enriched, thus lowering competition, bacterial species with low numbers may increase their percentage of the population and increase the chances of their identification.

AT THE BENCH

Materials

Cultures

Rich garden soil or compost sample.

Media

Per designated student group: Two Erlenmeyer flasks containing 20 ml of basal salts broth supplemented with 2 ml of 2.5% mandelic acid, two agar plates of the same composition as the broth, one Trypticase nitrate broth, one litmus milk, and one Trypticase soy agar slant.

Reagents

Crystal violet, Gram's iodine, 95% ethanol, safranin, Solution A (sulfanilic acid), Solution B (alpha-naphthylamine), and zinc powder. *Note: Solutions A and B are not Barritt's reagent.*

Equipment

Sterile 10-ml, 5-ml, and 1-ml pipettes, mechanical pipetting device, microspatula, Bunsen burner, staining tray, glass slides, lens paper, bibulous paper, inoculating loop, and glassware marking pencil.

Procedure Lab One

Primary Broth Culture Preparation

1. Inoculate an appropriately labeled Erlenmeyer flask containing the enrichment broth by adding an amount of the soil sample equivalent to the size of a pea with a microspatula. Gently swirl the flask to mix the culture.

2. Incubate the primary broth culture for 24 hours at 30°C.

Procedure Lab Two

Secondary Broth Culture Preparation

1. Examine the primary culture for presence of growth. If growth is not present, return the flask to the incubator for an additional 24 hours.
2. If growth is present, aseptically transfer 1 ml of the primary culture to an appropriately labeled Erlenmeyer flask containing fresh enrichment medium. Swirl the flask.
3. Incubate the secondary broth culture for 24 hours at 30°C.
4. Prepare and examine a Gram-stained smear from the primary culture. Record your observations of cellular morphology and Gram reaction in the Lab Report.
5. Refrigerate the primary broth culture.

Procedure Lab Three

Primary Agar Plate Preparation

1. If growth is present in the secondary broth culture, aseptically perform a four-way streak inoculation on the appropriately labeled agar plate of the enrichment medium (refer to Experiment 2).
2. Incubate the agar plate culture in an inverted position for 24 hours at 30°C.
3. Prepare and examine a Gram-stained smear of the secondary broth culture. Record your observations of cellular morphology and Gram reaction in the Lab Report.
4. Refrigerate the secondary broth culture.

Procedure Lab Four

Secondary Agar Plate Preparation

1. Examine the primary plate culture for the presence of discrete colonies. Record your observations of the cultural characteristics of these colonies in the Lab Report. Using a discrete colony:
   a. Aseptically prepare and examine a Gram-stained smear. Record your observations of cellular morphology and Gram reaction in the Lab Report.
   b. Aseptically perform a four-way streak inoculation on an appropriately labeled agar plate of the enrichment medium.
2. Incubate the secondary agar plate culture in an inverted position for 24 hours at 30°C.
3. Refrigerate the primary agar plate culture.

Procedure Lab Five

Pure Culture Isolation

1. Examine the secondary agar plate culture. Record your observations of the cultural characteristics of these colonies in the Lab Report. If the cultural characteristics of discrete colonies appear to be similar:
   a. Prepare and examine a Gram-stained smear from a discrete colony. Record your observations of cellular morphology and Gram reaction in the Lab Report.
   b. Pick a discrete colony and aseptically inoculate a Trypticase soy agar slant by means of a streak inoculation.
2. Incubate the agar slant culture for 24 to 48 hours at 30°C.
3. Refrigerate the secondary agar plate culture.

Procedure Lab Six

Genus Identification of Isolate

1. Prepare and examine a Gram-stained smear from the Trypticase agar slant culture. Record your observations of cellular morphology and Gram reaction in the Lab Report.
2. Using the Trypticase agar slant culture, aseptically inoculate the appropriately labeled tubes of Trypticase nitrate broth and litmus milk by means of a loop inoculation.
3. Incubate the litmus milk and Trypticase nitrate broth cultures for 24 to 48 hours at 30°C.

Procedure Lab Seven

1. Observe the litmus milk culture. Determine the type of reaction that has taken place (refer to Experiment 26), and record in the Lab Report.
2. Perform the nitrate reduction test on the Trypticase nitrate broth culture (refer to Experiment 27). Record your results in the Lab Report.

Name: _____

Date: _____ Section: _____

Observations and Results

Gram Reactions and Colony Characteristics

Culture	Gram Stain; Cellular Morphology	Cultural Characteristics
Primary broth culture		
Secondary broth culture		
Primary agar plate culture		
Secondary agar plate culture		
Trypticase soy agar slant culture		

Litmus Milk Reaction
Record the type of reaction below.

Nitrate Reduction Test
Record whether or not the organism was capable of nitrate reduction (+ or –) below.

Review Questions

1. A child is hospitalized with a severe gastroenteritis that is suspected to be food poisoning caused by a *Salmonella* species. Explain why the hospital laboratory supervisor uses an enrichment broth technique rather than selective media to confirm her suspicions.

2. A patient is afflicted with a disease that generates a large volume of gelatinous abdominal ascites. Drainage by surgical means is not successful. The use of a microbial enzyme capable of degrading this viscous ascites is suggested. Explain how you would go about isolating an organism that is enzymatically competent to act on this unusual substrate.

PART 13

Bacterial Genetics

Once you have completed the experiments in this section, you should be able to demonstrate the applicability of bacterial test systems in genetic-related studies. The procedures include

1. Enzyme induction.
2. Transfer of genetic material by means of conjugation.
3. Isolation of a streptomycin-resistant mutant.
4. Detection of potential chemical carcinogens.

Introduction

In recent years, bacteria have proved to be essential organisms in research into the structure and function of DNA, the universal genetic material. Their use is predicated on the following:

1. Their haploid genetic state, which allows the phenotypic, observable expression of a genetic trait in the presence of a single mutant gene.
2. Their rapid rate of growth, which permits observation of transmission of a trait through many generations.
3. The availability of large test populations, which allows isolation of spontaneous mutants and their induction by chemical and physical mutagenic agents.
4. Their low cost of maintenance and propagation, which make it possible to perform a large number of experimental procedures.

In the following experiments, bacterial test systems are used to demonstrate enzyme induction, screening for chemical carcinogens, and the genetic phenomena of mutation and genetic transfer. The last two mechanisms introduce genetic variability, which is essential for evolutionary survival in asexually reproducing bacterial populations.

Point mutations are permanent, sudden qualitative alterations in genetic material that arise as a result of the addition, deletion, or substitution of one or more bases in the region of a single gene. As a result, one or more amino acid substitutions occur during translation, and a protein that may be inactive, reduced in activity, or entirely different is synthesized. **Spontaneous mutations** are the result of the chemical and physical components in the organism's natural environment. The rate at which they occur is extremely low in all organisms. For example, in *Escherichia coli*, the spontaneous mutation rate at a single locus (specific site on the DNA) is estimated to be about 1×10^{-7} and the possibility of a mutation at any locus in the genome is approximately 1×10^{-4}. **Induced mutations** are genetic changes resulting from the organism's exposure to an artificial physical or chemical mutagen, that is, an agent capable of inducing a mutation. The resultant mutations are of the same type that occur spontaneously; however, their rate is increased, and in some cases dramatically so.

Transfer of genetic material and its subsequent incorporation into the bacterial genome is also a source of genetic variation in some bacteria. This transfer may occur by means of the following:

1. **Conjugation:** A mating process between "sexually" differentiated bacterial strains that allows unidirectional transfer of genetic material.

2. **Transduction:** A bacteriophage-mediated transfer of genetic material from one cell to another.

3. **Transformation:** A genetic alteration in a cell, resulting from the introduction of free DNA from the environment across the cell membrane.

Enzyme Induction

LEARNING OBJECTIVES

Once you have completed this experiment, you should be able to

1. Understand the mechanism of the lactose operon.

2. Understand the factors affecting the expression of the β-galactosidase gene.

Principle

Although bacteria possess a single chromosome, each cell is capable of synthesizing hundreds of different enzymes. Studies have shown that these enzymes are not present within the cells in equal concentrations. Some enzymes, called **constitutive enzymes**, are synthesized at a constant rate regardless of conditions in the cell's environment. Synthesis of other enzymes, called **adaptive enzymes**, occurs only when necessary, and it is subject to regulatory mechanisms that are dependent on the environment. One such mechanism, **induction**, requires the presence of a substrate, the inducer, in the environment to initiate synthesis of its specific enzyme, called an **inducible enzyme**. An extensively studied inducible enzyme in *E. coli* is **β-galactosidase**, which acts on the disaccharide lactose to yield the monosaccharides glucose and galactose. The gene for β-galactosidase is a member of a cluster of genes, called an **operon**, that is involved in the metabolism of lactose. The member genes of the lactose (*lac*) operon function as a unit, all being transcribed only when the inducer, lactose, is present in the surrounding medium. See **Figure 52.1**.

To illustrate β-galactosidase induction, two test strains of *E. coli* will be used: a prototrophic (wild type) strain (lactose-positive) and an auxotrophic (mutant) strain (lactose-negative), which carries a mutation in the gene for β-galactosidase

as well as a mutation in the lactose operon regulatory gene. Both test strains will be grown in the following media:

1. Inorganic synthetic medium lacking an organic carbon and energy source that is required by the heterotrophic *E. coli*.

2. Inorganic synthetic medium plus glucose, which can be utilized by both strains as a carbon and energy source.

3. Inorganic synthetic medium plus lactose, which can be utilized only by the prototrophic strain.

Orthonitrophenyl-β-d-galactoside (ONPG), a colorless analog of lactose, can serve as the substrate for the induction of β-galactosidase synthesis. As the inducer, it is hydrolyzed to galactose and a yellow nitrophenolate ion. Following a short incubation period, growth in all the cultures will be determined by spectrophotometry. Induction of β-galactosidase synthesis and activity will be indicated by the appearance of a yellow color in the medium following addition of ONPG, which occurs only in the presence of the nitrophenolate ion. Absence of this macroscopically visible color change indicates that enzyme induction in the lactose-negative strain did not occur.

CLINICAL APPLICATION

Enzyme Inducers and Cancer

Inducer molecules can include hormones produced by the body as well as toxins and drugs. Both enzyme induction and inhibition are used by the body to control a number of interactions that play a role in many cellular reactions from digestion to cell death. One important type of current research is the deliberate induction of human enzymes that protect against environmental carcinogens. Such intervention may provide advance protection against cell damage.

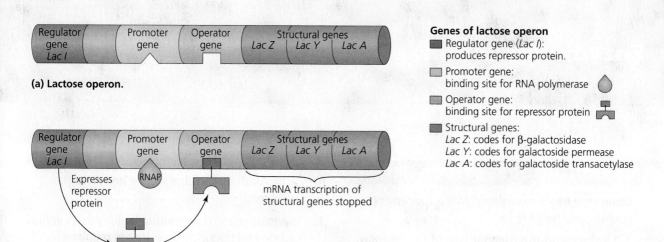

(a) Lactose operon.

Genes of lactose operon
- Regulator gene (*Lac I*): produces repressor protein.
- Promoter gene: binding site for RNA polymerase
- Operator gene: binding site for repressor protein
- Structural genes:
  Lac Z: codes for β-galactosidase
  Lac Y: codes for galactoside permease
  Lac A: codes for galactoside transacetylase

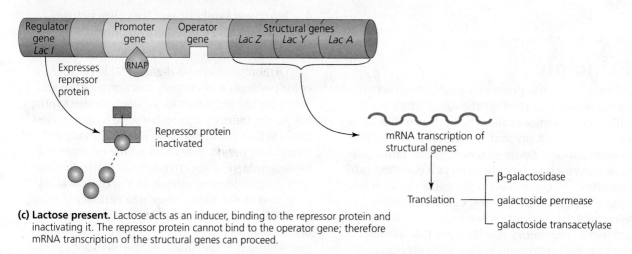

(b) **No lactose present.** The regulator gene (*Lac I*) expresses the repressor protein. Because no lactose is present, the repressor protein binds to the operator gene, blocking the RNA polymerase and stopping mRNA trascription of structural genes.

(c) **Lactose present.** Lactose acts as an inducer, binding to the repressor protein and inactivating it. The repressor protein cannot bind to the operator gene; therefore mRNA transcription of the structural genes can proceed.

Figure 52.1 Enzyme induction: The mechanism of operation of the lactose operon

Materials

Cultures

25-ml inorganic synthetic broth suspensions of 12-hour nutrient agar cultures of a lactose-positive *E. coli* strain (ATCC e 23725) and a lactose-negative *E. coli* strain (ATCC e 23735) adjusted to an absorbance of 0.1 at 600 nm.

Media

Per designated student group: dropper bottles of sterile 10% glucose, 10% lactose, and water.

Reagents

Dropper bottles of toluene and orthonitrophenyl -β-D-galactoside (ONPG).

Equipment

1-ml and 5-ml sterile pipettes, mechanical pipetting device, six sterile 13- × 100-mm test tubes, test tube racks, six sterile 25-ml Erlenmeyer flasks, Bausch & Lomb Spectronic 20 spectrophotometer, shaking water bath incubator, and glassware marking pencil.

Procedure

1. Label three sterile test tubes and three sterile 25-ml Erlenmeyer flasks as "Lac$^+$" (lactose-positive) and the name of the substrate to be added (glucose, lactose, or water). Similarly label three sterile tubes and flasks "Lac$^-$" (lactose-negative) for each test organism.

2. Using sterile 5-ml pipettes, aseptically transfer 5 ml of the Lac$^+$ and Lac$^-$ inorganic synthetic broth cultures to their respectively labeled test tubes.

3. Using a sterile 1-ml pipette, aseptically add 0.5 ml of the glucose and lactose solutions and 0.5 ml of sterile distilled water to the appropriately labeled tubes.

4. Determine the absorbance of all cultures at a wavelength of 600 nm. Record your results in the Lab Report.

5. Aseptically transfer each culture to its appropriately labeled flask. *(Note: If side-arm flasks are available, additions and absorbance readings may be made directly.)*

6. Incubate all flasks for 2 hours in a shaking water bath at 37°C and 100 strokes per minute.

7. Following incubation, transfer all cultures back to their appropriately labeled test tubes.

8. Determine and record in the Lab Report the absorbance for each culture at a wavelength of 600 nm. Based on your observations, indicate whether growth has occurred in each of the cultures.

9. To each culture, add 5 drops of toluene and shake vigorously (toluene ruptures the cells, releasing intact enzymes).

10. To each culture, add 5 drops of ONPG solution.

11. Incubate all cultures for 40 minutes at 37°C.

12. Following the addition of ONPG, observe the cultures for the presence of yellow coloration indicative of β-galactosidase synthesis and activity. In the Lab Report, record the colors of your cultures and the presence (+) or absence (–) of the β-galactosidase activity.

Name: _____

Date: _____ Section: _____

Observations and Results

| | ABSORBANCE AT 600 NM | | | | |
Cultures	Prior to Incubation	Following Incubation	Growth (+) or (−)	Color of Culture with ONPG	β-Galactosidase (+) or (−)
Lac⁺ E. coli Glucose					
Lactose					
Water					
Lac⁻ E. coli Glucose					
Lactose					
Water					

Explain the absence of growth in some of the cultures.

Review Questions

1. Distinguish between constitutive enzymes and inducible enzymes.

2. Explain what is meant by an operon.

3. Explain the purpose of the ONPG in the procedure.

4. Compare and contrast the methods for DNA transfer in microbial cells.

5. How can you explain why *Staphylococcus aureus*, which was initially sensitive to penicillin, is now resistant to this antibiotic?

Bacterial Conjugation

●

●

●

LEARNING OBJECTIVE

Once you have completed this experiment, you should be able to

1. Demonstrate genetic recombination in bacteria by the process of conjugation.

Principle

Genetic variability is essential for the evolutionary success of all organisms. In diploid eukaryotes, the processes of **crossing over** (exchange of genetic material between homologous chromosomes) and **meiosis** contribute to this variability. In haploid, asexually reproducing prokaryotic organisms, genetic recombination may occur by **conjugation**, **transduction**, and **transformation**. In this experiment, only the process of conjugation is considered.

Conjugation is a mating process during which a unidirectional transfer of genetic material occurs at physical contact between two "sexually" differentiated cell types. This differentiation, or existence of different mating strains in some bacteria, is determined by the presence of a **fertility factor**, or **F factor**, within the cell. Cells that lack the F factor are recipients (females) of the genetic material during conjugation and are designated as **F⁻**. Cells possessing the F factor have the ability to act as genetic donors (males) during mating. If this F factor is extrachromosomal (a **plasmid** or **episome**), the cells are designated as **F⁺**; most commonly only the F factor is transferred during conjugation. If this factor becomes incorporated into the bacterial chromosome, there is a transfer of chromosomal genes, although generally not involving the entire chromosome or the F factor. The resulting cells are designated **Hfr**, for **high-frequency recombinants**.

In this experiment, you will prepare a mixed culture representing a cross between an Hfr prototrophic (wild type) strain of *E. coli* that is streptomycin-sensitive and an F⁻ auxotrophic (mutant) *E. coli* strain that requires threonine (thr), leucine (leu), and thiamine (thi) and is streptomycin-resistant (Str-r). Following a short incubation period, isolation of only the threonine and leucine recombinants will be performed by plating the mixed culture on a minimal medium containing streptomycin and thiamine. The streptomycin is incorporated into the medium to inhibit the growth of the wild-type, streptomycin-sensitive (Str-s) parental Hfr cells. The thiamine is required as an essential growth factor for the thiamine-negative (thi⁻) recombinant cells. Because of its distant location on the chromosome, this marker will not be transferred during the short mating period. A genetic map denotes the time in minutes required for the transfer of a marker (operon) from the donor cell to the recipient cell. **Figure 53.1** is the genetic map showing the site of origin of transfer and location of relevant markers in this experiment.

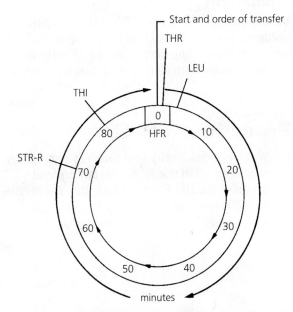

Figure 53.1 Genetic map of *Escherichia coli*

CLINICAL APPLICATION

Antibiotic Resistance

Conjugation is a major cause of the spread of antibiotic resistance and represents a serious problem in antibiotic therapy of immunosuppressed patients. Bacteria that carry several resistant genes are called multi-drug-resistant superbugs. The indiscriminate use of antibiotics both within the healthcare profession and the illegal sale of drugs without prescriptions are mainly responsible for the increased spread of antibiotic resistance.

AT THE BENCH

Materials

Cultures

12-hour nutrient broth cultures of F⁻ *E. coli* strain thr⁻, leu⁻, thi⁻, and Str-r (ATCC e 23724); and Hfr *E. coli* strain Str-s (ATCC e 23740).

Media

Per designated student group: three plates of minimal medium plus streptomycin and thiamine.

Equipment

Bunsen burner, beaker with 95% ethyl alcohol, L-shaped bent glass rod, 1-ml sterile pipettes, mechanical pipetting device, sterile 13 × 100-mm test tube, water bath shaker, and glassware marking pencil.

Procedure Lab One

1. With separate sterile 1-ml pipettes, aseptically transfer 1 ml of the F⁻ *E. coli* culture and 0.3 ml of the Hfr *E. coli* culture into the sterile 13 × 100-mm test tube.

2. Mix by gently rotating the culture between the palms of your hands.

3. Incubate the culture for 30 minutes at 37°C in a water bath shaker at the lowest speed setting.

4. Appropriately label two minimal plus streptomycin and thiamine agar plates, to be used for the control plates in Step 5.

5. To prepare control plates of the parental Hfr and F⁻ *E. coli* strains, aseptically add 0.1 ml of each *E. coli* strain to its appropriately labeled agar plate.

6. Use the spread-plate technique as shown in Figure 53.2 and as instructed below.

   a. Dip the bent glass rod into the beaker of 95% ethyl alcohol.

   b. Sterilize the glass rod by flaming with a Bunsen burner.

   c. Remove the glass rod from the Bunsen burner, allow flame to extinguish, and cool the glass rod.

   d. Spread the inoculum over the agar surface by rotating the plate.

7. Following incubation of the mixed culture, vigorously agitate it to terminate the genetic transfer.

8. Appropriately label a minimal plus streptomycin and thiamine plate. Aseptically add 0.1 ml of the mixed culture. Spread the inoculum over the entire surface with a sterile glass rod.

9. Incubate all plates in an inverted position for 48 hours at 37°C.

Procedure Lab Two

1. Observe all plates for growth of colonies.

2. Record your observations in the Lab Report.

PROCEDURE

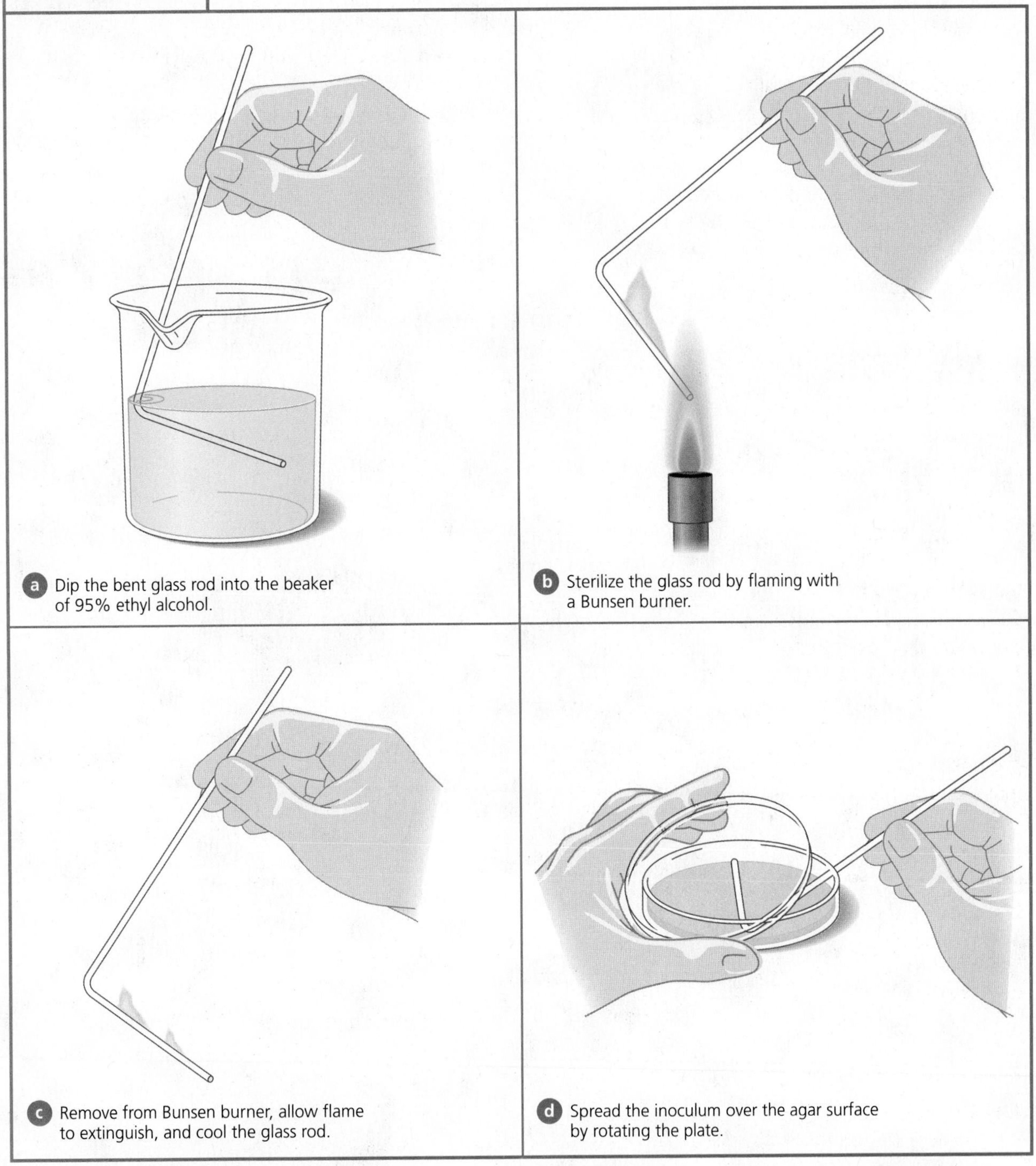

a Dip the bent glass rod into the beaker of 95% ethyl alcohol.
b Sterilize the glass rod by flaming with a Bunsen burner.
c Remove from Bunsen burner, allow flame to extinguish, and cool the glass rod.
d Spread the inoculum over the agar surface by rotating the plate.

Figure 53.2 Spread-plate technique

Name: _____

Date: _____   Section: _____

Observations and Results

1. Observe all plates for the presence (+) or absence (−) of colonies. Record your results in the chart.

	Hfr *E. coli* Plate	F⁻ *E. coli* Plate	Mixed-Culture Plate
Growth (+) or (−)			

2. Do you expect any growth to be present on the two parental *E. coli* minimal agar plates? Explain.

3. Did genetic recombination occur? Explain how your observations support your answer.

Review Questions

1. Explain how genetic variations may be introduced in eukaryotic and prokaryotic cells.

2. Explain the significance of the F factor.

3. Distinguish between F$^+$ and Hfr bacterial strains.

4. Explain the importance of the streptomycin marker in the parental *E. coli* strains.

Isolation of a Streptomycin-Resistant Mutant

LEARNING OBJECTIVE

Once you have completed this experiment, you should be able to

1. Isolate a streptomycin-resistant mutant in a prototrophic bacterial population by means of the gradient-plate technique.

Principle

Mutation, a change in the base sequence of a single gene, although infrequent, is one of the sources of genetic variability in cells. In some instances, these changes enable the cell to survive in an otherwise deleterious environment. An example of such a genetic adaptation is the development of **antibiotic resistance** in a small population of microorganisms prior to the advent and large-scale use of these agents. This microbial characteristic of antibiotic resistance is of major clinical importance because the number of drug-resistant microbial strains continues to increase. This occurs due to their extensive use and frequent misuse, which over the years have selected for the drug-resistant strains by their microbicidal effects on the sensitive cell forms. These agents select for the resistant mutant and do not act as inducers of the mutation.

In a drug-resistant organism, the mutated gene enables the cell to circumvent the antimicrobial

effect of the drug by a variety of mechanisms, including the following:

1. The production of an enzyme that alters the chemical structure of the antibiotic, as in penicillin resistance.

2. A change in the selective permeability of the cell membrane, as in streptomycin resistance.

3. A decrease in the sensitivity of the organism's enzymes to inhibiting mechanisms, as in the resistance to streptomycin, which interferes with the translation process at the ribosomes.

4. An overproduction of a natural substrate (metabolite) to compete effectively with the drug (antimetabolite), as in the resistance to sulfonamides, which produce their antimicrobial effect by competitive inhibition.

The following procedure is designed to allow you to isolate a streptomycin-resistant mutant from a prototrophic (wild type, streptomycin-sensitive) *Escherichia coli* culture by means of the **gradient-plate technique**. This requires preparation of a double-layered agar plate as illustrated in Figure 54.1. The lower, slanted agar-medium layer lacks streptomycin. When poured over the lower slanted layer, the molten agar medium containing the antibiotic will produce a streptomycin concentration gradient in the surface layer. Following a spread-plate inoculation of the prototrophic test culture and incubation, the appearance of colonies in a region of high streptomycin concentration is indicative of streptomycin-resistant mutants.

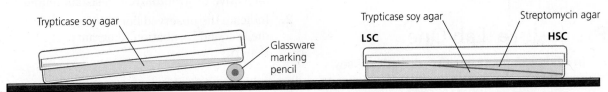

LSC = Low streptomycin concentration
HSC = High streptomycin concentration

Figure 54.1 Preparation of a streptomycin gradient plate

CLINICAL APPLICATION

Searching for Resistance Mutations

Once strains of bacteria that are known to be resistant are isolated, microbiologists attempt to find the gene or genes responsible. The *sasX gene,* found in methicillin-resistant *Staphylococcus aureus* (MRSA), has almost doubled in frequency over the past decade. It is located in a mobile genetic element that allows its easy transfer to other bacteria. This gene helps the bacterium to more effectively colonize nasal tissues and evade the host's immune system. Targeting this gene may provide a route for highly effective therapies.

AT THE BENCH

Materials

Cultures
24-hour nutrient broth culture of *E. coli*.

Media
Per designated student group: two 10-ml Trypticase soy agar deep tubes.

Reagent
Stock streptomycin solution (10 mg per 100 ml of sterile distilled water).

Equipment
Sterile Petri dish (100×15 mm) sterile 1-ml pipettes, mechanical pipetting device, inoculating loop, bent glass rod, beaker with 70% ethanol, water bath, and glassware marking pencil.

Procedure Lab One

1. In a hot water bath, melt two Trypticase soy agar tubes. Cool and maintain at 45°C.

2. Place a pencil under one end of a sterile Petri dish, pour in a sufficient amount of the molten agar medium to cover the entire bottom surface, and allow to solidify in the slanted position.

3. Using a sterile 1-ml pipette, add 0.1 ml of the streptomycin solution to a second tube of molten Trypticase soy agar. Mix by rotating the tube between the palms of your hands.

4. Place the dish in a horizontal position, pour in a sufficient amount of the molten agar medium containing streptomycin to cover the gradient agar layer, and allow to solidify.

5. With a sterile 1-ml pipette, add 0.2 ml of the *E. coli* test culture. With an alcohol-dipped and flamed bent glass rod, spread the culture over the entire agar surface as illustrated in Figure 53.2 on page 373.

6. Incubate the appropriately labeled culture in an inverted position for 48 hours at 37°C.

Procedure Lab Two

1. Observe the plate for the appearance of discrete colonies and indicate their positions in the "Initial Incubation" diagram in the Lab Report.

2. Select one or two isolated colonies present in the middle of the streptomycin concentration gradient. With a sterile inoculating loop, streak the selected colonies toward the high-concentration end of the plate.

3. Incubate the plate in an inverted position for 48 hours at 37°C.

Procedure Lab Three

1. Observe the plate for a line of growth from the streaked colonies into the area of high streptomycin concentration. Growth in this area is indicative of streptomycin-resistant mutants.

2. Indicate the observed line(s) of growth in the "Second Incubation" diagram in the Lab Report.

Name: _____

Date: _____   Section: _____

Observations and Results

1. Indicate the positions of discrete colonies in the diagram below (LSC = low streptomycin concentration; HSC = high streptomycin concentration).

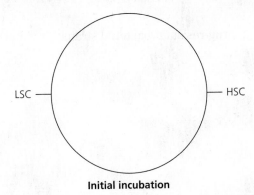

Initial incubation

2. Indicate the observed line(s) of growth in the diagram below.

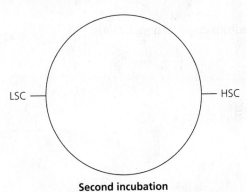

Second incubation

Review Questions

1. What mechanisms are responsible for antibiotic resistance?

2. Why is it necessary to use an antibiotic gradient-plate preparation for isolation of mutants?

3. Why has there been an increase in drug-resistant bacterial strains in recent years?

4. Does the streptomycin in the medium cause the mutations? Explain.

The Ames Test: A Bacterial Test System for Chemical Carcinogenicity

LEARNING OBJECTIVE

Once you have completed this experiment, you should be able to

1. Screen for potential chemical carcinogens using a bacterial test system.

Principle

Our exposure to a wide variety of chemical compounds has increased markedly over the past decades. Oncological epidemiologists strongly suspect that the intrusion of these chemicals in the form of industrial pollutants, pesticides, food additives, hair dyes, cigarette smoke, and the like may play a significant role in the induction of malignant transformations in humans. From a genetic aspect there is strong evidence linking **carcinogenicity** to **mutagenicity**. Research indicates that approximately 90% of the chemicals proved to be carcinogens are mutagens; they cause cancer by inducing mutations in somatic cells. These mutations are most frequently a result of base substitutions, the substitution of one base for another in the DNA molecule, and frameshift mutations, a shift in the reading frame of a gene because of the addition or deletion of a base.

In view of the rapid advent of new products and new industrial processes with their resultant pollutants, it is essential to determine their potential genetic hazards. Despite the fact that mammalian cell structure and human enzymatic pathways differ from those in bacteria, the chemical nature of DNA is common to all organisms; this permits the use of bacterial test systems for the rapid detection of possible mutagens and therefore possible carcinogens.

The **Ames test** is a simple and inexpensive procedure that uses a bacterial test organism to screen for mutagens. The test organism is a histidine-negative (his$^-$) and biotin-negative (bio$^-$) auxotrophic strain of *Salmonella typhimurium* that will not grow on a medium deficient in histidine unless a back mutation to his$^+$ (histidine-positive) has occurred. It is recognized that the mutagenic effect of a chemical is frequently influenced by the enzymatic pathways of an organism, whereby nonmutagens are transformed into mutagens and vice versa when introduced into human systems. In mammals, this toxification or detoxification frequently occurs in the liver. The Ames test generally requires the addition of a liver homogenate, S-9, which serves as a source of activating enzymes, to make this bacterial system more comparable to a mammalian test system.

In the Ames test, by means of the spot method, molten agar containing the test organism, S-9 mix, and a trace of histidine to allow the bacteria to undergo the several cell divisions necessary for mutation to occur is poured on a minimal agar plate. A disc impregnated with the test chemical is then placed in the center of the test plate. Following diffusion of the test compound from the disc, a concentration gradient of the chemical is established. Following incubation, a qualitative indication of the mutagenicity of the test chemical can be determined by noting the number of colonies present on the plate. Each colony represents a his$^-$ → his$^+$ revertant. A positive result, indicating mutagenicity, is obtained when an obvious increase in the number of colonies is evident compared to the number of spontaneous revertants on the negative control plate.

In the following procedure, you will perform a modified Ames test; you will not use the S-9 mix to test for the mutagenicity of nitro compounds, which, as in humans, are activated by the bacterial nitroreductases. Four minimal agar plates are inoculated with the *S. typhimurium* test organism. One plate, the negative control, is not exposed to a test chemical. Any colonies developing on this plate are representative of spontaneous his$^-$ → his$^+$ mutations. The second plate, the positive control, is exposed to a known nitrocarcinogen, 2-nitrofluorene. The remaining two plates are used to determine the mutagenicity of two commercial hair dyes.

CLINICAL APPLICATION

Testing for Cancer-Causing Chemicals

The Ames test is a procedure used for the identification of mutagenic chemical and physical agents. The test was named after Bruce Ames, who invented the test in the 1970s. While the Ames test does not detect all mutagenic chemicals, it is used in the pharmaceutical industry to test drugs prior to use in clinical trials and also in the cosmetic industry to check on the mutagenic potential of makeup. A positive Ames test results in the rejection of the drug or agent for further development and testing.

AT THE BENCH

Materials

Cultures

24-hour nutrient broth cultures of *S. typhimurium* BSL-2 , Strain TA 1538 (ATCC e 29631).

Media

Per designated student group: four minimal agar plates and four 2-ml top agar tubes.

Reagents

Sterile biotin-histidine solution, 2-nitrofluorene dissolved in alcohol, and two commercial hair dyes.

Equipment

1-ml sterile pipettes, mechanical pipetting device, sterile discs, forceps, water bath, Bunsen burner, and glassware marking pencil.

Procedure Lab One

⚠ **Wear disposable gloves and a laboratory coat when handling 2-nitrofluorene. For disposal of this chemical, place excess into a sealable container and put it inside a fume hood for subsequent removal according to your institution's policy for disposal of hazardous materials.**

Refer to Figure 55.1 as you read the following instructions.

1. Label three minimal agar plates with the name of the test chemical to be used. Label the fourth plate as a negative control.

2. Melt four tubes of top agar in a hot water bath and maintain the molten agar at 45°C.

3. To each molten top agar tube, aseptically add 0.2 ml of the sterile biotin-histidine solution and 0.1 ml of the *S. typhimurium* BSL-2 test culture. Mix by rotating the test tube between the palms of your hands.

4. Aseptically pour the top agar cultures onto the minimal agar plates and allow to solidify.

5. Using sterile forceps, dip each disc into its respective test chemical solution and drain by touching the side of the container.

6. Place the chemical-impregnated discs in the center of the respectively labeled minimal agar plates. Place a sterile disc on the plate labeled negative control. With the sterile forceps, *gently* press down on the discs so that they adhere to the surface of the agar.

7. Incubate all plates in an inverted position for 24 hours at 37°C.

Procedure Lab Two

1. Count the number of large colonies present on each plate and record on the chart in the Lab Report.

2. Determine and record the number of chemically induced mutations by subtracting the number of colonies on the negative control plate, representative of spontaneous mutations, from the number of colonies on each test plate.

3. Determine and record in the Lab Report the relative mutagenicity of the test compounds on the basis of the number of induced mutations: If below 10, (−); if greater than 10, (1+); if greater than 100, (2+); and if greater than 500, (3+).

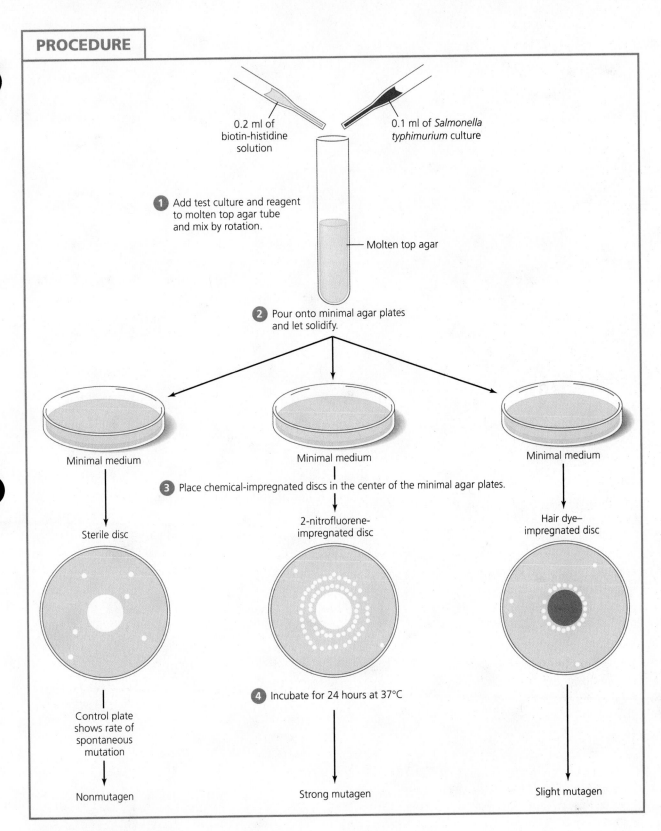

0.2 ml of biotin-histidine solution

0.1 ml of *Salmonella typhimurium* culture

1 Add test culture and reagent to molten top agar tube and mix by rotation.

Molten top agar

2 Pour onto minimal agar plates and let solidify.

Minimal medium

Minimal medium

Minimal medium

3 Place chemical-impregnated discs in the center of the minimal agar plates.

Sterile disc

2-nitrofluorene-impregnated disc

Hair dye–impregnated disc

4 Incubate for 24 hours at 37°C

Control plate shows rate of spontaneous mutation

Nonmutagen

Strong mutagen

Slight mutagen

Figure 55.1 **The Ames test**

Name: _____

Date: _____ Section: _____

Observations and Results

Test Chemical	Number of Colonies	Number of Induced Mutations	Degree of Mutagenicity (−), (1+), (2+), or (3+)*
Negative control			
2-Nitrofluorene			
Hair dye 1			
Hair dye 2			

*If below 10, (−); if greater than 10, (1+); if greater than 100, (2+); and if greater than 500, (3+).

Review Questions

1. What is the purpose of the S-9 in the Ames test?

2. What is the purpose of the biotin-histidine solution in the Ames test?

3. What is the relationship between chemical carcinogenicity and mutagenicity?

4. What are the advantages of using bacterial systems instead of mammalian systems to test for chemical carcinogenicity?

5. What are the disadvantages of using bacterial systems instead of mammalian systems to test for chemical carcinogenicity?

PART 14

Biotechnology

LEARNING OBJECTIVES

Once you have completed the experiments in this section, you should be familiar with the

1. Process of bacterial transformation.

2. Isolation of bacterial plasmids.

3. Use of restriction endonucleases to cut (digest) DNA molecules into small fragments.

4. Use of agarose gel electrophoresis to separate DNA fragments and determine their size.

Introduction

In the past, microbiologists focused on macromolecular cellular components. However, in recent years, this focus has rapidly shifted, and biological studies can now be easily carried out at the molecular level. **Molecular biology** is the study of life at the molecular level. This rapidly expanding field is based on the ability to manipulate the genes of an organism. As technology has expanded, so too have researchers' abilities to modify an organism's genetic structure by removing the genetic material from one organism and combining it with that of another organism. Commonly, this is referred to as **genetic engineering** or **recombinant DNA technology**.

Most often, genomic manipulation begins with isolating plasmid DNA and foreign DNA and cutting them with the same restriction enzyme. The foreign DNA inserts itself into the plasmid and the recombinant plasmid is introduced into a bacterium. The bacteria are then grown and screened to select the desired gene. An example of the use of this technology is the exploration of human genetic structure to identify and ameliorate genetic diseases, particularly through prenatal diagnosis. Scientists are also interested in the genetic engineering of hormones (e.g., pancreatic insulin), which are normally obtained only in their natural state. Genetically engineered organisms also help produce vaccines and antibiotic substances and remediate environmental toxins that threaten human and environmental health. The role of genetic engineering technology in clinical and forensic science is commonplace today. For example, DNA profiling is used to resolve paternity disputes, and it was used to identify the bones of Czar Nicholas, the last Russian czar, and his family. This technology is also used extensively in criminology as a tool to help establish guilt or innocence of individuals involved in criminal cases.

The many applications of recombinant DNA technology have brought about significant advances in medicine, pharmacology, basic research, industry, and agriculture. In medicine, advances in biotechnology now allow insulin-deficient diabetics to use different types of genetically engineered insulin (e.g., rapid acting, moderate acting, or slow acting) to meet diverse medical needs, rather than relying on the slower-acting insulin obtained from the pancreas of slaughtered animals. Other proteins, such as somatostatin, human growth hormone, interferon, interleukin-2—a regulator of the immune system—and blood clotting factor VIII, have all been cloned using genetic engineering and are in clinical use. Biotechnology plays a prominent role in agriculture as well. Plants have been made resistant to herbicides, such as glyphosate, and pathogens, such as the European corn borer. These few examples illustrate the amazing achievements made by the use of recombinant DNA technology. More fascinating biotechnological advances are anticipated in the years ahead.

In the experiments that follow, you will explore some of the basic techniques in biotechnology. You will transform an antibiotic-susceptible cell into an antibiotic-resistant one by means of a plasmid. You will isolate plasmid DNA from bacterial cells and cut it into small fragments using restriction endonucleases. Finally, using agarose gel electrophoresis, you will separate these DNA fragments and determine their size.

Bacterial Transformation

LEARNING OBJECTIVES

Once you have completed this experiment, you should be able to

1. Transform a competent ampicillin-susceptible strain of *Escherichia coli* into one that is ampicillin-resistant by means of a DNA plasmid.

2. Visualize transformed cells using a color marker gene carried in the plasmid.

3. Calculate the efficiency of transformation.

Principle

Transformation is a process whereby small pieces of host cell genomic DNA are able to enter a recipient cell and become incorporated into a homologous area on the recipient cell's genome. Historically, transformation had its origin in the pioneering experiments of Fred Griffith in London in the late 1920s. Working with *Streptococcus pneumoniae*, he noted that when an encapsulated smooth (S) strain that was lethal for mice was heat-killed and mixed with a living culture of an avirulent nonencapsulated rough (R) strain and then injected into mice, the result was fatal (Figure 56.1). Subsequent isolation of the organisms from the tissues of the dead mice revealed that the rough avirulent strain had been converted to a smooth, encapsulated, and lethal strain of *S. pneumoniae*.

This unusual experiment by Griffith, done long before DNA was determined to be the genetic basis for life, was simply termed by him as a "transformation." Today, in retrospect, we realize that this experiment proved to be the first indication of gene activity and the first demonstration of genetic recombination in bacteria. Later, Avery, McLeod, and McCarty, research scientists at the then Rockefeller Institute, were able to show that the transforming factor in Griffith's experiment was not a protein, as had been previously suspected, but a little-studied organic chemical called deoxyribonucleic acid (DNA).

During the **transformation** process, the donor cells forcibly lyse, releasing small segments of DNA containing 10 to 20 genes. These small segments have the ability to pass through the cell wall and cell membrane of a **competent cell** (a cell that is able to take up DNA from its environment). During naturally occurring transformations, a double-stranded DNA segment passes through the cell wall and into the cell's cytoplasm, and if there is sufficient sequence similarity, the foreign DNA undergoes homologous recombination with the recipient chromosome. The genome of the recipient cell has now been modified to contain DNA with genetic characteristics of the donor cell. Naturally occurring transformations are of great interest medically because they may serve as a vehicle for genetic exchange among pathogenic organisms. Interestingly, it appears that a larger percentage of pathogenic bacteria, such as *Streptococcus pneumoniae*, *Neisseria gonorrhoeae*, and *Haemophilus influenzae*, is capable of natural transformation than the nonpathogenic bacteria. This raises the intriguing possibility that the exchange of genetic material allows pathogenic cells to acquire the ability to evade a host's bodily defenses.

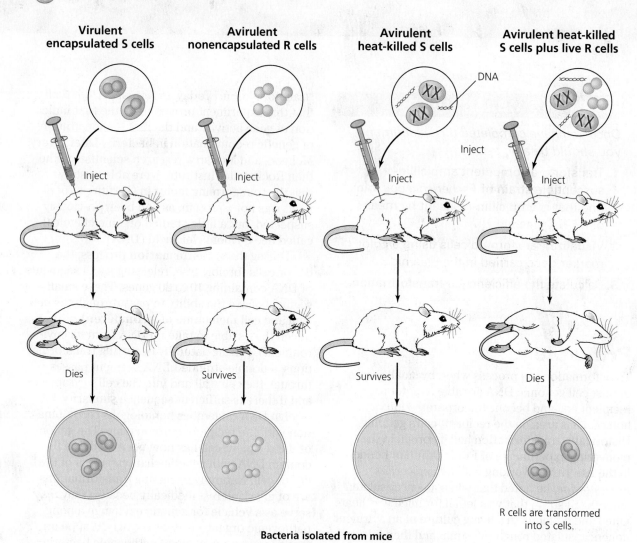

○○ Encapsulated smooth strain (S cell)
○○ Nonencapsulated rough strain (R cell)
ⓧⓧ Heat-killed S cell

Virulent encapsulated S cells

Avirulent nonencapsulated R cells

Avirulent heat-killed S cells

Avirulent heat-killed S cells plus live R cells

DNA

Inject

Inject

Inject

Inject

Dies

Survives

Survives

Dies

R cells are transformed into S cells.

Bacteria isolated from mice

Figure 56.1 Griffith's experiment on transformation. An avirulent rough strain of Streptococcus pneumoniae becomes lethal when transformed after mixing with the DNA of a heat-killed virulent smooth strain of S. pneumoniae.

Not all bacteria are naturally transformable, however, and methods have been developed to produce competency in various types of cells and transform those cells artificially. This process was initiated in the 1970s, when it was shown that treating a recipient cell with a cold calcium chloride ($CaCl_2$) solution allows the passage of donor DNA into the cell. The porosity of the cell wall is already almost sufficient to allow the passage of intact DNA; it is the cell membrane that is the true barrier, and its permeability is altered by this drastic treatment with $CaCl_2$, allowing DNA to pass through the membrane and into the cell. With our rapidly advancing knowledge in the field of molecular genetics, it is now possible to artificially induce transformations by the use of **plasmids**.

Plasmids are small, circular pieces of extrachromosomal DNA with a length of 5,000 to 100,000 base pairs (bp), capable of autonomous replication in the bacterial cytoplasm. Another membrane-altering method is **electroporation**. In this method, cells are suspended in a DNA solution and subjected to high-voltage electric impulses that destabilize the cell membrane, resulting in increased permeability and enabling DNA to pass into the cells. **Transduction** is a method of horizontal passage of genetic material from one bacterial cell to another by means of a bacteriophage. **Conjugation** occurs when bacterial DNA is transferred from one cell to another via the formation of a protoplasmic bridge, called a conjugative, or sex, pilus.

In the following experiment, the transformation of *E. coli* is artificial because the cells must be treated with a salt concentration and temperature shocks, an environment that is not natural for these cells. You will use a competent strain of *E. coli* in which the *lac* operon has been deleted, leaving it devoid of a β-galactosidase gene. The procedure directs you to suspend the *E. coli* in cold $CaCl_2$ and then introduce an ampicillin-resistant plasmid (pBLU®). The plasmid confers ampicillin resistance because it carries the β-lactamase gene amp^r, as shown in **Figure 56.2**. This suspension is incubated in ice and then heated. The cold shock and heat shock in the presence of $CaCl_2$ alters the permeability of the outer surfaces of the cell and facilitates the passage of the DNA into the cellular cytoplasm.

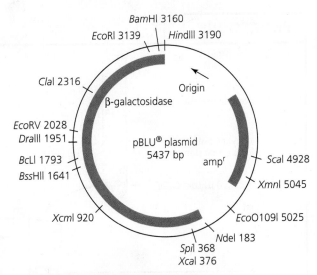

Figure 56.2 **Plasmid map of pBLU**

![AT THE BENCH]

Materials

Cultures

18- to 24-hour Luria-Bertani (LB) agar base streak plate cultures of *Escherichia coli*.

Media

Per designated student group: two LB agar base plates, three LB agar base plates plus ampicillin (Amp), three LB agar base plates plus ampicillin/X-Gal, and one tube of LB broth. *Note: X-Gal is a colorimetric analog of lactose that is cleaved by β-galactosidase to yield blue-colored colonies.*

Reagent

50 mM $CaCl_2$ solution.

Plasmid pBLU is 5437 bp long and has the gene for β-lactamase (penicillinase) and the gene for β-galactosidase.

Equipment

Sterile plastic 13- × 100-mm test tubes or plastic 1.5-ml centrifuge tubes, adjustable micropipette (0.5 to 100 μl) with sterile plastic micropipette

---

CLINICAL APPLICATION

Transferring Genes Between Bacteria

Bacterial transformation is the exchange of genetic material (DNA) between strains of bacteria by the transfer of a fragment of naked DNA from a donor cell to a recipient cell, followed by recombination in the recipient chromosome. Called "horizontal gene transfer," this allows for the transfer of genes between bacteria of the same or different species. Human and other genes are routinely transferred into bacteria in order to synthesize products for medical and commercial use, such as human insulin, human growth hormone, and vaccines.

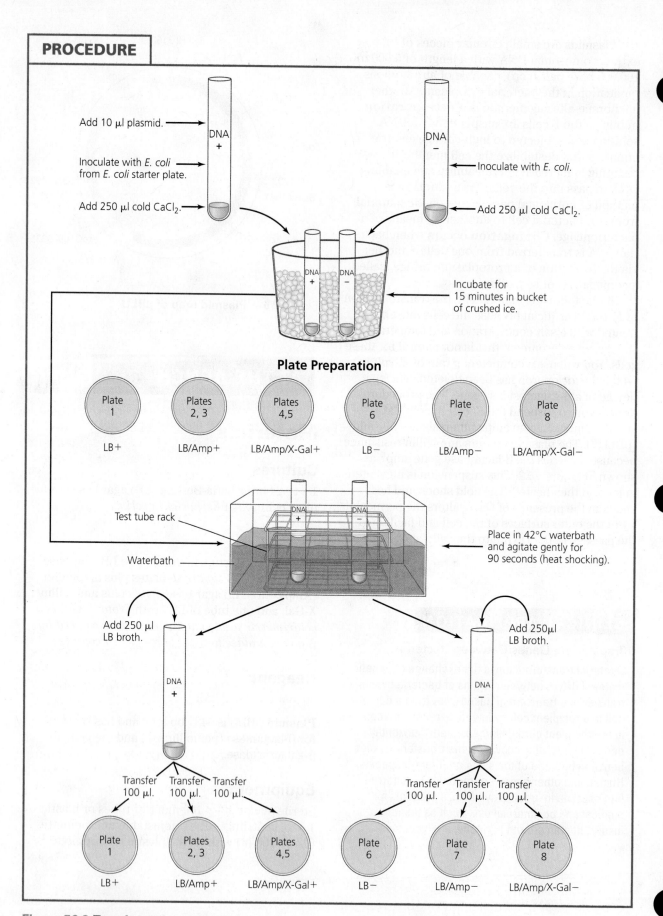

Add 10 µl plasmid.

Inoculate with *E. coli* from *E. coli* starter plate.

Add 250 µl cold $CaCl_2$.

Inoculate with *E. coli*.

Add 250 µl cold $CaCl_2$.

DNA +

DNA −

Incubate for 15 minutes in bucket of crushed ice.

Plate Preparation

Plate 1	Plates 2, 3	Plates 4,5	Plate 6	Plate 7	Plate 8
LB+	LB/Amp+	LB/Amp/X-Gal+	LB−	LB/Amp−	LB/Amp/X-Gal−

Test tube rack

Waterbath

Place in 42°C waterbath and agitate gently for 90 seconds (heat shocking).

Add 250 µl LB broth.

Add 250µl LB broth.

DNA +

DNA −

Transfer 100 µl.   Transfer 100 µl.   Transfer 100 µl.

Transfer 100 µl.   Transfer 100 µl.   Transfer 100 µl.

Plate 1	Plates 2, 3	Plates 4,5	Plate 6	Plate 7	Plate 8
LB+	LB/Amp+	LB/Amp/X-Gal+	LB−	LB/Amp−	LB/Amp/X-Gal−

Figure 56.3 **Transformation procedure**

tips (10 to 100 µl) (or 1.0-ml graduated, individually wrapped, disposable plastic transfer pipettes), glass beads (6-mm diameter), glassware marking pencil, disposable plastic inoculating loops (standard wire loops may be used), Bunsen burner, water bath, 500-ml beaker of crushed ice, 500-ml beaker labeled "waste," 500-ml beaker containing disinfectant solution, and a Quebec colony counter (or permanent marker); if the spread-plate method is used, bent glass rod, beaker of alcohol, and turntable.

Procedure Lab One

Refer to Figure 56.3 as you read the numbered instructions.

Note: If using a plastic transfer pipette (Figure 56.4), it is essential that you calibrate it to deliver a volume of 100 µl (0.1 ml), required for plasmid transfer. Once calibrated, it should be marked with a permanent glassware marker and retained to be used as a guide in the transformation experiment. Ask your instructor for help if needed.

1. With a glassware marking pencil, label two 13 × 100-mm test tubes, one as "DNA+" and the other as "DNA−." The DNA+ tube will receive the plasmid.

2. Using a sterile pipette, transfer 250 µl (0.25 ml) of ice-cold CaCl₂ solution into each tube.

3. Place both tubes in a 500-ml beaker of crushed ice.

4. Using a sterile inoculation loop, obtain a large mass of cells approximately 5 mm in size (about the size of a pencil eraser) from the *E. coli* starter plate, and inoculate the tube labeled "DNA+." *Note: Be sure to immerse the loop directly into the CaCl₂ and shake the loop vigorously to dislodge the inoculum. Discard the plastic loops in the beaker labeled "waste" or sterilize the wire loop by flaming it.*

5. Disperse the cells by gently tapping the tube with your finger until a uniform milky-white translucent cell suspension is obtained.

6. Repeat Steps 4 and 5 to inoculate the tube marked "DNA−," using an equal amount of inoculum and a sterile inoculating loop.

7. Using a sterile pipette, deliver 10 µl (0.01 ml) of pBLU plasmid into the DNA+ tube. Tap the tube several times with your finger to ensure complete mixing of the plasmid and cell suspension. *Note: Discard the plastic tip or*

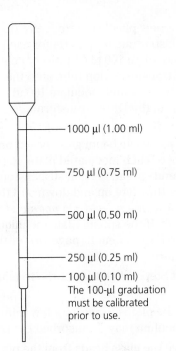

1000 µl (1.00 ml)

750 µl (0.75 ml)

500 µl (0.50 ml)

250 µl (0.25 ml)

100 µl (0.10 ml)
The 100-µl graduation must be calibrated prior to use.

Figure 56.4 Plastic transfer pipette

disposable pipette into a beaker containing disinfectant solution.

8. Return the DNA+ tube to the crushed-ice beaker and incubate for 15 minutes. During this time, label your agar plates as described in Step 9.

9. Label the 8 LB agar plates as follows:

   Plate 1: LB+

   Plates 2, 3: LB/Amp+

   Plates 4, 5: LB/Amp/X-Gal+

   Plate 6: LB−

   Plate 7: LB/Amp−

   Plate 8: LB/Amp/X-Gal−

10. Remove both tubes from ice after 15 minutes; place them in a test tube rack and immediately into a 42°C water bath with *gentle agitation* for 90 seconds (heat shocking).

11. Return both tubes to the crushed-ice beaker for 1 minute.

12. With a sterile pipette, add 250 µl (0.25 ml) of LB broth to both the DNA+ and DNA− tubes. Tap the tubes with your finger to achieve uniform cell suspension. (These are the transformation tubes.)

13. Incubate both tubes in a test tube rack for 10 minutes. *Note: This is the **recovery period**, when the cells convert their newly modified genotype into a functionally ampicillin-resistant phenotype.*

14. Using a new plastic micropipette tip or disposable plastic transfer pipette for each inoculation, inoculate 100 µl (0.1 ml) of cells from the DNA+ transformation tube onto the surface of LB plates 1–5, and inoculate 100 µl (0.1 ml) of cells from the DNA– transformation tube onto plates 6–8.

15. Place six sterile 6-mm glass beads on the surface of each inoculated plate. Replace the cover and spread the cell suspension by gently moving the plate up and down and then side to side a few times. *Note: Do not swirl or rotate the plate.* If the spread-plate technique is used, refer to Experiment 2 (page 15) for the correct procedure.

16. Repeat Steps 14 and 15 for the remaining plates.

17. Allow the plates to set for a few minutes so the inoculum may be absorbed by the agar.

18. Remove the glass beads from the plate by lifting the cover slightly while holding the plate vertically over the beaker of disinfectant, allowing the beads to leave the plate. *Note: This step may be eliminated if the spread-plate procedure is used.*

19. Incubate all plates at 37°C for 24 to 36 hours or at room temperature for 48 to 72 hours.

Procedure Lab Two

1. In the Lab Report, predict whether each plate will experience growth or no growth. Use a plus (+) sign for growth and a minus (–) for no growth.

2. Without removing the cover of the Petri plates, observe the colonies through the bottom of each plate.

3. Perform a colony count on each plate using an electronic colony counter if available, or use a permanent marker to mark each colony on the bottom of the plate as it is counted. Plates with more than 300 colonies should be designated as **TNTC (too numerous to count)**; plates with fewer than 30 colonies are designated as **TFTC (too few to count)**. Record your results in the Lab Report.

4. For each plate, did transformation occur? Record your results in the Lab Report.

Name: _____

Date: _____   Section: _____

Observations and Results

Plate Number	Designation	Growth + or −	Transformation Yes or No	Number of Colonies
1	LB+			
2	LB/Amp+			
3	LB/Amp+			
4	LB/Amp/X-Gal+			
5	LB/Amp/X-Gal+			
6	LB−			
7	LB/Amp−			
8	LB/Amp/X-Gal−			

Determine the sum of the colonies on the two experimental plates for LB/AMP+ and the two LB/AMP/X-Gal+ plates. Determine the averages and use these figures to calculate the transformation efficiency, using the following protocol.

Avg. LB/Amp+ colonies _____   Avg. LB/Amp/X-Gal+ colonies _____

a. The concentration of the plasmid used in this experiment was 0.005 μg/μl. Keep in mind that you used 10 μl.

Total mass of plasmid = volume × concentration

Total mass of plasmid = _____

b. Calculate the total volume of the cell suspension.

Total volume = volume of $CaCl_2$ solution + volume of plasmid + volume of LB broth

Total volume = _____

c. Calculate the fraction of cell suspension spread on each plate.

Fraction spread = volume of suspension spread on plate/total volume

Fraction spread = _____

d. Calculate the mass of DNA plasmid in the cell suspension spread on each plate.

Mass of DNA plasmid spread = total mass of plasmid × fraction spread

Mass of DNA plasmid spread = _____

e. Calculate the transformation efficiency (the number of colonies per µg of plasmid DNA).

Transformation efficiency = average of colonies counted/mass of plasmid spread

Express your results using scientific notation, as found in Appendix 1.

TRANSFORMATION EFFICIENCY	
LB/Amp+	**LB/Amp/X-Gal+**

Review Questions

1. Why do you think that it is necessary to autoclave transformed cells immediately after the termination of the experiment?

2. How would you explain to an untrained neighbor the process scientists use to change the genes (genotype) of an organism to give it useful new traits?

3. At a neighboring lab bench, students find that there are a few white colonies scattered among a few hundred blue colonies on their LB/Amp/X-Gal+ plates.

   a. How would you explain this result?

   b. To confirm your hypothesis, you ask to see their LB/Amp/X-Gal– plate. What do you see that supports your hypothesis?

Isolation of Bacterial Plasmids

LEARNING OBJECTIVES

Once you have completed this experiment, you should be able to

1. Isolate plasmids from plasmid-bearing bacteria.

2. Separate plasmids using agarose gel electrophoresis.

3. Compare electrophoretic mobilities of plasmids.

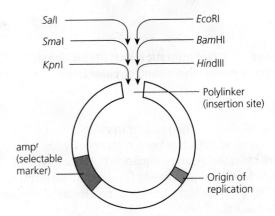

Figure 57.1 Illustration of a plasmid, showing a selectable marker and a multiple cloning site

Principle

Isolating pure DNA is a necessary step in studies that incorporate cloning, gene sequencing, gene mapping, or any other recombinant DNA technique. Many microorganisms contain small pieces of circular DNA called **plasmids** that exist separately from the host-cell genome. In studies that use recombinant DNA techniques, plasmid DNA is often preferred over chromosomal DNA because it is smaller and easier to manipulate. Plasmids that are commonly used as cloning vectors possess three important elements: an origin of replication, which allows the plasmid to be replicated independently of the cell's chromosome; a selectable marker, so the presence of the plasmid in the cell can be detected; and a cloning site into which a gene can be inserted. Although a selectable marker is not a required element of a plasmid, it can be useful in order to signal that the plasmid has been incorporated into the host cell. Plasmid genes that code for resistance to antibiotics are able to confer this resistance to the host cell, and a test of the host cell reveals that the transfer has occurred. In fact, antibiotic-resistance genes are among the most commonly used selectable markers.

Because plasmids are circular and capable of self-replication, they are able to serve as vectors for transportation of cloned fragments of DNA

into other cells for genetic engineering purposes. To do this, plasmids must have a multiple cloning site, or **polylinker**, which is a DNA segment with several unique insertion sites for restriction endonucleases located next to each other, as shown in Figure 57.1.

Gel electrophoresis is a technique used to separate different sizes of DNA fragments from a sample of DNA. Because DNA is negatively charged, when a sample is loaded into a porous agarose gel and subjected to an electric current with a positive charge at one end of the gel and a negative charge at the other, DNA fragments will migrate through the pores in the gel, toward the positively charged end. Different-sized fragments of linear and circular DNA move through the gel at different speeds, thus traveling different distances in the gel over a set time period. Larger, longer pieces snake their way through the pores more slowly, while shorter, smaller pieces move more quickly and travel further toward the positively charged end of the gel.

There are two types of circular DNA: closed and nicked. Closed circular DNA has all of its nucleotides linked with phosphodiester bonds and is supercoiled. Nicked circular DNA has at least one broken phosphodiester linkage. Nicked DNA is sometimes referred to as "relaxed" because

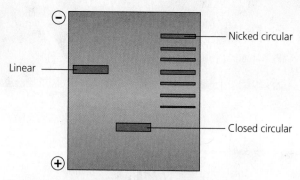

Figure 57.2 **Gel migration pattern for linear, nicked circular, and closed circular DNA**

Labels on figure: (−), Linear, Nicked circular, Closed circular, (+)

some of the tension present in covalently coiled and twisted DNA has been released. **Figure 57.2** illustrates the relative distance that each type of DNA described previously (linear, closed circular, and nicked circular) travels in an electrophoresed agarose gel.

Isolating plasmids is a multistep process, which involves rupturing a plasmid-bearing bacterium, using a variety of reagents to remove cellular components, and suspending plasmid DNA in an aqueous solution. After a plasmid-bearing organism is cultured, cells are lysed using alkali to release the plasmid DNA. The cellular debris is then precipitated by using a detergent and potassium acetate. Following centrifugation, the pellet that forms is removed, and alcohol is added to the supernatant to precipitate the DNA. The DNA precipitate is resuspended in Tris-EDTA buffer.

In the following experiment, two different strains of plasmid-bearing *Escherichia coli* will be used—*E. coli*-1 and *E. coli*-2. Half of the class will isolate *E. coli*-1 plasmid DNA, and the other half will isolate *E. coli*-2 plasmid DNA.

CLINICAL APPLICATION

Plasmids and Genetic Engineering

Plasmids are mostly found in bacteria and are used in recombinant DNA research to transfer genes between cells. Plasmids that confer antibiotic resistance (R Plasmids) have been of special interest because of their medical importance, and also because of their significant role in genetic engineering.

AT THE BENCH

Materials

Cultures

24-hour Luria-Bertani broth, 50 µg/ml of ampicillin, cultures of plasmid-bearing *Escherichia coli* ATCC 39991 (plasmid designation pIVEV) and *Escherichia coli* ATCC 53100 (plasmid designation pDGR-2).

Reagents

Glucose-Tris-ethylenediaminetetraacetic acid (EDTA) buffer, Tris-EDTA buffer, Tris-acetate-EDTA buffer, 5M potassium acetate (KOAc), sodium hydroxide containing 1% sodium dodecyl-sulfate (NaOH/SDS), 95% ethanol at 0°C, 70% ethanol, molten agarose at 55°C, gel electrophoresis running dye, Carolina Blu stain or 0.025% methylene blue stain, and *Hind*III-cut bacteriophage lambda (λ) DNA (used as the standard for comparing fragment sizes).

Note: The formulation for some of these buffers and reagents may be found in Appendix 4. Although it is cheaper to prepare your own solutions, your instructor may have ordered a kit containing premade solutions.

Equipment

Microcentrifuge, 2-ml microcentrifuge tubes, digital micropipette—10, 100, and 200 µl, small and large micropipette tips, water bath, rubber micro test tube racks capable of floating, glassware marking pencil, ice bucket, crushed ice, a light box or overhead projector, millimeter ruler, agarose gel casting tray, staining tray, plastic sandwich size bags, and an electrophoretic apparatus.

Procedure
Using a Micropipette

Before the start of the experiment, familiarize yourself with the use of a micropipette, the function of which is to accurately deliver microliter volumes of solution. Not all micropipettes work the same way. Some are designed to deliver a fixed volume, while others can deliver variable volumes. Your instructor will demonstrate the proper handling and use of these expensive instruments.

Using samples of colored water, practice using a micropipette, attaching different-sized micropipette tips and delivering various sample volumes to digestion tubes.

1. Set the scale on the pipette to the volume you wish to deliver.

2. Place a tip on the micropipette by pushing it firmly onto the pipette.

3. Depress the plunger to the first stop. This is necessary to remove all of the air from the tip.

4. To load the pipette, dip the pipette tip into the solution and release the plunger slowly to draw up the solution.

5. Touch the end of the tip to the side of the tube to remove any excess solution.

6. To deliver the solution, touch the side of the micropipette tip to the inside of the tube receiving the solution to produce a capillary.

7. Depress the plunger to the *first stop* and then continue depressing the plunger to the *second stop* to deliver the full volume of sample, blowing out the last bit in the tip.

8. Continue depressing the plunger while you remove the pipette tip from the tube.
Note: Releasing the plunger before removing the tip of the pipette from the tube will cause fluid to suck back into the tip.

Before the laboratory session, *E. coli*-1 and *E. coli*-2 were inoculated in their medium and grown overnight. Before the start of the lab, your instructor dispensed 1 ml of culture into a microcentrifuge tube and spun it for 1 minute in a centrifuge. The supernatant was discarded, and the pellet retained. Another 1 ml of culture was added to the tube, and the process was repeated.

Isolating the Plasmid

Obtain a microcentrifuge tube from your instructor with a retained pellet labeled "EC-1" or "EC-2." With a glassware marking pencil, label the tube with your group name or number to identify it later. Refer to **Figure 57.3** as you complete the following steps.

1. Add 100 μl of GTE (glucose, Tris, and EDTA) buffer to your tube and resuspend the pelleted cells by tapping with your finger or mixing by vortex. *Note: The EDTA in the buffer chelates the divalent metal ions, Ca^{2+} and Mg^{2+}, which destabilizes the cell membrane and*

PROCEDURE

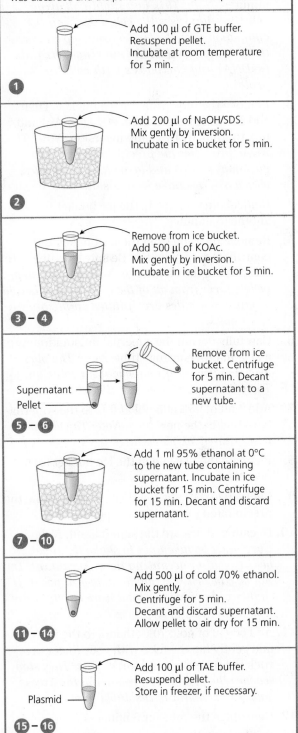

Prior to the lab, your instructor centrifuged 1 ml of *E. coli* culture for 1 minute. The supernatant was discarded and the pellet retained. Another 1 ml of *E. coli* culture was added to the pellet. It was centrifuged for 1 minute. The supernatant was discarded and the pellet retained for this experiment.

1. Add 100 μl of GTE buffer. Resuspend pellet. Incubate at room temperature for 5 min.

2. Add 200 μl of NaOH/SDS. Mix gently by inversion. Incubate in ice bucket for 5 min.

3 – 4. Remove from ice bucket. Add 500 μl of KOAc. Mix gently by inversion. Incubate in ice bucket for 5 min.

5 – 6. Supernatant / Pellet. Remove from ice bucket. Centrifuge for 5 min. Decant supernatant to a new tube.

7 – 10. Add 1 ml 95% ethanol at 0°C to the new tube containing supernatant. Incubate in ice bucket for 15 min. Centrifuge for 15 min. Decant and discard supernatant.

11 – 14. Add 500 μl of cold 70% ethanol. Mix gently. Centrifuge for 5 min. Decant and discard supernatant. Allow pellet to air dry for 15 min.

15 – 16. Plasmid. Add 100 μl of TAE buffer. Resuspend pellet. Store in freezer, if necessary.

Figure 57.3 Procedure for isolating bacterial plasmid DNA

inhibits the activity of DNases. The glucose maintains the osmolarity, preventing the buffer from bursting the cell.

2. Add 200 μl of NaOH/SDS solution and mix gently by inversion about four to five times. Incubate the tube in an ice bucket for 5 minutes. *Note: This is a highly alkaline solution that lyses the cell, releasing the cytoplasm into the buffer, and separates the chromosomal DNA into single strands (ssDNA) and complexes with cellular proteins.*

3. Remove the tube from the ice bucket. Then add 500 μl of potassium acetate (KOAc) and mix thoroughly by gentle inversion. *Note: The KOAc promotes the precipitation of chromosomal ssDNA and large RNA molecules, which are insoluble in this salt.*

4. Reincubate the tube in the ice bucket for another 5 minutes.

5. Remove the tube from the ice bucket and centrifuge for 5 minutes. Be sure the tubes are balanced in the centrifuge. *Note: In this step, pellets form from all of the cellular debris and organic molecules precipitated in the previous steps.*

6. Carefully decant the supernatant solution into a new microcentrifuge tube. *Note: The plasmid remains in the supernatant solution. The pellet and the tube are discarded.*

7. Add 1 ml of 95% ethanol at 0°C to the supernatant fluid in the new tube. *Note: The ethanol precipitates the plasmid.*

8. Incubate the plasmid in the ice bucket for 15 minutes.

9. Centrifuge the tube for 15 minutes to make the precipitated plasmid form a pellet.

10. Decant and discard the supernatant. *Note: Care must be taken not to shake the tube before or after decanting the supernatant. Do not be concerned if you do not see a pellet. It is there, provided that you were careful during the decanting step.*

11. Add 500 μl of cold 70% ethanol to the pellet and gently tap the tube with your finger or rock the tube back and forth. *Note: This step washes the plasmid by removing the excess salt. The plasmid is insoluble in ethanol.*

12. Centrifuge the tube for 5 minutes.

13. Decant and discard the supernatant fluid.

14. Allow the pellet to dry for about 15 minutes until you no longer smell alcohol.

15. Add 100 μl of TAE (Tris-acetate-EDTA) buffer to resuspend the pellet.

16. The plasmid may be placed in the freezer until the next lab class, or you may proceed to the electrophoresis step. *Note: If the electrophoresis is to be done during this class period, practice loading and casting the gel, which is described next.*

Casting the Agarose Gel

Note: *Not all casting trays are the same. Your instructor will indicate which type will be used and whether there are special considerations during the setup.*

Your instructor prepared the 0.8% agarose gel in a 1X TAE buffer solution before class and maintained it at 55°C in a water bath. It is ready to pour. One or two drops of Carolina Blu stain was added to the agarose buffer solution to give a small tinge of blue to the gel. At this concentration, the pores that form the gel lattice are such that they allow the free migration of the cut DNA fragments between 0.5 and 10 kilobases (kb).

1. Place the casting tray inside the casting tray box on a level surface.

2. Close off the ends of the tray with the rubber dams by tightening the knob on the top of the casting tray box.

3. Place a well-forming comb in the first notch at the end of the casting tray.

4. Pour 60 ml to 70 ml of agarose solution that has been cooled to 55°C into the tray. Use a toothpick or applicator stick to move the bubbles to the edge of the gel before it solidifies.

5. Allow the gel to solidify completely. It should be firm to the touch after 20 minutes.

6. Slowly remove the rubber dams and *very gently* remove the well-forming comb by pulling it straight up. *Note: Use extreme care not to damage or tear the wells.*

7. Place the gel on the platform in the electrophoresis box so that the formed wells are properly oriented toward the anode (negative pole with black cord). Because DNA is negatively charged, the cut DNA fragments will migrate to the cathode (positive pole with the red cord). Refer to **Figure 57.4** to see the proper setup of an electrophoretic apparatus.

8. Fill the electrophoresis box with TAE buffer to a level that just covers the gel, about 2 mm. Make sure that all of the wells are filled with the buffer.

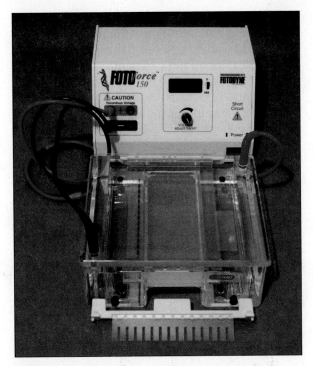

Figure 57.4 **Setup of agarose gel unit for DNA electrophoresis**

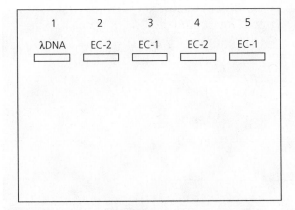

Figure 57.5 **Gel loading scheme**

Practicing Gel Loading

Before loading your sample into the wells of the agarose gel, practice this challenging technique. Your instructor will demonstrate the proper method for loading the wells. Each student should practice on a gel that has been prepared earlier by your instructor, not on the gel to be used for running the samples.

1. Load the pipette with 22 μl of loading gel.

2. Hold the pipette with both hands and dip the tip slightly through the buffer covering the gel, with the tip barely in the well.

3. Slowly discharge the contents of the pipette. *Note: The loading gel contains sucrose, which is heavier than the DNA and will pull the sample into the well.*

4. Practice the technique until you are comfortable with it.

Electrophoresing of the Plasmids

1. Add 18 μl of plasmid in the 1X TAE buffer to a new microcentrifuge tube. Then add 4 μl of gel electrophoresis running dye to the tube.

2. Add 18 μl of *Hind*III-cut lambda (λ) DNA and 4 μl of the gel electrophoresis running dye. With your glass marking pencil, label this tube

*Hind*III. This DNA has been cut into six linear fragments with the *Hind*III restriction enzyme. The fragments (bands) are various sizes: 23 kb, 9.4 kb, 6.6 kb, 4.4 kb, 2.3 kb, and 2.0 kb.

3. When the wells are ready to be loaded, make a diagram so that you will know the position of your sample in the agarose gel.

4. Fill the wells by designating the EC-1 samples as odd-numbered groups and EC-2 samples as even-numbered groups, as shown in Figure 57.5.

5. After the samples are loaded into the wells, place the lid on the electrophoresis gel box. Check that the power switch is turned to the "off" position and then attach the electrical leads (red to red and black to black) from the power supply to the box.

6. Turn the power pack on and adjust the rheostat dial to 110V.

7. Electrophorese the gel for 30 to 40 minutes or until the leading edge of the bromphenol blue dye (the dye in the loading gel) has traveled roughly three-fourths of the distance to the edge of the gel.

8. Turn the rheostat to zero and turn off the power. Disconnect the leads and remove the cover from the gel box.

Staining the Gel

1. Put on a pair of disposable laboratory gloves.

2. Lift the gel tray out of the electrophoresis box, and slide the gel into a staining tray containing approximately 100 ml of Carolina Blu stain or 0.025% of methylene blue stain.

3. Allow the gel to remain in the stain for 30 to 40 minutes.

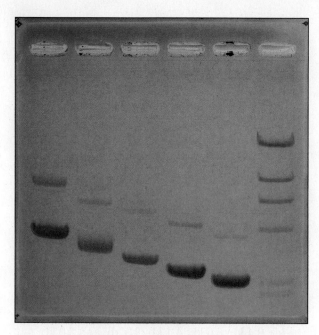

Figure 57.6 **Agarose gel stained with methylene blue following the electrophoresis of plasmid DNA**

4. Pour off the stain into a waste beaker. Transfer the gel to a staining tray containing 100 ml of distilled water and allow the gel to decolorize (destain) for another 30 minutes. Frequent destaining with fresh distilled water for 2 minutes increases the intensity of the bands. For best results, let the gel destain overnight in a small volume of water. *Note: If the gel is left overnight in a large volume of water, it may destain too much.*

5. Pour off the water, carefully remove the gel from the staining tray, and place it in a plastic sandwich-size bag or wrap it in a piece of clear plastic wrap. *Note: Be careful to keep the gel flat as you place it in the bag or plastic wrap.*

6. The gel can be placed in the refrigerator until the next lab period.

Figure 57.6 shows an agarose gel stain with methylene blue following electrophoresis of plasmid DNA.

Name: _____

Date: _____ Section: _____

Observations and Results

1. Tape a millimeter ruler to your light box or to the glass on your overhead projector. If you prefer, you may use a millimeter ruler to measure the migration of the plasmid.

2. Align your gel so that the front end of the well is set at the zero point on the ruler.

3. Measure the migration distances from the front of the well to the front edge of the band and record the distance in the following chart.

4. Prepare a standard curve on the semilog paper provided on page 405 by plotting the distance traveled in millimeters on the x-axis versus the size of the fragment of *Hind*III-cut λ DNA in kilobases. Record your results in the following chart.

   Note: 0.8% agarose gel has pore sizes that will allow the free movement of nucleic acids between 0.5 and 10 kb. Therefore, draw the best-fit straight line for all bands except the 23-kb band.

Migration Distances of Lambda DNA						
Kilobases	23	9.4	6.6	4.4	2.3	2.0
Millimeters						

5. Determine the number of bands in each plasmid and use a ruler to measure the migration distance in centimeters.

Band #	1	2	3	4	5	Total
E. coli-1						
E. coli-2						

6. Draw a diagram of your agarose gel and indicate which bands are linear, closed circular, or nicked circular.

7. Determine the size of linear DNA segments that would migrate the same distance as the various forms of the plasmids, using the standard curve. Record your results below:

	Plasmid DNA	
Linear DNA	E. coli-1	E. coli-2

Review Questions

1. Why is plasmid DNA preferred for genetic engineering studies?

2. What are selectable markers and why are they important to cloning vectors?

3. What is the rationale for using the following solutions for the isolation of plasmids?

   a. EDTA:

   b. Sodium dodecylsulfate (SDS):

   c. Potassium acetate:

(Questions continue on page 406.)

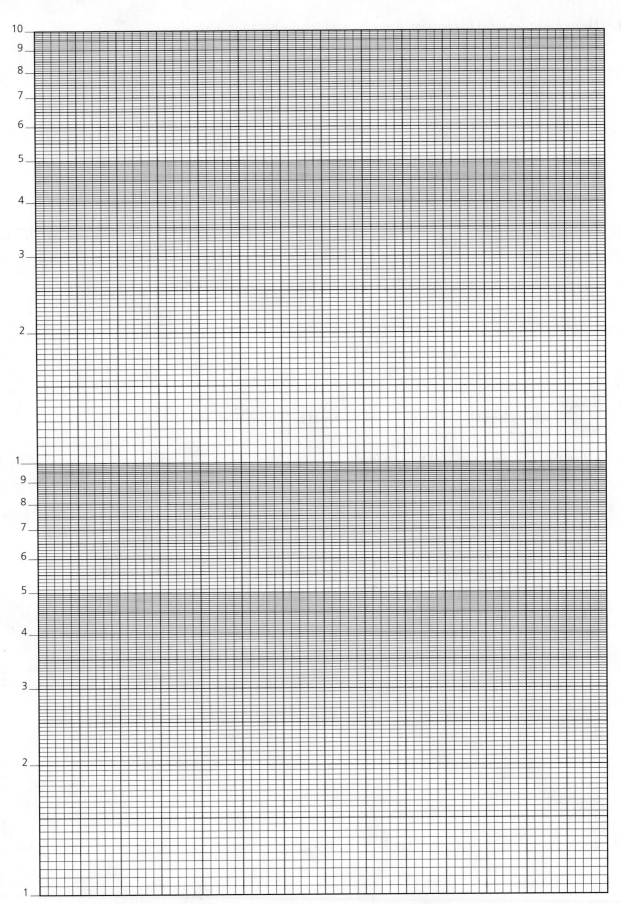

Migration Distance (mm) versus Fragment Size (kb)

4. Alcohol is considered to be a significant reagent for the isolation of nucleic acids (RNA and DNA). Why is this so?

5. What is nicked circular DNA and why is it termed "relaxed"?

6. When might you not be able to use a standard curve to determine the size of a plasmid?

7. When plasmids are isolated from bacterial cells, they may exist in a number of forms.

a. List the different forms that may be found.

b. Which do you think would migrate the fastest and farthest in an electrophoresis experiment and why?

Restriction Analysis and Electrophoretic Separation of Bacteriophage Lambda DNA

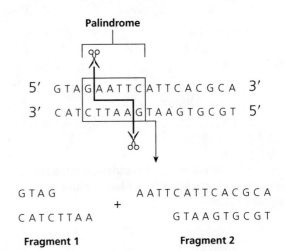

Figure 58.1 Palindrome for EcoRI endonuclease

LEARNING OBJECTIVES

Once you have completed this experiment, you should be able to

1. Cut DNA into small fragments by using restriction endonucleases.
2. Separate DNA fragments through agarose gel electrophoresis.
3. Determine the length of DNA fragments in kilobases.

Principle

Through their pioneering work, Werner Arber and Hamilton Smith discovered that bacteria possess enzymes that can act as molecular scissors and cut DNA molecules into smaller fragments. These enzymes, called **endonucleases**, are able to differentiate between DNA endogenous to an organism and foreign DNA, including infecting bacteriophages. Endonucleases can cut foreign DNA, rendering it nonfunctional, which spares the bacterium from infection. For scientists, the discovery of restriction endonucleases has been vital in advancing research over the past 40 years because small DNA fragments are much easier to manipulate than longer DNA strands. Now scientists can accurately map a gene's location on a chromosome and describe its base-pair sequence. **Restriction endonucleases** are also being used to develop DNA recombinants for commercial use, detect genetic defects, map restriction sites on plasmids, and create DNA profiles for use in medicine and forensics.

Endonucleases recognize palindromic sequences, four to six base pairs long, on DNA molecules. In everyday usage, a palindrome is a word that is spelled the same way forward and backward. For example, the word "racecar" is a common palindrome. On a double-stranded DNA molecule, a palindrome is a sequence of base pairs that reads the same on one strand 5′ to 3′ as

it does on the other strand 5′ to 3′. Keep in mind that reading 5′ to 3′ on one strand is the opposite direction of reading 5′ to 3′ on the other. Each endonuclease has its own unique restriction site. Figure 58.1 shows an example of a palindromic base-pair sequence and the cutting site for the restriction enzyme *Eco*RI. In the figure, *Eco*RI cuts the molecule between guanine and adenine, producing two fragments with staggered ends.

The key property of endonucleases is that they recognize and digest, or cut, one specific sequence of nucleotides on a DNA molecule and cut this same sequence every time. Several endonucleases make staggered cuts in the double-stranded molecule, producing single strands of DNA with cohesive, or sticky, ends that allows them to combine with complementary single-stranded DNA. Other endonucleases cut DNA sequences straight through both strands, producing blunt ends. Figure 58.2 illustrates the restriction sites of some commonly used endonucleases. The arrows indicate the cutting sites on each strand. The endonucleases that produce sticky, staggered ends are clearly distinguishable in Figure 58.2 from those that produce blunt ends.

DNA fragments cut with the same restriction enzyme can pair with one another. The sticky ends of different strands will join together because of

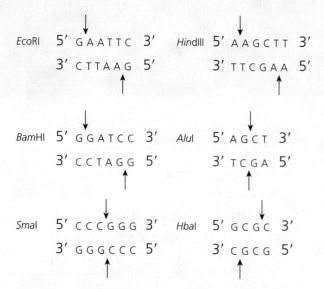

Figure 58.2 Illustration of restriction sites for common endonucleases that leave blunt and staggered ends

the formation of hydrogen bonds between complementary bases. However, joined fragments lack phosphodiester bonds between guanine and adenine, and nicks form as a result. These nicks are annealed by DNA ligase enzymes. Under optimum environmental conditions (salt concentration, pH, and temperature), restriction endonucleases will cut a strand of DNA into a number of varying-sized fragments. The exact number and size of the fragments depends on the location and number of restriction sites for the enzyme.

Restriction enzymes are named based on the genus and species of bacteria from which they are obtained. The first letter of the genus name is followed by the first two letters of the species name. For example, an endonuclease from *Escherichia coli* is named *Eco*. If a bacterium produces more than one restriction enzyme, each endonuclease is differentiated by Roman numerals. If the enzyme is coded for on a resistance factor, it is further designated with an "R." Thus *Eco*RI is one of several endonucleases produced by *E. coli* and is coded for on a restriction site. Other widely used endonucleases are obtained from *Haemophilus influenzae D* (*Hind*III, which cuts between adenine bases) and *Bacillus amyloliquefaciens H* (*Bam*HI, which cuts between two guanine bases).

In the following experiment, you will use endonucleases to cut bacteriophage lambda (λ) DNA, containing 48,502 base pairs (48.5 kb), into fragments. You will separate the fragments by using agarose gel electrophoresis, and you will determine the size of each.

CLINICAL APPLICATION

Enzyme Digestion to Isolate Genes of Interest

It was quickly recognized that restriction enzymes would provide a new tool for the investigation of gene organization, function, and expression. Current medical research is examining how restriction enzymes can be used to insert the genes required for insulin production into diabetic patients.

AT THE BENCH

Materials

DNA Source
Bacteriophage λ (200 µl).

Restriction Endonucleases
*Eco*RI, *Hind*III, and *Bam*HI.

Reagents
Tris-acetate buffer, type-specific buffers for *Eco*RI, *Hind*III, and *Bam*HI, electrophoresis loading dye, Carolina Blu or 0.025% methylene blue stain, and 0.8% agarose in 1X TAE buffer. *Note: Formulas for the preparation of type-specific buffers may be found in Appendix 4.*

Equipment
Plastic 1.5-ml microcentrifuge tubes, microcentrifuge, adjustable micropipettes (0.5 µl to 10 µl), (5 µl to 10 µl), (10 µl to 100 µl), large and small fine-point micropipette tips, water bath, ice bucket, crushed ice, staining tray, disposable gloves, glassware marking pencil, Bunsen burner or hot plate, 250-ml Erlenmeyer flask, beaker for waste, micro test tube racks, electrophoretic apparatus, millimeter ruler, and a light box or overhead projector.

Procedure
Note: In the following procedure, the steps for practicing micropipette use, casting a gel, and loading samples were covered in Experiment 57: Isolation of Bacterial Plasmids. If you completed these practice steps in Experiment 57, you may want to skip them here. If you are uncomfortable with any of the techniques, practice them again.

TABLE 58.1 Microliters (µl) per Digestion Tube

| TUBE | LAMBDA DNA | DH₂O | 10X RESTRICTION BUFFER | ENZYMES | | | TOTAL VOLUME |
				EcoRI	HinDIII	BamHI	
B	6	10	2	0	0	2	20
E	6	10	2	2	0	0	20
H	6	10	2	0	2	0	20
L	6	12	2	0	0	0	20

Using a Micropipette

Before the start of the experiment, familiarize yourself with the use of a micropipette, the function of which is to accurately deliver microliter volumes of solution. Not all micropipettes work the same way. Some are designed to deliver a fixed volume, while others can deliver variable volumes. Your instructor will demonstrate the proper handling and use of these expensive instruments.

Using samples of colored water, practice using a micropipette, attaching different-sized micropipette tips and delivering various sample volumes to digestion tubes.

1. Set the scale on the pipette to the volume you wish to deliver.

2. Place a tip on the micropipette by pushing it firmly onto the pipette.

3. Depress the plunger to the first stop. This is necessary to remove all of the air from the tip.

4. To load the pipette, dip the pipette tip into the solution and release the plunger slowly to draw up the solution.

5. Touch the end of the tip to the side of the tube to remove any excess solution.

6. To deliver the solution, touch the side of the micropipette tip to the inside of the tube receiving the solution to produce a capillary.

7. Depress the plunger to the *first stop* and then continue depressing the plunger to the *second stop* to deliver the full volume of sample, blowing out the last bit in the tip.

8. Continue depressing the plunger while you remove the pipette tip from the tube. *Note: Releasing the plunger before removing the tip of the pipette from the tube will cause fluid to suck back into the tip.*

Digesting of Lambda (λ) DNA

⚠ Be sure to wear gloves, as enzymes on your skin degrade DNA in the experiment.

1. Obtain a sample of lambda DNA from the instructor.

2. With a glassware marking pencil, label four microcentrifuge tubes with your name or group number followed by an "L" for the uncut DNA, "E" for EcoRI, "H" for HindIII, and "B" for BamHI.

3. Using a new pipette tip for each reagent, add the reagents to the digestion tubes in the following order:

   a. Lambda DNA

   b. Deionized or distilled water (dH₂O)

   c. Restriction enzyme buffer 10X

   d. Restriction endonucleases (10 units/µl)

   Note: The restriction enzyme must be added last to the digestion tubes. Addition of the endonucleases before DNA or buffer may inactivate the endonuclease. Each reagent is added with a new pipette tip to avoid contaminating the digestion tubes.

4. The addition of the reagents to each tube may be made following the scheme in Table 58.1.

5. Pulse centrifuge or tap your finger on each tube several times to mix the reagents.

6. Place all tubes in a foam rubber test tube rack or a suitable microcentrifuge rack and incubate them in the water bath at 37°C for 60 minutes.

7. The digestion tubes may be stored in the refrigerator until the next class period. If you are continuing with the experiment now, place the tubes in an ice bucket and proceed to the next step.

Casting the Agarose Gel

Note: *Not all casting trays are the same. Your instructor will indicate which type will be used and whether there are special considerations during the setup.*

Your instructor prepared the 0.8% agarose gel in a 1× TAE buffer solution before class and maintained it at 55°C in a water bath. It is ready to pour. One or two drops of Carolina Blu stain was added to the agarose buffer solution to give a small tinge of blue to the gel. At this concentration, the pores that form the gel lattice are such that they allow the free migration of the cut DNA fragments between 0.5 and 10 kb.

Refer to Figure 57.4 on page 401 to see the proper setup of an electrophoretic apparatus.

1. Place the casting tray inside the casting tray box on a level surface.
2. Close off the ends of the tray with the rubber dams by tightening the knob on the top of the casting tray box.
3. Place a well-forming comb in the first notch at the end of the casting tray.
4. Pour 60 ml to 70 ml of agarose solution that has been cooled to 55°C into the tray. Use a toothpick or applicator stick to move the bubbles to the edge of the gel before it solidifies.
5. Allow the gel to solidify completely. It should be firm to the touch after 20 minutes.
6. Slowly remove the rubber dams and *very gently* remove the well-forming comb by pulling it straight up. *Note: Use extreme care not to damage or tear the wells.*
7. Place the gel on the platform in the electrophoresis box so that the formed wells are properly oriented toward the anode (negative pole with black cord). Because DNA is negatively charged, the cut DNA fragments will migrate to the cathode (positive pole with the red cord).
8. Fill the electrophoresis box with TAE buffer to a level that just covers the gel, about 2 mm. Make sure that all of the wells are filled with the buffer.

Practicing Gel Loading

Before loading your sample into the wells of the agarose gel, practice this challenging technique. Your instructor will demonstrate the proper method for loading the wells. Each student should practice on a gel that has been prepared earlier by your instructor, not on the gel to be used for running the samples.

1. Load the pipette with 22 µl of loading gel.
2. Hold the pipette with both hands and dip the tip slightly through the buffer covering the gel, with the tip barely in the well.
3. Slowly discharge the contents of the pipette. *Note: The loading gel contains sucrose, which is heavier than the DNA and will pull the sample into the well.*
4. Practice the technique until you are comfortable with it.

Loading the DNA Digests into the Wells and Electrophoresing the Samples

1. Remove the digestion tubes from the ice bucket and add 4 µl of 6X loading dye to each tube.
2. Pulse centrifuge or tap your finger on each tube several times so that the contents of the tube move to the bottom.
3. Set the dial on the micropipette to deliver 24 µl (20 µl of restriction digests plus 4 µl of loading dye).
4. Deliver each of the four enzyme digests to separate wells in the agarose gel.
5. Remember the order of your samples and the position of each in the agarose gel. Because the gel cannot be marked, you should draw a diagram of the gel and label the position of your samples as shown in Figure 58.3.
6. After the samples are loaded into the wells, place the lid on the electrophoresis gel box. Check that the power switch is turned to the "off" position and then attach the electrical leads (red to red and black to black) from the power supply to the box.
7. Turn the power pack on and adjust the rheostat dial to 110V.
8. Electrophorese the gel for 30 to 40 minutes or until the leading edge of the bromphenol blue dye (the dye in the loading gel) has traveled

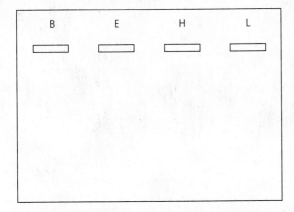

B E H L

Figure 58.3 **Example of gel loading scheme**

roughly three-fourths of the distance to the edge of the gel.

9. Turn the rheostat to zero and turn off the power. Disconnect the leads and remove the cover from the gel box.

Staining the Gel

1. Put on a pair of disposable laboratory gloves.

2. Lift the gel tray out of the electrophoresis box, and slide the gel into a staining tray containing approximately 100 ml of Carolina Blu stain.

3. Allow the gel to remain in the stain for 30 to 40 minutes.

4. Pour off the stain into a waste beaker. Transfer the gel to a staining tray containing 100 ml of distilled water and allow the gel to decolorize (destain) for another 30 minutes. Frequent destaining with fresh distilled water for 2 minutes increases the intensity of the bands. For best results, let the gel destain overnight in a small volume of water. *Note: If the gel is left overnight in a large volume of water, it may destain too much.*

5. Pour off the water, carefully remove the gel from the staining tray, and place it in a plastic sandwich-size bag or wrap it in a piece of clear plastic wrap. *Note: Be very careful to keep the gel flat as you place it in the bag or plastic wrap.*

6. The gel can be placed in the refrigerator until the next lab period. Refer to Figure 57.6, page 402, for a photo of an electrophoresed and stained gel.

Name: _____

Date: _____ Section: _____

Observations and Results

1. Tape a millimeter ruler to the surface of a light box or to the glass surface of an overhead projector.

2. Place the stained gel (inside plastic bag) next to the zero point on the ruler and measure the distances that each fragment (band) migrated. Measure the distance from the front of the well to the front of the band. Record your results in the following chart.

	Migration Distances of Fragments (mm)					
Uncut λ DNA						
*Bam*HI cut λ DNA						
*Eco*RI cut λ DNA						
*Hin*dIII cut λ DNA						

Linear DNA fragments migrate at rates inversely proportional to the $\log_{10}$ of their molecular weight and base-pair length.

3. A graph (standard curve) can be constructed by plotting known kilobase-pair fragments versus distances migrated from the wells to the front of the fragment. The six kilobase-pair fragment sizes for *Hin*dIII are well established and can be used to plot a standard curve.

4. Once the fragment sizes are measured and distances traveled are plotted on semilog paper found on page 415, a best-fit straight line can be drawn. The size of each unknown fragment can be determined by drawing a vertical line from the migration distance (mm) on the x-axis up to the point on the curve that intersects that straight line. From there, draw a horizontal line to the fragment size on the y-axis.

5. In the following table, the kilobase lengths of *Hind*III are provided. From the standard curve, use the migration distance you have measured to determine the base-pair lengths for the three restriction enzymes. Record your results in the table.

*Hind*III*		*Bam*HI		*Eco*RI		λ DNA	
Distance (mm)	Actual kb	Distance (mm)	Calculated kb	Distance (mm)	Calculated kb	Distance (mm)	Calculated kb
	27.4*						
	23.1*						
	9.4						
	6.6						
	4.4						
	2.3						
	2.0						

*Note: Remember that 0.8% agarose allows the free migration of DNA in the range of 0.5 to 10 kb. Therefore, the 27.4- and 23.1-kb fragments will not be detected.

6. Calculate the fragment lengths of *Eco*RI and *Bam*HI from the standard curve and compare them to the actual kilobase lengths listed in the following chart.

*Hind*III		*Bam*HI			*Eco*RI		
Actual kb	Distance (mm)	Actual kb	Distance (mm)	Calculated Length	Actual kb	Distance (mm)	Calculated Length
27.4*		16.8*			24.6*		
23.1*		12.3			21.2*		
9.4		7.2			7.4		
6.6		6.7*			5.8*		
4.3		6.5*			5.6*		
2.3		5.6*			4.9		
2.0		5.5*			3.5		

*Note: These fragments appear as a single band.

7. Compare and contrast your calculated kilobase pair from the standard curve with the actual kilobase pair for the restriction endonucleases.

a. List those that were most accurate.

b. List those that were least accurate.

(See Review Questions on page 416.)

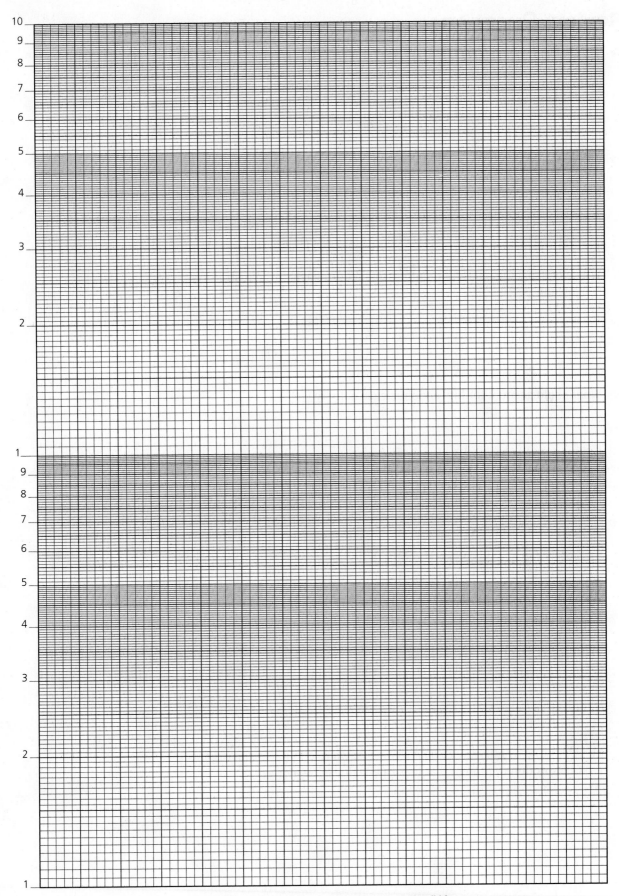

10
9
8
7
6
5
4
3
2
1
9
8
7
6
5
4
3
2
1

Migration Distance (mm) versus Fragment Size (kb)

Review Questions

1. Why were the DNA digestions carried out at 37°C?

2. Would any or all of these endonucleases cut the DNA of another bacteriophage or bacterium?

3. Why were specific restriction buffers needed for each of the restriction enzyme digests?

4. What could account for low endonuclease activity?

5. Why are the restriction enzymes added last to the digestion mixtures?

6. Assume you have one organism with a gene for ampicillin resistance and another organism with a gene for luciferinase. How would you isolate the gene from one organism and connect it with the gene of the other organism?

7. How would restriction enzymes play a role in having an organism produce a protein that it normally doesn't make?

PART 15

Medical Microbiology

LEARNING OBJECTIVES

Once you have completed the experiments in this section, you should be familiar with the

1. Characteristics and methodology for isolating and identifying selected pathogenic microorganisms.
2. Indigenous microbial flora of selected human anatomical sites.

Introduction

Although microorganisms are ubiquitous and their benefits to humans have been recorded, a small group of organisms remains a focus of concern: They are the pathogens, whose existence makes medical or clinical microbiology an especially important science.

That living agents are capable of inducing infections (*contagium vivum*) was first put forward by the monk Fracastoro in Verona about 500 years ago. In 1659, Kircher reported the presence of minute motile organisms in the blood of plague victims. Two hundred years after Fracastoro developed his initial concept, the germ theory of disease was formulated by Plenciz based on Leeuwenhoek's revolutionary microscopic observation of microorganisms. Perhaps the most important contributions to microbiology were made by Pasteur, Koch, and Lister during the **Golden Era of Microbiology** from 1870 to 1920. These investigators and their students recorded the observations and discoveries that cemented the cornerstone of medical microbiology. The body of knowledge that has accrued since these early years has made clinical microbiology a major

component of laboratory or diagnostic medicine. The major responsibility of this science is isolating and identifying infectious pathogens to enable physicians to treat patients with infectious disease prudently, intelligently, and rapidly.

Many of the experiments so far described have application in the field of clinical microbiology. Among these are isolation and identification of unknown cultures, the use of selective and differential media, and biochemical tests used to separate and identify various microorganisms. Although studying all of the bacterial pathogens responsible for human illness is not possible here, routine experiments for isolating and identifying some of the most frequently encountered infectious organisms and microorganisms that constitute the indigenous flora of the human body are included. The pathogens chosen are pyogenic cocci, members of the genera *Staphylococcus* and *Streptococcus*, the *Enterobacteriaceae*, and the organisms suspected in formation of dental caries. Experimental procedures designed for the detection and presumptive identification of microorganisms in blood and urine, which are normally sterile body fluids, have also been incorporated into this section. Organisms that naturally reside in or on body surfaces and constitute the body's **normal flora** are also examined.

The need for the expeditious detection and identification of pathogens has led to the development of rapid testing methods. These are microbiologically and immunologically based and can be performed quickly and without the need for sophisticated and expensive equipment. Some prototypic experiments using these rapid methods are included along with the traditional procedures.

⚠ Many of the organisms that are used, although attenuated by having been subcultured on artificial complex media for many generations, must be viewed as potential pathogens and therefore handled with respect. At this point in your training, your manipulative skills should be sufficiently developed, allowing you to perform aseptically in any medical, hospital, or clinical laboratory setting to prevent infection of yourself and others.

LEARNING OBJECTIVES·

Once you have completed this experiment, you should be

1. Familiar with the organisms responsible for dental caries.

2. Able to perform experiments that demonstrate the host's susceptibility to formation of caries.

Principle

A variety of microorganisms are known to be involved in the formation of dental caries, including *Lactobacillus acidophilus*, *Streptococcus mutans*, and *Actinomyces odontolyticus*. These organisms in the oral flora produce organic acids, particularly lactic acid, by fermenting carbohydrates that adhere to the surface of the teeth. In the continued presence of lactic acid, dental enamel undergoes decalcification and softening, which result in the formation of tiny perforations called dental caries.

The actual mechanism of action of these organisms is still unclear. However, it has been noted that *S. mutans* excretes an enzyme called **dextransucrase** (glycosyl transferase), which is capable of polymerizing sucroses into a large polymer, dextran, plus the monosaccharide fructose. This polysaccharide clings tenaciously to the teeth and forms dental plaque, in which streptococci reside and ferment fructose with the formation of lactic acid (Figure 59.1).

Similarly, *L. acidophilus* produces lactic acid as an end product of carbohydrate fermentation. Oral lactobacilli are capable of metabolizing glucose found in the mouth, producing organic acids that reduce the oral acid concentration to a pH of less than 5. At this pH, decalcification occurs and dental decay begins.

One of the best microbiological methods for determining susceptibility to dental caries is the **Snyder test**. This test measures the amount of acid produced by the action of the lactobacilli on glucose. The test employs a differential medium, Snyder agar (pH 4.7), which contains glucose and the pH indicator bromcresol green that gives the medium a green color.

Following incubation, Snyder agar cultures containing lactobacilli from the saliva will show glucose fermentation with the production of acid, which tends to lower the pH to 4.4, the level of acidity at which dental caries form. At this pH the green medium turns yellow. A culture demonstrating a yellow color within 24 to 48 hours is suggestive of the host's susceptibility to the formation of dental caries. A culture that does not change color is indicative of lower susceptibility.

CLINICAL APPLICATION

Preventing Dental Caries

Factors that help control the development of dental caries are proper oral hygiene, consumption of adequate fluoride, and moderation in the consumption of foods that cause decay. Foods likely to lead to decay are sticky, highly processed, and high in fermentable carbohydrates, such as breads, muffins and dried fruits. Also, the use of products to control oral pH might be helpful to ensure that bacteria that cause caries will not flourish.

AT THE BENCH

Materials

Cultures

Organisms of the normal oral flora present in saliva.

Media

Per designated student group: two Snyder test agar deep tubes.

Equipment

Microincinerator or Bunsen burner, ice-water bath, 1-in. square blocks of paraffin, sterile 1-ml pipettes, mechanical pipetting device, sterile test tubes, and glassware marking pencil.

Figure 59.1

Sucrose →(Sucrase)→ Glucose + Fructose

Sucrose **Glucose** **Fructose**

Glycosyl transferase

Dextran

Lactic acid

Figure 59.1 **Degradation of sucrose and subsequent conversion of glucose into dextran by** *Streptococcus mutans*

Procedure Lab One

1. Melt two appropriately labeled Snyder agar deep tubes and cool to 45°C.

2. Chew one square of paraffin for 3 minutes *without swallowing the saliva*. As saliva develops, collect it in a sterile test tube.

3. Vigorously shake the collected saliva sample and transfer 0.2 ml of saliva with a sterile pipette into one of the Snyder test medium tubes that have been cooled to 45°C. *Note: Don't let the pipette touch the sides of the tubes or the agar.*

4. Mix the contents of the tube thoroughly by rolling the tube between the palms of your hands or by tapping it with your finger.

5. Rapidly cool the inoculated tube of Snyder agar in an ice-water bath.

6. Repeat Steps 3 through 5 to inoculate the second tube.

7. Incubate both tubes for 72 hours at 37°C. Observe cultures at 24, 48, and 72 hours.

Procedure Lab Two

1. Examine the Snyder test cultures daily during the 72-hour incubation period for a change in the color of the culture medium. Use an uninoculated tube of the medium as a control. Figure 59.2 shows a positive and a negative Snyder test.

2. Record the color of the cultures in the Lab Report.

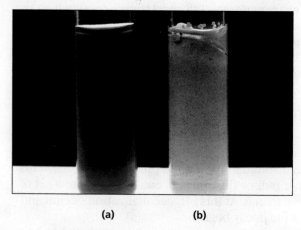

(a) **(b)**

Figure 59.2 **Snyder test. (a)** No change in the color indicates a negative result. **(b)** The color change to yellow indicates a positive result.

Name: _____

Date: _____ Section: _____

Observations and Results

Using Table 59.1 to interpret your observations, record your findings about susceptibility to caries in the chart below.

TABLE 59.1 Assessment of Susceptibility to Dental Caries

CARIES ACTIVITY	HOURS OF INCUBATION		
	24	48	72
Marked	Positive	...	...
Moderate	Negative	Positive	...
Slight	Negative	Negative	Positive
Negative	Negative	Negative	Negative

Source: Courtesy of Difco Laboratories, Inc., Detroit, Michigan.

Positive: Complete color change; green is no longer dominant.

Negative: No color change or a slight color change; medium retains green color throughout.

Tube Number	COLOR OF SNYDER TEST CULTURES			Caries Susceptibility (Yes or No)
	24 hr	48 hr	72 hr	

Review Questions

1. How would you explain the differential nature of the Snyder agar medium as used for the detection of dental caries?

2. How would you explain the mechanism responsible for the formation of dental caries by resident microorganisms?

3. What is the function of the paraffin in this procedure?

4. Based on your results, what is your tendency to form dental caries?

Is this result consistent with your dental history?

5. Are all members of the resident flora of the mouth capable of initiating dental caries? Explain.

6. What is the ideal time of day to perform this procedure? Why?

Normal Microbial Flora of the Throat and Skin

LEARNING OBJECTIVE

Once you have completed this experiment, you should be able to:

1. Identify microorganisms that normally reside in the throat and skin.

Principle

The normal flora are regularly found in specific areas of the body. This specificity is far from arbitrary and depends on environmental factors such as pH, oxygen concentration, amount of moisture present, and types of secretions associated with each anatomical site. Native microbial flora are broadly located as follows:

1. **Skin:** Staphylococci (predominantly *Staphylococcus epidermidis*), streptococci (alpha-hemolytic, nonhemolytic), enterococci, diphtheroid bacilli, yeasts, and fungi.

2. **Eye conjunctiva:** Staphylococci, streptococci, diphtheroids, and neisseriae.

3. **Upper respiratory tract:** Staphylococci, streptococci (alpha-hemolytic, nonhemolytic, and *Streptococcus pneumoniae*), enterocci, diphtheroids, spirochetes, and members of the genera *Moraxella* (formerly called *Branhamella*), *Neisseria*, and *Haemophilus*.

4. **Mouth and teeth:** Anaerobic spirochetes and vibrios, fusiform bacteria, staphylococci, and anaerobic levan-producing and dextran-producing streptococci responsible for dental caries.

5. **Intestinal tract:** In the upper intestine, predominantly lactobacilli and enterococci. In the lower intestine and colon, 96% to 99% is composed of anaerobes, such as members of the genera *Bacteroides*, *Lactobacillus*, *Clostridium*, and *Streptococcus*, and 1% to 4% is composed of aerobes, including coliforms, enterococci, and a small number of *Proteus*, *Pseudomonas*, and *Candida* species.

6. **Genitourinary tract:** Staphylococci, streptococci, lactobacilli, gram-negative enteric bacilli, clostridia, spirochetes, yeasts, and protozoa, such as *Trichomonas* species.

PART A   Isolation of Microbial Flora

In this exercise, you will study the resident flora of the throat and skin. Since these sites represent sources of mixed microbial populations, you will perform streak-plate inoculations, as outlined in Experiment 2 to effect their separations. The discrete colonies thus formed can be studied morphologically, biochemically, and microscopically to identify the individual genera of these mixed flora.

The procedure used to identify the native flora of the throat involves the following steps:

1. A *blood agar plate* is inoculated to demonstrate the alpha-hemolytic and beta-hemolytic reactions of some streptococci and staphylococci. Hemolytic reactions on blood agar are shown in Figure 60.1. A distinction between these two genera can be made based on their colonial and microscopic appearances. The streptococci typically form pinpoint colonies on blood agar, whereas the staphylococci form larger pinhead colonies that might show a golden coloration. When viewed under a microscope, the streptococcal cells form chains of varying lengths, whereas the staphylococci are arranged in clusters.

2. A *chocolate agar plate* is inoculated to detect *Neisseria* spp. by means of the oxidase test. Members of this genus are recognized when the colonies develop coloration that is pink to dark purple on addition of *p*-aminodimethylaniline oxalate following incubation. Figure 60.2 shows colonies growing on chocolate agar from a throat culture.

3. A *Mueller-Hinton tellurite* or *Tinsdale agar plate* is inoculated to demonstrate the presence of diphtheroids, which appear as black, pinpoint colonies on this medium (Figure 60.3).

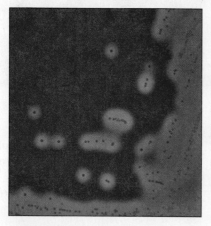
(a) Beta hemolysis

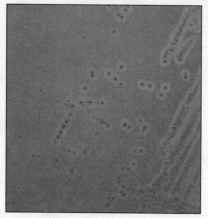
(b) Alpha hemolysis

Figure 60.1 **Beta- and alpha-hemolytic reactions on blood agar**

Figure 60.2 **Colony growth on chocolate agar from a throat culture**

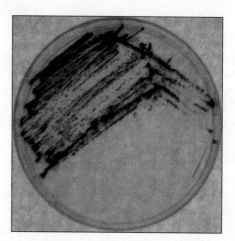

Figure 60.3 **Mueller-Hinton tellurite agar plate.** Growth of black, pinpoint colonies indicates the presence of diphtheroids.

This coloration is due to the diffusion of the tellurite ions into the bacterial cells and their subsequent reduction to tellurium metal, which precipitates inside the cells.

The procedure used to identify the native flora of the skin involves the following steps:

1. A *blood agar plate* inoculated to determine the presence of hemolytic microorganisms, specifically the staphylococci and streptococci: Differentiation between these two genera may be made as previously described.

2. A *mannitol salt agar plate* inoculated for the isolation of the staphylococci: The generally avirulent staphylococcal species can be differentiated from the pathogenic *Staphylococcus aureus* because the latter is able to ferment mannitol, causing yellow coloration of this

medium surrounding the growth. Figure 60.4 shows a fermenter and a nonfermenter organism on a mannitol salt agar plate.

3. A *Sabouraud agar plate* inoculated to detect yeasts and molds: Yeast cells will develop pigmented or nonpigmented colonies that are elevated, moist, and glistening. Mold colonies will appear as fuzzy, powdery growths arising from a mycelial mat in the agar medium. Figure 60.5 shows yeast colonies and a mold colony.

4. *Chocolate agar plate* inoculated to detect *Neisseria* spp. The presence of *Neisseria* spp. produces pink to purple to black colonies on this medium.

5. *Mueller-Hinton tellurite* or *Tinsdale media* inoculated to detect *Corynebacterium* spp. (diphtheroids). These colonies are black in appearance.

CLINICAL APPLICATION

Skin Flora and Acne

The bacterial population on a single human's skin is about 10^{12} organisms. A normal flora of microorganisms colonizes the human skin at birth as it passes through the birth canal, and typically inhabits the superficial layers of the epidermis and upper parts of the hair follicles. They consist mainly of *Staphylococcus epidermidis*, *Micrococcus*, and corynebacteria such as *Propionibacterium*. *Propionibacterium acnes* is normally found in low concentrations, but overgrows in the anaerobic environment of a blocked hair follicle, producing acne.

AT THE BENCH

Materials

Media

Per designated student group: two blood agar plates, two mannitol salt agar plates, one chocolate agar plate, one Mueller-Hinton tellurite or Tinsdale agar plate, one Sabouraud agar plate, and two 5-ml sterile saline tubes.

Reagents

Crystal violet, Gram's iodine, safranin, 1% *p*-aminodimethylaniline oxalate, and lactophenol–cotton-blue.

Equipment

Sterile cotton swabs, tongue depressors, desiccator jar with candle, microscope, glass slides, microincinerator or Bunsen burner, glassware marking pencil, and disposable gloves.

Procedure Lab One

⚠ **You must wear disposable gloves in Steps 1–3.**

Fermenter   Nonfermenter

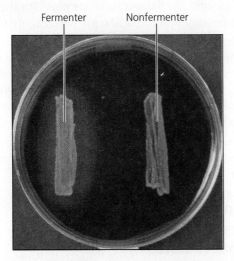

Figure 60.4 Mannitol salt agar plate showing a fermenter and a nonfermenter organism

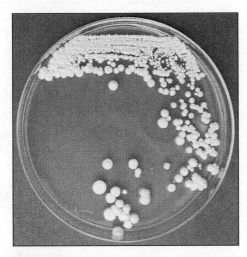

(a) Yeast colonies

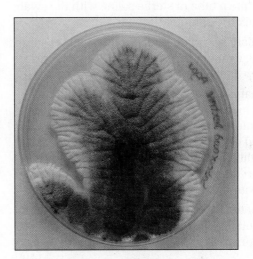

(b) Mold colony

Figure 60.5 Sabouraud agar plate. (a) Yeast colonies have an elevated, moist, and glistening appearance. **(b)** A mold colony shows fuzzy, powdery growth.

1. Place a tongue depressor on the extended tongue and with a sterile cotton swab, obtain a specimen from the palatine tonsil by rotating the swab vigorously over its surface without touching the tongue, as illustrated.

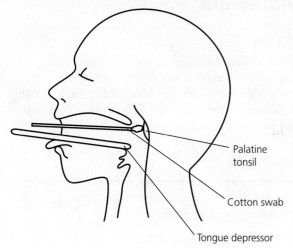

Palatine tonsil

Cotton swab

Tongue depressor

2. Inoculate a tube of sterile saline with the swab and mix until you have a uniform suspension.

3. Using a sterile inoculating loop, inoculate one plate each of blood agar, chocolate agar, mannitol salt agar, and Mueller-Hinton tellurite or Tinsdale agar, all previously labeled with the source of the specimen, by means of a four-way streak inoculation as described in Experiment 2.

4. Using a sterile cotton swab moistened in sterile saline, obtain a specimen from the skin by rubbing the swab vigorously against the palm of your hand.

5. Inoculate a tube of sterile saline with the swab and mix the solution.

6. Inoculate one plate each of blood agar, mannitol salt agar, and Sabouraud agar, as described in Step 3.

7. Incubate the inverted chocolate agar plate in a CO_2 incubator, in a CO_2 incubation bag, or in a candle jar. If you use the candle jar, place a lighted candle in a desiccator jar and cover the jar tightly to effect a 5% to 10% CO_2 environment required for the growth of the *Neisseria*. Incubate the jar for 48 hours at 37°C.

8. Incubate the inverted Sabouraud agar plate for 48 hours at 25°C and the remaining plates for 48 hours at 37°C.

Procedure Lab Two

Selection and Differentiation of Skin and Throat Isolates

1. Examine the blood agar plate cultures for zones of hemolysis (refer to Figure 60.1 and Experiment 13).

2. Add the *p*-aminodimethylaniline oxalate to the surface of the growth on the chocolate agar plate. Observe for the appearance of a pink-to-purple-to-black color on the surface of any of the colonies (Figure 60.2).

3. Examine the Mueller-Hinton tellurite or Tinsdale agar plate for the presence of black colonies (Figure 60.3).

4. Examine the Sabouraud agar plate for the appearance of mold-like growth (Figure 60.5).

5. Examine the mannitol salt agar plate for the presence of growth that is indicative of staphylococci. Then examine the color of the medium surrounding the growth. A yellow color is indicative of *S. aureus* BSL-2 (refer to Figure 60.4 and Experiment 13).

6. Record your observations in the Lab Report and indicate the types of organisms that may be present in each specimen.

Staining and Morphological Characteristics of Skin and Throat Isolates

1. Prepare two Gram-stained smears from each of the blood agar cultures, choosing well-isolated colonies that differ in their cultural appearances and demonstrate hemolytic activity. Observe microscopically for the Gram reaction and the size, shape, and arrangement of the cells. Record your observations in the Lab Report and attempt to identify each isolate.

2. Prepare two lactophenol–cotton-blue–stained smears of organisms obtained from discrete colonies that differ in appearance on the Sabouraud agar culture (refer to Experiment 34). Observe microscopically, draw a representative field in the Lab Report, and attempt to identify the fungi by referring to Experiment 36.

LEARNING OBJECTIVES

Once you have completed this experiment, you should understand

1. The difference between the residential flora and transient flora found on skin surfaces.
2. The effect of handwashing on the reduction of organisms on the skin.
3. The effectiveness of using soap alone or soap accompanied by surgical brushing.

Principle

Each day our hands come in contact with numerous objects and surfaces that are contaminated with microorganisms. These may include door handles, light switches, shopping carts, sinks, toilet seats, books, or even things like compost piles or body fluids, to name a few. The lack of adequate handwashing is a major vehicle in the transmission of microbial infection and disease.

The skin of a human being is sterile while *in utero* and first becomes colonized by a normal microbial flora at birth as it is passed through the birth canal. By the time you reach adulthood, your skin is calculated to contain 10^{12} (1,000,000,000,000), or one trillion, bacteria, most of which are found in the superficial layers of the epidermis and upper hair follicles. This normal flora of microorganisms is called the **resident flora**, the presence of which does not cause negative effects in healthy individuals. In fact, it forms a symbiotic relationship with your skin, which is vital to your health. This beneficial relationship can change in patients who are immunocompromised, or when residential flora accidently gains entrance to the host via inoculating needles, indwelling catheters, lacerations, and the like. Microorganisms that are less permanent and present for only short periods are termed **transient flora**. This latter flora can be removed with good handwashing techniques. The resident flora is more difficult to remove because they are found in the hair follicles and covered by hair, oil, and dead skin cells that obstruct their removal by simple handwashing with soap. Surgical scrubbing is the best means for removal of these organisms from the skin.

Surgical handwashing was introduced into medical practice in the mid-19th century by the Hungarian physician Ignatz Semmelweis while working at an obstetric hospital in Vienna. He observed that the incidence of puerperal fever (child birth fever) was very high, with a death rate of about 20%. He further observed that medical students examining patients and assisting in deliveries came directly from cadaver (autopsy) laboratories without stopping to wash their hands. On his insistence, medical students and all medical personnel were required to wash their hands in a chloride of lime (bleach) solution before and after all patient contact. The incidence of death from puerperal fever dropped drastically to around 1%. Semmelweis's effort was responsible for the development of routine surgical scrubbing by surgeons, which has become essential practice for all surgical procedures in modern medicine.

CLINICAL APPLICATION

Preventing Nosocomial Infections

Nosocomial (hospital-acquired) infections are mainly transmitted from the unwashed hands of healthcare providers. Transient and residential flora on health care providers' skin can infect hospital patients whose immune systems are compromised. The cornerstone for the prevention of nosocomial infections is the meticulous handwashing and scrubbing of healthcare personnel. In the laboratory setting, your normal flora may contaminate patient samples and skew your result, leading to a misdiagnosis. It is important for everyone in the lab to correctly wash their hands before and after handling biological materials.

AT THE BENCH

Materials

Media

Per designated student group: four nutrient agar plates

Equipment

Liquid antibacterial soap, 8 sterile cotton swabs, 2 test tubes of sterile saline, microincinerator or Bunsen burner, glass marking pencil, surgical hand brush, Quebec colony counter, stopwatch.

Procedure Lab One

1. One student will become the washer and the other student the assistant. **The washer must not wash hands before coming to the lab.**

2. Use the glass marking pencil to label the bottoms of all agar plates; one set of plates as "Water" and the second set of plates as "Soap" and draw a line down the middle of each plate to divide each plate in half. For the "Water" plates, label the halves as R1, R2, R3, and R4. For the "Soap" plates, label the halves as L1, L2, L3, and L4.

3. Aseptically dip a sterile cotton swab into the first test tube of sterile saline then rub the moistened cotton swab on the pad of the washer's **right** thumb.

4. Aseptically inoculate the half of the nutrient agar plate labeled R1 by streaking the far edge of the plate several times then making a zig zag streak only on the half labeled R1.

5. The assistant will turn on the tap on the lab sink, so that the washer can wash the right hand under warm running water, **without soap**, concentrating on the thumb (rubbing the thumb over the right index and middle finger) for one minute. The assistant will turn off the tap. The washer will shake off the excess water from the hand, but not blot dry. The assistant, using a new, dry (not moistened with saline) sterile cotton swab, will obtain a sample from the right thumb pad and inoculate the section of the nutrient agar plate labeled R2 in the same way that R1 was inoculated.

6. Repeat step 5 two more times, washing the thumb for 2 minutes and then 3 minutes, respectively. The assistant will use a new, dry sterile cotton swab each time, and will aseptically inoculate R3 and R4, respectively. See Table 60.1.

7. The assistant and washer will now move to the left hand. The assistant will aseptically dip the sterile cotton swab into the second test tube of sterile saline (following the process from Step 3) and will rub the moistened cotton swab over the pad of the left thumb and aseptically inoculate L1.

8. The assistant will turn on the tap of the lab's sink so that the washer can wet the thumb and index finger of the left hand under warm running water. The assistant will apply one or two drops of liquid soap to the thumb and index finger and the washer will wash for 1 minute by rubbing the thumb over the index finger. Rinse well. Shake off water from the hand but do not blot dry. The assistant will turn off the tap. The assistant will then use a dry, sterile cotton swab to obtain a sample from the washed thumb pad and inoculate L2.

9. Repeat step 8 two more times, not only using soap but also scrubbing the thumb with a surgical brush, for 2 minutes and then 3 minutes, respectively. The washer will obtain the surgical brush and the assistant will add saline to the brush to dampen it, and then add one or two drops of soap to the thumb and also the brush. *Caution: Place the brush bristles up on a dry paper towel between washings.* The assistant will use a new, dry sterile cotton swab each time, and will aseptically inoculate L3 and L4, respectively. Refer back to Table 60.1.

10. Incubate all plates in an inverted position at 37°C for 24 to 48 hours.

TABLE 60.1	Inoculation of Agar Plates		
	WATER—RIGHT THUMB		**SOAP—LEFT THUMB**
R1	No wash, damp cotton swab	L1	No wash, damp cotton swab
R2	Wash 1 minute, dry cotton swab	L2	Wash with soap 1 minute, dry cotton swab
R3	Wash 2 minutes, dry cotton swab	L3	Soap and surgical brush 2 minutes, dry cotton swab
R4	Wash 3 minutes, dry cotton swab	L4	Soap and surgical brush 3 minutes, dry cotton swab

Procedure Lab Two

Examine and record the amount of growth found on each nutrient agar plate. Results may be determined by two methods.

1. Macroscopically. Visually observe the presence of growth on the surface of each agar plate in each section. Record your results in your Lab Report as 0 no growth, 1+ = slight growth, 2+ = moderate growth, 3+ = heavy growth, and 4+ = maximum growth.

2. **Percent Growth Reduction.**

   a. Count the colonies that appear in each section of the agar plates using a Quebec colony counter. If more than 300 colonies are present, label it as "too numerous to count (TNTC)," if fewer than 30 colonies are present, label it as "too few to count (TFTC)."

   b. For sections R2, R3, R4 and L2, L3, L4, calculate the percent growth reduction from the first section, using the following equation:

   Percent reduction = [Colonies (section 1) − Colonies (section X)] ÷ Colonies (section 1)

   X = sections 2, 3, 4 for each hand

Name: _____

Date: _____ Section: _____

Observations and Results
Part A: Isolation of Microbial Flora
Selection and Differentiation of Skin and Throat Isolates

Cultures	Throat Specimen	Skin Specimen
Blood agar: *Staphylococcus* spp. *Streptococcus* spp. Type of hemolysis: alpha 　　　　　　　beta		
Chocolate agar: *Neisseria* spp. (+) or (−) pink to purple to black		
Mueller-Hinton tellurite or Tinsdale: *Corynebacterium* spp. (+) or (−) black colonies		
Sabouraud agar: Fungal colonies (+) or (−)		
Mannitol salt agar: *Staphylococcus aureus* Other *Staphylococcus* spp. (*S. epidermidis, S saprophyticus*) 　　　(+) or (−) growth Color of medium		
Types of organisms present		

Staining and Morphological Characteristics of Skin and Throat Isolates

Skin Specimen	Isolate 1	Isolate 2
Draw a representative field.		
Gram reaction		
Morphology		
Organism		
Throat Specimen	Isolate 1	Isolate 2
Draw a representative field.		
Gram reaction		
Morphology		
Organism		
Sabouraud Agar Colonies Specimen	Isolate 1	Isolate 2
Draw a representative field.		
Morphology		
Organism		

Part B: Effectiveness of Handwashing

1. Record the macroscopic observations in the chart below.

Section (Water—Right Thumb)	Time (mins)	Growth (0 = none, 1+ = slight, 2+ = moderate, 3+ = heavy, 4+ = maximum)	Section (Soap—Left Thumb)	Time (mins)	Growth (0 = none, 1+ = slight, 2+ = moderate, 3+ = heavy, 4+ = maximum)
R1	0		L1	0	
R2	1		L2	1	
R3	2		L3	2	
R4	3		L4	3	

2. Record the percent growth reduction in the chart below.

Section (Water—Right Thumb)	Time (mins)	Number of Colonies	Percent Reduction	Section (Soap—Left Thumb)	Time (mins)	Number of Colonies	Percent Reduction
R1	0		—	L1	0		—
R2	1			L2	1		
R3	2			L3	2		
R4	3			L4	3		

Review Questions

1. How does the presence of residential flora influence the infectious process?

2. Why are some microorganisms termed "normal flora," and of what value are they to the well-being of the host?

3. A 6-year-old female is taken to her pediatrician for a checkup. As the doctor takes the child's history, her mother reports that the child had a severe sore throat several weeks earlier that regressed without treatment. Upon examination the pediatrician notes that the child has a systolic heart murmur consistent with mitral insufficiency and suspects that she has rheumatic fever.
   a. How was the earlier pharyngitis related to the subsequent development of rheumatic fever?

b. Rheumatic fever is diagnosed on clinical and serological findings. What test should be done to diagnose rheumatic fever?

c. How are rheumatic fever patients treated?

4. A 35-year-old female underwent serious abdominal surgery involving extensive bowel resection. She was maintained postoperatively on a regimen of intravenous broad-spectrum antibiotics. Three days postoperative she spiked a fever without a clear source. She complains of vaginal discomfort. Blood cultures reveal the presence of an ovoid cell that reproduced by budding.
   a. Based on this observation, what do you think this organism is?

   b. Is it part of the normal flora in humans?

   c. How did the treatment with broad-spectrum antibiotics predispose the patient to infection with this organism?

5. Compare the effectiveness of handwashing with water, with soap, and with soap and surgical scrubbing.

6. How does handwashing affect residential versus transient flora?

LEARNING OBJECTIVES

*Once you have completed this experiment,
you should understand*

1. The medical significance of the
   staphylococci.

2. Selected laboratory procedures designed
   to differentiate among the major
   staphylococcal species.

Principle

The genus *Staphylococcus* is composed of both
pathogenic and nonpathogenic organisms. The
three major species are *S. aureus, S. saprophyticus,* and *S. epidermidis.* Strains of the last two
species are generally avirulent; however, under
special circumstances in which a suitable portal of
entry is provided, S. epidermidis may be the etiological agent for skin lesions and endocarditis, and
S. saprophyticus has been implicated in some urinary tract infections. **Figure 61.1** is a streak-plate
culture of *Staphylococcus aureus.*

Infections are primarily associated with *S.
aureus* pathogenic strains that are often responsible for the formation of abscesses, localized pus-
producing lesions. These lesions most commonly
occur in the skin and its associated structures,
resulting in boils, carbuncles, acne, and impetigo.
Infections of internal organs and tissues are not
uncommon however, and include pneumonia,
osteomyelitis (abscesses in bone and bone mar-
row), endocarditis (inflammation of the endocar-
dium), cystitis (inflammation of the urinary bladder), pyelonephritis (inflammation of the kidneys),
staphylococcal enteritis due to enterotoxin contamination of foods, and, on occasion, septicemia.
Strains of *S. aureus* produce a variety of metabolic end products, some of which may play roles
in the organisms' pathogenicity. Included among
these are coagulase, which causes clot formation;
leukocidin, which lyses white blood cells; hemolysins, which are active against red blood cells; and

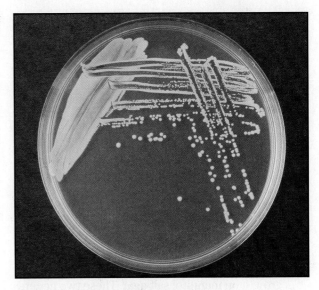

**Figure 61.1 Streak-plate culture of
Staphylococcus aureus.** Produces colonies that are
circular, convex, smooth, and cream-colored to
golden yellow in appearance.

enterotoxin, which is responsible for a type of gastroenteritis. Additional metabolites of a nontoxic
nature are DNAse, lipase, gelatinase, and the fibrinolysin staphylokinase.

When there is a possibility of staphylococcal infection, isolation of *S. aureus* is of clinical
importance. These virulent strains can be differentiated from other staphylococci and identified
by a variety of laboratory tests, some of which are
illustrated in **Table 61.1**.

In this exercise you will distinguish among the
Staphylococcus species by performing traditional
test procedures, a computer-assisted multitest procedure, or a newer rapid latex agglutination test.

Traditional Procedures

The traditional procedures involve the following
steps:

1. Mannitol salt agar: This medium is selective for
   salt-tolerant organisms such as staphylococci.
   Differentiation among the staphylococci is

TABLE 61.1 Laboratory Tests for Differentiation of Staphylococcal Species

TEST	S. AUREUS	S. EPIDERMIDIS	S. SAPROPHYTICUS
Mannitol salt agar			
Growth	+	+	+
Fermentation	+	–	–
Colonial pigmentation	Generally golden yellow	White	White
Coagulase	+	–	–
DNase	+	–	–
Hemolysis	Generally beta	–	–
Novobiocin sensitivity	Sensitive	Sensitive	Resistant

predicated on their ability to ferment mannitol. Following incubation, mannitol-fermenting organisms, typically *S. aureus* strains, exhibit a yellow halo surrounding their growth, and non-fermenting strains do not. (Refer to Figure 60.4.) It should be noted that other salt-tolerant micro-organisms, such as enterococci, are capable of growth on mannitol salt agar. These two genera are easily differentiated by performing a cata-lase test. *Staphylococcus* will grow in the presence of catalase while the enterococci will not.

2. Coagulase test: Production of coagulase is indicative of an *S. aureus* strain. The enzyme acts within host tissues to convert fibrinogen to fibrin. It is theorized that the fibrin meshwork that is formed by this conversion surrounds the bacterial cells or infected tissues, protecting the organism from nonspecific host resistance mechanisms such as phagocytosis and the antistaphylococcal activity of normal serum. In the coagulase tube test for bound and free coagulase, a suspension of the test organism in citrated plasma is prepared and the inocu-lated plasma is then periodically examined for fibrin formation, or coagulation. Clot formation within 4 hours is interpreted as a positive result and indicative of a virulent S. *aureus* strain. The absence of coagulation after 24 hours of incubation is a negative result, indicative of an avirulent strain (Figure 61.2).

3. Deoxyribonuclease (DNase) test: Generally, coagulase-positive staphylococci also produce the hydrolytic enzyme DNase; thus this test can be used to reconfirm the identification of S. aureus. The test organism is grown on an agar medium containing DNA. Following incubation,

DNase activity is determined by the addition of 0.1% toluidine blue to the surface of the agar. DNase-positive cultures capable of DNA hydro-lysis will show a rose-pink halo around the area of growth. The absence of this halo is indica-tive of a negative result and the inability of the organism to produce DNase (Figure 61.3).

4. Novobiocin sensitivity: This test is used to dis-tinguish *S. epidermidis* from *S. saprophyticus*. The Mueller-Hinton agar plate is heavily seeded with the test organism to produce a confluent growth on the agar surface. After the seeding, a 30-µg novobiocin antibiotic disc is applied to the agar surface. Following incubation, the sen-sitivity of an organism to the antibiotic is deter-mined by the Kirby-Bauer method as described in Experiment 42 and as shown in Figure 61.4.

STAPH-IDENT® System Procedure

A computer-assisted procedure is the API® (Analytical Profile Index) STAPH-IDENT® system (developed by Analytab Products, Division of Sherwood Medical, Plainview, New York). STAPH-IDENT is a rapid, computer-based micromethod for the separation and identification of the newly proposed 13 species of staphylococci. The system consists of 10 microcupules containing dehydrated substrates for the performance of conventional and chromogenic tests. The addition of a suspen-sion of the test organism serves to hydrate the media and to initiate the biochemical reactions. The identification of the staphylococcal species is made with the aid of the differential charts or the STAPH-IDENT Profile Register that is part of the system (Table 61.2 on page 439), or both.

(a) Positive coagulase test

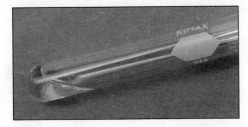

(b) Negative coagulase test

Figure 61.2 Coagulase test. (a) Clot formation indicates a positive result; **(b)** the absence of coagulation is a negative result.

Latex Agglutination Procedure

The latex agglutination test as a rapid diagnostic slide test for Staphylococcus aureus. The Remel BactiStaph® diagnostic kit (Fisher Health Care) uses protein-coated latex particles that are able to detect the clumping factor (bound coagulase and protein A) that causes the *S. aureus* to adhere to the black latex particles, producing a visible agglutination.

CLINICAL APPLICATION

Staphyloxanthin

Staphylococcus aureus is one of the most common species of staphylococci to cause human disease, producing many types of skin infections as well as life-threatening diseases like meningitis, osteomyelitis, endocarditis, and toxic shock syndrome. Its pathogenic success is due to its immune-evasive properties, mainly through the production of its yellow pigment staphyloxanthin. This pigment behaves as a virulence factor that helps the organism evade the immune system of the host. Blocking the synthesis of staphyloxanthin may present a unique and vital target for antimicrobials.

AT THE BENCH

Materials

Cultures

24-hour Trypticase soy agar slant cultures of *Staphylococcus epidermidis, Staphylococcus saprophyticus* (ATCC 15305), and *Staphylococcus aureus* (ATCC 27660) BSL-2 . Number-coded, 24-hour blood agar cultures of the above organisms for the STAPH-IDENT system.

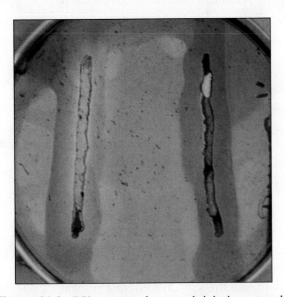

Figure 61.3 DNase test. A rose-pink halo around the area of growth on the left side of the plate indicates a positive result, while the absence of a halo on the right is a negative result.

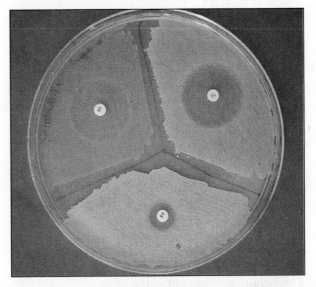

Figure 61.4 Novobiocin test. *Staphylococcus aureus* and *Staphylococcus epidermidis* (on top) are sensitive to the antibiotic, while *Staphylococcus saprophyticus* (on bottom) is resistant.

Media

Per designated student group: three mannitol salt agar plates, one DNA agar plate, three Mueller-Hinton agar plates, and the STAPH-IDENT system.

Reagents

Citrated human or rabbit plasma, 0.1% toluidine blue, 0.85% saline (pH 5.5–7.0), McFarland barium sulfate standards, and BactiStaph diagnostic kit (latex agglutination test).

Equipment

Microincinerator or Bunsen burner, inoculating loop, 13 × 100-mm test tubes, 15 × 150-mm test tubes, sterile Pasteur pipettes, 1-ml sterile pipettes, mechanical pipetting device, sterile cotton swabs, 30-μg novobiocin antibiotic discs, glassware marking pencil, metric ruler, forceps, and beaker with 95% ethyl alcohol.

Procedure Lab One
Traditional Procedures

1. Preparation of DNA agar plate culture:

   a. With a glassware marking pencil, divide the bottom of the plate into three sections. Label each section with the name of the organism to be inoculated.

   b. Aseptically make a single line of inoculation of each test organism in its respective sector on the agar plate.

2. Preparation of agar plate cultures for novobiocin sensitivity determination:

   a. Label the three Mueller-Hinton agar plates with the name of the test organism to be inoculated. Inoculate each plate with its respective organism according to the Kirby-Bauer procedure as outlined in Experiment 42.

   b. Using alcohol-dipped and flamed forceps, aseptically apply a novobiocin antibiotic disc to the surface of each inoculated plate. Gently press the discs down with sterile forceps to ensure that they adhere to the agar surface.

3. Preparation of mannitol salt agar plate cultures: Aseptically make a single line of inoculation of each test organism in the center of the appropriately labeled agar plates.

4. Incubation of all plate cultures: Incubate them in an inverted position for 24 to 48 hours at 37°C.

5. Coagulase test procedure:

   a. Label three 13 × 100-mm test tubes with the name of the organism to be inoculated.

   b. Aseptically add 0.5 ml of a 1:4 dilution of citrated rabbit or human plasma and 0.1 ml of each test culture to its appropriately labeled test tube.

   c. Examine the bacterial plasma suspensions for clot formation at 5 minutes, 20 minutes, 1 hour, and 4 hours after inoculation by holding the test tubes in a slanted position. Record your observations and results in the Lab Report.

   d. At the end of the laboratory session, place all tubes that are coagulase-negative in an incubator for 20 hours at 37°C.

STAPH-IDENT System Procedure

1. Prepare strip:

   a. Dispense 5 ml of tap water into incubation tray.

   b. Place API strip in incubation tray.

2. Prepare inoculum:

   a. Add 2 ml of 0.85% saline (pH 5.5–7.0) to a sterile 15 × 150-mm test tube.

   b. Using a sterile swab, pick up a sufficient amount of inoculum to prepare a saline suspension with a final turbidity that is equivalent to a No. 3 McFarland (BaSO₄) turbidity standard. Note: Be sure to use suspension within 15 minutes of preparation.

3. With a sterile Pasteur pipette, add 2 or 3 drops of the inoculum to each microcupule.

4. Place plastic lid on tray and incubate for 5 hours at 37°C.

Latex Agglutination Procedure

1. Label three of the provided slides (cards) with the name of the organism to be inoculated.

2. Place one drop of *Staphylococcus* latex reagent in the center of the circle on the provided slide.

3. Using an applicator stick or sterile needle, spread one colony of each organism in the reagent of its respective slide.

4. Spread the mixture over the entire circle.

5. Rotate the slide in a circular motion for 60 seconds.

6. Observe all slides for the presence or absence of agglutination. A positive agglutination

reaction usually occurs in 15 seconds and is indicated by a clumping together of the black latex suspension, followed by the loss of the black background. A negative reaction results in little or no agglutination and no loss of the black background within 60 seconds.

7. Record your results as positive (+) or as negative (–) in the chart provided in the Lab Report.

Procedure Lab Two

Traditional Procedures

1. Examine the coagulase-negative tubes, and record your observations in the Lab Report.

2. Examine the mannitol salt agar plate. Note and record the following in the Lab Report.

   a. Presence (+) or absence (–) of growth of each test organism.

   b. Color of the medium surrounding the growth of each test organism.

   c. Whether each test organism is a mannitol fermenter (+) or non–mannitol fermenter (–).

3. Flood the DNA agar plate with 0.1% toluidine blue. Observe for the delayed development of a rose-pink coloration surrounding the growth of each test organism. Record your color observation and indicate the presence (+) or absence (–) of DNase activity in the Lab Report.

4. With a metric ruler, measure the size of the zone of inhibition, if present, surrounding each of the novobiocin discs on the agar plates. A zone of inhibition of 17 mm or less is indicative of novobiocin resistance, whereas a zone greater than 17 mm indicates that the organism is sensitive to this antibiotic. Record the susceptibility of each test organism to novobiocin as sensitive (S) or resistant (R) in the Lab Report.

STAPH-IDENT System Procedure

1. Interpret your STAPH-IDENT system reactions on the basis of the observed color changes in each of the microcupules described in the chart in the Lab Report. Report your color observations and results as (+) or (–) for each test in the Lab Report.

2. Construct a four-digit profile for your unknown organisms using the guidelines provided in the Lab Report.

TABLE 61.2 API STAPH-IDENT Profile Register

PROFILE	IDENTIFICATION		PROFILE	IDENTIFICATION	
0 040	STAPH CAPITIS		2 000	STAPH SAPROPHYTICUS	NOVO R
0 060	STAPH HAEMOLYTICUS			STAPH HOMINIS	NOVO S
0 100	STAPH CAPITIS		2 001	STAPH SAPROPHYTICUS	
0 140	STAPH CAPITIS		2 040	STAPH SAPROPHYTICUS	NOVO R
0 200	STAPH COHNII			STAPH HOMINIS	NOVO S
0 240	STAPH CAPITIS		2 041	STAPH SIMULANS	
0 300	STAPH CAPITIS		2 061	STAPH SIMULANS	
0 340	STAPH CAPITIS		2 141	STAPH SIMULANS	
0 440	STAPH HAEMOLYTICUS		2 161	STAPH SIMULANS	
0 460	STAPH HAEMOLYTICUS		2 201	STAPH SAPROPHYTICUS	
0 600	STAPH COHNII		2 241	STAPH SIMULANS	
0 620	STAPH HAEMOLYTICUS		2 261	STAPH SIMULANS	
0 640	STAPH HAEMOLYTICUS		2 341	STAPH SIMULANS	
0 660	STAPH HAEMOLYTICUS		2 361	STAPH SIMULANS	
			2 400	STAPH HOMINIS	NOVO S
1 000	STAPH EPIDERMIDIS			STAPH SAPROPHYTICUS	NOVO R
1 040	STAPH EPIDERMIDIS		2 401	STAPH SAPROPHYTICUS	
1 300	STAPH AUREUS		2 421	STAPH SIMULANS	
1 540	STAPH HYICUS (An)		2 441	STAPH SIMULANS	
1 560	STAPH HYICUS (An)		2 461	STAPH SIMULANS	
2 541	STAPH SIMULANS		6 101	STAPH XYLOSUS	
2 561	STAPH SIMULANS		6 121	STAPH XYLOSUS	
2 601	STAPH SAPROPHYTICUS		6 221	STAPH XYLOSUS	

TABLE 61.2 API STAPH-IDENT Profile Register (continued)

PROFILE	IDENTIFICATION		PROFILE	IDENTIFICATION	
2 611	STAPH SAPROPHYTICUS		6 300	STAPH AUREUS	
2 661	STAPH SIMULANS		6 301	STAPH XYLOSUS	
2 721	STAPH COHNII (SSP1)		6 311	STAPH XYLOSUS	
2 741	STAPH SIMULANS		6 321	STAPH XYLOSUS	
2 761	STAPH SIMULANS		6 340	STAPH AUREUS	COAG+
				STAPH WARNERI	COAG−
3 000	STAPH EPIDERMIDIS		6 400	STAPH WARNERI	
3 040	STAPH EPIDERMIDIS		6 401	STAPH XYLOSUS	XYL + ARA+
3 140	STAPH EPIDERMIDIS			STAPH SAPROPHYTICUS	XYL − ARA−
3 540	STAPH HYICUS (An)		6 421	STAPH XYLOSUS	
3 541	STAPH INTERMEDIUS (An)		6 460	STAPH WARNERI	
3 560	STAPH HYICUS (An)		6 501	STAPH XYLOSUS	
3 601	STAPH SIMULANS	NOVO	6 521	STAPH XYLOSUS	
	STAPH SAPROPHYTICUS	NOVO R	6 600	STAPH WARNERI	
			6 601	STAPH SAPROPHYTICUS	XYL − ARA−
4 060	STAPH HAEMOLYTICUS			STAPH XYLOSUS	XYL + ARA+
4 210	STAPH SCIURI		6 611	STAPH XYLOSUS	
4 310	STAPH SCIURI		6 621	STAPH XYLOSUS	
4 420	STAPH HAEMOLYTICUS		6 700	STAPH AUREUS	
4 440	STAPH HAEMOLYTICUS		6 701	STAPH XYLOSUS	
4 460	STAPH HAEMOLYTICUS		6 721	STAPH XYLOSUS	
4 610	STAPH SCIURI		6 731	STAPH XYLOSUS	
4 620	STAPH HAEMOLYTICUS				
4 660	STAPH HAEMOLYTICUS		7 000	STAPH EPIDERMIDIS	
4 700	STAPH AUREUS	COAG+	7 021	STAPH XYLOSUS	
	STAPH SCIURI	COAG−	7 040	STAPH EPIDERMIDIS	
4 710	STAPH SCIURI		7 141	STAPH INTERMEDIUS (An)	
			7 300	STAPH AUREUS	
5 040	STAPH EPIDERMIDIS		7 321	STAPH XYLOSUS	
5 200	STAPH SCIURI		7 340	STAPH AUREUS	COAG−
5 210	STAPH SCIURI	COAG+	7 401	STAPH XYLOSUS	
5 300	STAPH AUREUS	COAG−	7 421	STAPH XYLOSUS	
	STAPH SCIURI		7 501	STAPH INTERMEDIUS (An)	COAG+
5 310	STAPH SCIURI			STAPH XYLOSUS	COAG−
5 600	STAPH SCIURI		7 521	STAPH XYLOSUS	
5 610	STAPH SCIURI	COAG+	7 541	STAPH INTERMEDIUS (An)	
5 700	STAPH AUREUS	COAG−	7 560	STAPH HYICUS (An)	
	STAPH SCIURI		7 601	STAPH XYLOSUS	
5 710	STAPH SCIURI		7 621	STAPH XYLOSUS	
5 740	STAPH AUREUS		7 631	STAPH XYLOSUS	
			7 700	STAPH AUREUS	
6 001	STAPH XYLOSUS	XYL + ARA+	7 701	STAPH XYLOSUS	
	STAPH SAPROPHYTICUS	XYL − ARA−	7 721	STAPH XYLOSUS	
6 011	STAPH XYLOSUS		7 740	STAPH AUREUS	
6 021	STAPH XYLOSUS				

Note: The API STAPH-IDENT Profile Register uses "STAPH" as an abbreviation for *Staphylococcus*. The correct scientific abbreviation is *S. aureus*, for example.

Source: STAPH-IDENT, Analytab Products, Division of Sherwood Medical, Plainview, New York.

Name: _____

Date: _____   Section: _____

Observations and Results

Traditional Procedures

Staphylococcal Species	APPEARANCE OF PLASMA: CLOTTED (+) OR UNCLOTTED (−)					Coagulase (+) or (−)
	5 min	20 min	1 hr	4 hr	24 hr	
S. aureus						
S. epidermidis						
S. saprophyticus						

Procedure	S. aureus	S. epidermidis	S. saprophyticus
Mannitol salt agar:			
Growth			
Color of medium			
Fermentation			
DNA agar:			
Color of medium			
DNase activity			
Novobiocin sensitivity:			
Growth inhibition in mm			
Susceptibility—(R) or (S)			

STAPH-IDENT System Procedure

MICROCUPULE			INTERPRETATION OF REACTIONS		REACTION RESULTS	
No.	Substrate		Positive	Negative	Color	(+) or (−)
1	PHS	p-Nitrophenyl-phosphate, disodium salt	Yellow	Clear or straw-colored		
2	URE	Urea	Purple to red-orange	Yellow or yellow-orange		
3	GLS	p-Nitrophenyl- β-d-glucopyranoside	Yellow	Clear or straw-colored		

STAPH-IDENT System Procedure (continued)

MICROCUPULE		INTERPRETATION OF REACTIONS		REACTION RESULTS	
No.	Substrate	Positive	Negative	Color	(+) or (−)
4	MNE Mannose	Yellow or yellow-orange	Red or orange		
5	MAN Mannitol				
6	TRE Trehalose				
7	SAL Salicin				
8	GLC p-Nitrophenyl-β-d-glucuronide	Yellow	Clear or straw-colored		
9	ARG Arginine	Purple to red-orange	Yellow or yellow-orange		
10	NGP 2-Naphthyl-β-d-galactopyranoside	Add 1–2 drops of STAPH-IDENT reagent			
		Plum-purple (mauve)	Yellow or colorless		

Construct a four-digit profile for your unknown organism as follows: A four-digit profile is derived from the results obtained with STAPH-IDENT. The 10 biochemical tests are divided into four groups, as follows:

PHS	MNE	SAL	NGP
URE	MAN	GLC	
GLS	TRE	ARG	

Only positive reactions are assigned a numerical value. The value depends on the location within the group.

A value of 1 for the first biochemical in each group (e.g., PHS, MNE)
A value of 2 for the second biochemical in each group (e.g., URE, MAN)
A value of 4 for the third biochemical in each group (e.g., GLS, TRE)
A value of 0 for all negative reactions

A four-digit number is obtained by totaling the values of each of the groups.

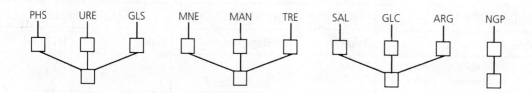

Using Table 61.2 and your four-digit profile number, identify your organism.

Unknown organism: _____

Latex Agglutination Procedure

Record the presence of agglutination as (+), and the absence of agglutination as (−).

	S. aureus	S. epidermidis	S. saprophyticus
Agglutination			
No agglutination			

Identification of Human Streptococcal Pathogens

LEARNING OBJECTIVES

Once you have completed this experiment, you should understand

1. The medical significance of streptococci.

2. Selected laboratory procedures designed to differentiate streptococci on the basis of hemolytic activity and biochemical patterns associated with the Lancefield group classifications.

Principle

Members of the genus *Streptococcus* are perhaps responsible for a greater number of infectious diseases than any other group of microorganisms. Morphologically, they are cocci that divide in a single plane, forming chains. They form circular, translucent to opaque, pinpoint colonies on solid media. All members of this group are gram-positive, and many are nutritionally fastidious, requiring enriched media such as blood for growth.

The streptococci are classified by means of two major methods: (1) their **hemolytic activity**, and (2) the **serologic classification of Lancefield**. The observed hemolytic reactions on blood agar are of the following three types:

1. **(α) Alpha hemolysis,** an incomplete form of hemolysis, produces a green zone around the colony. α-Hemolytic streptococci, the *Streptococcus viridans* species, are usually nonpathogenic opportunists. In some instances, however, they are capable of inducing human infections such as **subacute endocarditis**, which may precipitate valvular damage and heart failure if untreated. *Streptococcus pneumoniae*, the causative agent of **lobar pneumonia**, will be studied in a separate experiment.

2. **(β) Beta hemolysis,** a complete destruction of red blood cells, exhibits a clear zone of approximately 2- to 4-times the diameter of the colony. The streptococci capable of producing β -hemolysins are most frequently associated with pathogenicity.

3. **(γ) Gamma hemolysis** is indicative of the absence of any hemolysis around the colony. Most commonly, γ-hemolytic streptococci are avirulent.

These hemolytic reactions are shown in **Figure 62.1.**

Lancefield classified the streptococci into 20 **serogroups**, designated A through V, omitting I and J, based on the presence of an antigenic group-specific hapten called the **C-substance.** This method of classification generally implicates the members of Groups A, B, C, and D in human infectious processes.

β-Hemolytic streptococci belonging to **Group A,** and collectively referred to as *Streptococcus pyogenes*, are the human pathogens of prime importance. Members of this group are the main etiological agents of human respiratory infections such as **tonsillitis, bronchopneumonia**, and **scarlet fever**, as well as skin disorders such as **erysipelas** and **cellulitis.** In addition, these organisms are responsible for the development of complicating infections, namely **glomerulonephritis** and **rheumatic fever,** which may surface when primary streptococcal infections either go untreated or are not completely eradicated by antibiotics. The β-hemolytic streptococci found in **Group B** are indigenous to the vaginal mucosa and have been shown to be responsible for **puerperal fever** (childbirth fever), a sometimes-fatal **neonatal meningitis**, and **endocarditis.** Members of **Group C** are also β-hemolytic and have been implicated in **erysipelas, puerperal fever,** and **throat infections.** The enterococci formerly classified as Group D streptococci have been reclassified and are now considered a separate genus. The enterococci differ significantly from other members of Group D, such as *S. bovis*, which may be the etiological agent of urinary tract infections. Enterococci such as *Enterococcus faecalis* may cause infections to the lungs, urinary

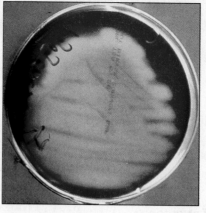

(a) Alpha hemolysis   **(b)** Beta hemolysis   **(c)** Gamma hemolysis

Figure 62.1 **Types of hemolytic reactions on blood agar**

tract infections, or bloodstream through an intestinal laceration or poor personal hygiene. The enterococci tend to be antibiotic-resistant, particularly to penicillin and more recently to vancomycin.

The virulence of the streptococci is associated with their ability to produce a wide variety of extracellular metabolites. Included among these are the **hemolysins** (α and β), **leukocidins** that destroy phagocytes, and the **erythrogenic toxin** responsible for the rash of scarlet fever. Also of medical significance are three metabolic end products that facilitate the spread of the organisms, thereby initiating secondary sites of streptococcal infection. These metabolites are **hyaluronidase** (the spreading factor), which hydrolyzes the tissue cement hyaluronic acid; **streptokinase,** a fibrinolysin; and the **nucleases,** ribonuclease and deoxyribonuclease, which destroy viscous tissue debris.

Although the different groups of streptococci have similar colonial morphology and microscopic appearance, they can be separated and identified by the performance of a variety of laboratory tests. Toward this end, you will perform laboratory procedures to differentiate among the medically significant streptococci on the basis of their Lancefield group classification and their hemolytic patterns. Table 62.1 will aid in this separation.

Identification of Group A streptococci involves the following procedures:

1. **Bacitracin sensitivity test:** A filter-paper disc impregnated with 0.04 unit of bacitracin

TABLE 62.1	Laboratory Differentiation of Streptococci					
GROUP:	A	B	C	D	K, H, N	
ORGANISMS:	*S. pyogenes*	*S. agalactiae*	*S. equi*	S. bovis NON-ENTEROCOCCI	S. salivarius S. sanguis S. mitis	*E. Faecalis* ENTEROCOCCI
Hemolysis	β	B	β	α → γ	A	α → γ
Bacitracin sensitivity	S	R	R	R	R	R
CAMP test	–	+	–	–	–	–
Bile esculin hydrolysis	–	–	–	+	–	+
6.5% NaCl medium	NG	NG	NG	NG	NG	G
Growth at 10°C	NG	NG	NG	NG	NG	G
Growth at 45°C	NG	NG	NG	NG or G	NG	G

NG = no growth; G = growth; S = sensitive; R = resistant

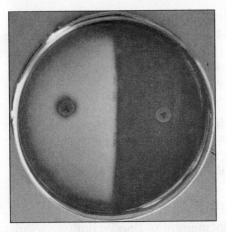

Figure 62.2 Bacitracin sensitivity test. Positive for beta-hemolytic Group A streptococci on the left; negative on the right.

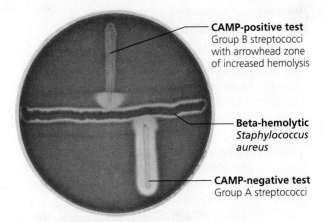

CAMP-positive test
Group B streptococci with arrowhead zone of increased hemolysis

Beta-hemolytic
Staphylococcus aureus

CAMP-negative test
Group A streptococci

Figure 62.3 CAMP reactions

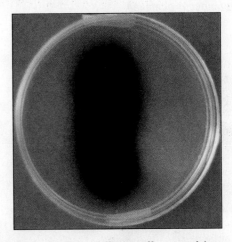

Figure 62.4 Positive bile esculin test. A brown-to-black coloration of the medium indicates positive identification of Group D streptococci.

is applied to the surface of a blood agar plate previously streaked with the organism to be identified. Following incubation, the appearance of a zone of growth inhibition surrounding the disc is indicative of Group A streptococci. Absence of this zone suggests a non–Group A organism. **Figure 62.2** shows the result of a bacitracin sensitivity test.

2. **Directigen™ test:** A rapid, non–growth-dependent immunological procedure for the detection of the Group A antigen, developed by Becton Dickinson and Company. In this test, a clinical specimen is subjected to reagents designed to extract the Group A antigen, which is then mixed with a reactive and a negative control latex. Agglutination with the reactive latex is indicative of Group A streptococci.

Group B streptococci are identified with the **CAMP test** (named for Christie, Atkins, and Munch-Petersen). Group B streptococci produce a peptide, the CAMP substance, that acts in concert with the β-hemolysins produced by some strains of *Staphylococcus aureus*, causing an increased hemolytic effect. Following inoculation and incubation, the resultant effect appears as an arrow-shaped zone of hemolysis adjacent to the central streak of *S. aureus* growth. The non–Group B streptococci do not produce this reaction. **Figure 62.3** illustrates the CAMP reactions.

Identification of Group D streptococci involves the following:

1. **Bile esculin test:** In the presence of bile, Group D streptococci hydrolyze the glycoside esculin to 6,7-dihydroxy coumarin that reacts

with the iron salts in the medium to produce a brown-to-black coloration of the medium following incubation (**Figure 62.4**). Lack of this dark coloration is indicative of a non–Group D organism.

2. **6.5% sodium chloride broth:** The enterococci can be separated from the non-enterococci by the ability of the former to grow in this medium. This reaction is shown in **Figure 62.5**.

Hemolytic activity is identified with a blood agar medium. The pathogenic streptococci, primarily the β-hemolytic, can be separated from the generally avirulent α- and γ-hemolytic streptococci by the type of hemolysis produced on blood agar, as previously described.

CLINICAL APPLICATION

Streptococci Infections

Medically the streptococci are of significant importance because they are responsible for a wide variety of infections, many of which are pyogenic (pus-producing). *Streptococcus agalactiae* (Lancefield group B) may colonize the vagina as well as the upper respiratory tract of humans, and is the most frequent cause of neonatal pneumonia in the United States. Meanwhile, *Streptococcus pyogenes* (Lancefield group A) causes necrotizing fasciitis, a rare but devastating infection that destroys skin, muscle, and underlying tissue. The CAMP test is used to identify Group *A Streptococcus pyogenes* from Group B *Streptococcus agalatiae*.

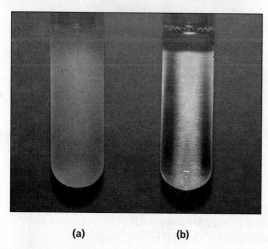

(a)   (b)

Figure 62.5 65% sodium chloride test. (a) Growth indicates the presence of Group D enterococci. **(b)** The absence of growth indicates the presence of Group D non-enterococci.

AT THE BENCH

Materials

Cultures

24-hour blood agar slant cultures of *Streptococcus pyogenes* (ATCC 12385) BSL-2 , *Enterococcus faecalis* BSL-2 , *Streptococcus bovis*, *Streptococcus agalactiae* BSL-2 , *Streptococcus mitis* BSL-2 , and *Staphylococcus aureus* (ATCC 25923) BSL-2 .

Media

Per designated student group: five blood agar plates, three bile esculin agar plates, and three 6.5% sodium chloride broths.

Reagents

Directigen Rapid Group A Strep Test (Becton Dickinson and Company), crystal violet, Gram's iodine, ethyl alcohol, safranin, and Taxo™ A discs (0.04 unit of bacitracin).

Equipment

Microincinerator or Bunsen burner, inoculating loop, staining tray, lens paper, bibulous paper, microscope, sterile cotton swabs, glassware marking pencil, sterile 12 × 75-mm test tubes, sterile Pasteur pipettes, sterile applicators, 95% ethyl alcohol in beaker, forceps, and mechanical rotator.

Procedure Lab One

1. Prepare a Gram-stained preparation of each streptococcal culture and observe under oil immersion. Record in the Lab Report your observations of cell morphology and Gram reaction.

2. Prepare the blood agar plate cultures to identify the type of hemolysis as follows:

    a. With a glassware marking pencil, divide the bottoms of two blood agar plates to accommodate the five test organisms. Label each section with the name of the culture to be inoculated.

    b. Using aseptic inoculating technique, make a single line streak of inoculation of each organism in its respective sector on the blood plates.

3. Prepare the blood agar plate cultures for the **bacitracin** test as follows:

    a. With a glassware marking pencil, label the covers of two blood agar plates with the names of the organisms to be inoculated, *S. pyogenes* BSL-2 and *S. agalactiae* BSL-2 .

    b. Using a sterile cotton swab, inoculate the agar surface of each plate with its respective test organism by streaking first in a horizontal direction, then vertically to ensure a heavy growth over the entire surface.

    c. Using alcohol-dipped and flamed forceps, apply a single 0.04-unit bacitracin disc to the surface of each plate. Gently touch each disc to ensure its adherence to the agar surface.

4. Prepare a blood agar plate culture for the **CAMP test** as follows:

   a. Using a sterile inoculating loop, make a single line of inoculation along the center of the plate using the *S. aureus* `BSL-2` culture.

   b. With a sterile loop, inoculate *S. pyogenes* `BSL-2` on one side and perpendicular to the central *S. aureus* `BSL-2` streak, starting about 5 mm from the central streak and extending toward the periphery of the agar plate.

   c. On the opposite side of the central streak, but not directly opposite the *S. pyogenes* `BSL-2` line of inoculation, repeat Step 4b using *S. agalactiae* `BSL-2`.

5. Prepare the bile esculin agar plate cultures as follows:

   a. Label the three bile esculin plates with the names of the organisms to be inoculated, *S. bovis, S. mitis* `BSL-2`, and *E. faecalis* `BSL-2`.

   b. Aseptically inoculate each plate with its test organism by making several lines of inoculation on the agar surface.

6. Prepare 6.5% sodium chloride broth cultures as follows:

   a. Label three tubes of 6.5% sodium chloride broth with the names of the organisms to be inoculated, *S. bovis, E. faecalis* `BSL-2`, and *S. mitis* `BSL-2`.

   b. With a sterile loop, inoculate each tube with its organism.

7. Conduct the **Directigen test** procedure as follows:

   a. Label two sterile 12 × 75-mm test tubes as *S. pyogenes* `BSL-2` and *S. agalactiae* `BSL-2`.

   b. Add 0.3 ml of Reagent 1 to both test tubes.

   c. Using a sterile cotton swab, transfer the test organisms into their respectively labeled test tubes. *Note: These samples will emulate the throat swabs obtained in a clinical solution.*

   d. Add 1 drop of Reagent 2 to each test tube. Mix by rotating the swab against the side of the tube. Allow the swabs to remain in the test tubes for 3 minutes.

   e. Add 1 drop of Reagent 3 to both tubes and mix.

   f. Remove swabs after extracting as much liquid as possible by rolling them against the sides of the tubes.

   g. Place 1 drop of negative antigen control on both circles in Column A of test slide.

   h. Place 1 drop of positive antigen control on both circles in Column B of test slide.

   i. Dispense 1 drop of each streptococcal sample on both circles in Columns C and D, respectively.

   j. Using a new sterile applicator for each specimen, spread each specimen within the confines of both circles in Columns A, B, C, and D.

   k. Add 1 drop of reactive latex to the top row of circles.

   l. Add 1 drop of control latex to the bottom row of circles.

   m. Place the slide on a mechanical rotator for 4 minutes under a moistened humidifying cover.

   n. Compare the agglutination seen in the upper "reactive latex" circles with the consistency of the latex in the bottom "control latex" circles. Any agglutination in the top circles distinct from any background granules seen in the bottom circles indicates Group A streptococci.

8. Incubate all tubes and plates in an inverted position for 24 hours at 37°C.

Procedure Lab Two

1. Examine the two blood agar plates for bacitracin activity. Record in the Lab Report your observations of the presence (+) or absence (−) of a zone of inhibition of any size surrounding the discs.

2. Examine the blood agar plate for the CAMP reaction. Record your observations of the presence (+) or absence (−) of increased arrow-shaped hemolysis.

3. Examine the bile esculin plates for the presence (+) or absence (−) of a brown-black coloration in the medium and record your observations.

4. Observe the 6.5% sodium chloride broth cultures for the presence (+) or absence (–) of growth and record your observations.

5. Examine the two blood agar plates for the presence and type of hemolysis produced by each of the test organisms. Record your observations of the appearance of the medium surrounding the growth and the type of hemolytic reaction that has occurred—α, β, or γ.

6. Observe the Directigen test slide for the presence (+) or absence (–) of agglutination in the reactive and control latex circles. Based on your observations, indicate the Lancefield group classification of each test organism. Record your results.

7. Based on your observations, classify each test organism according to its Lancefield group.

8. Check that all of your observations have been recorded in the Lab Report.

Name: _____

Date: _____   Section: _____

Observations and Results

Procedure	S. pyogenes	S. agalactiae	S. bovis	E. faecalis	S. mitis
Gram stain:					
Morphology	_____	_____	_____	_____	_____
Reaction	_____	_____	_____	_____	_____
Bacitracin test:					
Zone of inhibition	_____	_____	_____	_____	_____
CAMP test:					
Increased hemolysis	_____	_____	_____	_____	_____
Bile esculin test:					
Color of medium	_____	_____	_____	_____	_____
Result: (+) or (−)	_____	_____	_____	_____	_____
6.5% NaCl broth:					
Growth	_____	_____	_____	_____	_____
Hemolytic activity:					
Appearance of medium	_____	_____	_____	_____	_____
Type of hemolysis	_____	_____	_____	_____	_____
Directigen test:					
Agglutination (+) or (−) in:					
Reactive circle	_____	_____	_____	_____	_____
Control ircle	_____	_____	_____	_____	_____
Lancefield group	_____	_____	_____	_____	_____
Group classification					

Review Questions

1. How do the purposes of the bacitracin and CAMP tests differ?

2. What is the mechanism of the bile esculin test?

3. Why is it important medically to distinguish between the enterococci and the non-enterococci?

4. Why can some streptococci produce secondary sites of infection?

5. The streptococci are known to be fastidious organisms that require an enriched medium for growth. How would you account for the fact that a medium enriched with blood (blood agar) is the medium of preference for growth of these organisms?

Identification of *Streptococcus pneumoniae*

LEARNING OBJECTIVE

Once you have completed this experiment, you should understand

1. Laboratory procedures to differentiate between *Streptococcus pneumoniae* and other α-hemolytic streptococci.

Principle

The pneumococcus *Streptococcus pneumoniae* is the major α-hemolytic, streptococcal pathogen in humans. It serves as an etiological agent of **lobar pneumonia,** an infection characterized by acute inflammation of the bronchial and alveolar membranes. These organisms are gram-positive cocci, tapered or lancet-shaped at their edges, and occur in pairs or as short, tight chains. The large, thick **capsules** formed *in vivo* are responsible for antiphagocytic activity, which is believed to enhance the organisms' virulence. In addition, the pneumococci produce α-**hemolysis** on blood agar plates. **Figure 63.1** shows the effects of *Streptococcus pneumoniae* on blood agar. Because of these properties (short-chain formation, α-hemolysis, and failure of the capsule to stain on Gram staining), the organisms closely resemble *Streptococcus viridans* species. The *S. pneumoniae* can be differentiated from other α-hemolytic streptococci on the basis of the following laboratory tests:

Test	S. pneumoniae	S. mitis
Hemolysis	α	α
Bile solubility	+	−
Optochin sensitivity	+	−
Inulin fermentation	+	−
Quellung reaction	+	−
Mouse virulence	+	−

Brief descriptions of the tests and their mechanisms follow:

1. **Bile solubility test:** In the presence of surface-active agents such as **bile** and **bile salts** (sodium desoxycholate or sodium dodecyl sulfate), the cell wall of the pneumococcus undergoes lysis. Other members of the α-hemolytic streptococci will not be lysed by these agents and are bile-insoluble. Following incubation, bile-soluble cultures will appear clear, and bile-insoluble cultures will be turbid.

2. **Optochin sensitivity test:** This is a growth inhibition test in which 6-mm filter-paper discs impregnated with 5 mg of **ethylhydrocupreine hydrochloride** (optochin) and called Taxo P discs are applied to the surface of a blood agar plate streaked with the test organisms. The *S. pneumoniae*, being sensitive to this surface-active agent, are lysed with the resultant formation of a zone of inhibition greater than 15 mm surrounding the P disc. Nonpneumococcal α-hemolytic streptococci are resistant to optochin and fail to show a zone of inhibition or produce a zone less than 15 mm. Sensitivity to optochin is illustrated in **Figure 63.2.**

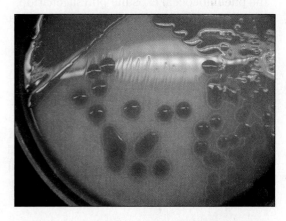

Figure 63.1 *Streptococcus pnuemoniae* **forms large thick capsules and produces alpha hemolysis on blood agar**

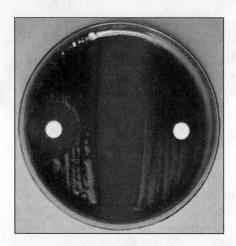

Figure 63.2 Optochin sensitivity test. The formation of a zone of inhibition greater than 15 mm on the left indicates the presence of alpha-hemolytic *S. pneumoniae*. No zone of inhibition on the right indicates the presence of other alpha-hemolytic streptococcal species.

3. **Inulin fermentation:** The pneumococci are capable of fermenting inulin, while most other α-hemolytic streptococci are non–inulin fermenters. Following incubation, the **acid** resulting from inulin fermentation will change the color of the culture from red to yellow. Cultures that are not capable of fermenting inulin will not exhibit a color change, which is a negative test result.

4. **Quellung** (Neufeld) **reaction:** This **capsular swelling** reaction is a sensitive and accurate method of determining the presence of *S. pneumoniae* in sputum. The reaction of the pneumococcal capsular polysaccharide, a hapten antigen, with an omnivalent capsular antiserum (Abcam, Inc.) produces a microscopically visible swollen capsule surrounding the *S. pneumoniae* organisms.

5. **Mouse virulence test:** Laboratory white mice are highly susceptible to infection by *S. pneumoniae* and resistant to other streptococcal infections. Intraperitoneal injection of 0.1 ml of pneumococcus-infected sputum will kill the mouse. Examination of the peritoneal fluid by Gram stain and culture will reveal the presence of *S. pneumoniae*.

In the following experiment, you will use hemolytic patterns, bile solubility, the Quellung reaction, the optochin test, and the inulin fermentation test for laboratory differentiation of *S. pneumoniae* from other α-hemolytic streptococci.

CLINICAL APPLICATION

Pneumococcus Infections

Streptococcus pneumoniae, formerly called *Diplococcus pneumoniae,* appears as a lancet-shaped diplococcus and is unlike all other cocci. The pneumococcus, as it is called, is the causative agent of lobar pneumonia (lung), otitis media (middle ear), and meningitis (meninges) infections. It is currently the leading invasive bacterial disease in children and the elderly. Presently a vaccine is available for people who are designated as high risk for infection with this organism.

AT THE BENCH

Materials

Cultures

24-hour blood agar slant cultures of *Streptococcus pneumoniae* `BSL-2` and *Streptococcus mitis* `BSL-2` .

Media

Per designated student group: one blood agar plate, two phenol red inulin broth tubes, and four 13 × 75-mm tubes containing 1 ml of nutrient broth.

Reagents

Crystal violet, Gram's iodine, ethyl alcohol, safranin, methylene blue, 10% sodium desoxycholate, commercially available Taxo P discs (5 mg of optochin), and omnivalent pneumococcal antiserum.

Equipment

Microincinerator or Bunsen burner, waterbath, inoculating loop, glass slides, coverslips, sterile cotton swabs, sterile 1-ml serological pipettes, mechanical pipetting device, 95% ethyl alcohol in beaker, forceps, and glassware marking pencil.

Procedure Lab One

1. **Bile solubility test**

   a. Label two nutrient broth tubes *S. pneumoniae* BSL-2 and two other tubes *S. mitis* BSL-2.

   b. Aseptically add 2 loop-fulls of the test organisms to the appropriately labeled sterile test tubes to effect a heavy suspension.

   c. Aseptically add 0.5 ml of sodium desoxycholate to one tube of each test culture. The remaining two cultures will serve as controls.

   d. Incubate the tubes in a waterbath at 37°C for 1 hour.

   e. After incubation, examine the tubes for the presence or absence of turbidity in each culture. Record your observations of the appearance (clear or turbid) and bile solubility of each test organism in the Lab Report.

2. **Optochin test**

   a. With a glassware marking pencil, divide the bottom of a blood agar plate into two equal sections and label one section *S. pneumoniae* BSL-2 and the other *S. mitis* BSL-2.

   b. Using a sterile cotton swab, heavily inoculate the surface of each section with its respective test organism in a horizontal and then vertical direction, being careful to stay within the limits of each section.

   c. Using alcohol-dipped and flamed forceps, apply a single Taxo P disc (optochin) to the surface of the agar in each section of the inoculated plate. Touch each disc slightly to ensure its adherence to the agar surface.

   d. Incubate the plate in an inverted position for 24 to 48 hours at 37°C.

3. **Inulin fermentation test**

   a. Label two phenol red inulin broth tubes with the name of each test organism to be inoculated.

   b. Using aseptic technique and loop inoculation, inoculate each experimental organism in its appropriately labeled tube of medium.

   c. Incubate the tube cultures for 24 to 48 hours at 37°C.

4. **Quellung reaction**

   a. Spread a loop-full of each test culture on a separate labeled clean glass slide and allow the slides to air-dry.

   b. Place a loop-full of the omnivalent capsular antiserum and a loop-full of methylene blue on each of two coverslips.

   c. Place the coverslips over the dried bacterial smears. Prepare a Gram-stained preparation of each test organism and observe under oil immersion. Record your observations of cell morphology and Gram reaction in the Lab Report.

Procedure Lab Two

1. Examine blood agar plates for the presence of hemolysis and optochin activity by measuring the zone of inhibition, if any, surrounding the disc. Record the measurement in the Lab Report and indicate whether each organism is optochin-sensitive (zone of inhibition greater than 15 mm) or optochin-resistant (no zone or less than 15 mm).

2. Observe the inulin fermentation broth cultures containing phenol red and record the color of each culture and whether it is indicative of a positive (+) or negative (–) result in the Lab Report.

3. Examine slides of the Quellung reaction under oil immersion and indicate in the Lab Report the presence (+) or absence (–) of capsular swelling surrounding the blue-stained cells.

Name: _____

Date: _____ Section: _____

Observations and Results

Procedure	*S. pneumoniae*	*S. mitis*

Bile solubility test:
    Appearance of culture
 	
    Bile solubility
	_____
	
	_____
		_____
		
		_____

Gram stain:
    Morphology
 	
    Reaction
	_____
	
	_____
		_____
		
		_____

Optochin test:
    Zone of inhibition in mm
 	
    Resistant or sensitive
	_____
	
	_____
		_____
		
		_____

Inulin fermentation:
    Color of medium
 	
    Fermentation (+) or (−)
	_____
	
	_____
		_____
		
		_____

Quellung reaction: Capsular swelling (+) or (−)	_____	_____

Review Questions

1. Why is it clinically important to distinguish *S. pneumoniae* from other α-hemolytic streptococci?

2. How would you separate *S. pneumoniae* from other α-hemolytic streptococci?

3. What are secondary pneumonias? Why do they develop most frequently following viral infections?

4. Why did it require many years of research to develop an effective, long-term pneumococcal vaccine?

Identification of Enteric Microorganisms Using Computer-Assisted Multitest Microsystems

LEARNING OBJECTIVES

Once you have completed this experiment, you should be familiar with

1. The members of the family Enterobacteriaceae.

2. Laboratory procedures designed to identify enteric pathogens using commercial multitest microsystems.

Principle

Enterobacteriaceae is a significant group of bacteria that is endogenous to the intestinal tract or that may gain access to this site via a host's ingestion of contaminated food and water. The family consists of a number of genera whose members vary in their capacity to produce disease. *Salmonella* and *Shigella* are considered to be pathogenic. Members of other genera, particularly *Escherichia* and *Enterobacter*, and to a lesser extent *Klebsiella* and *Proteus*, constitute the natural flora of the intestines and are generally considered to be avirulent. Remember, however, that all can produce disease under appropriate conditions.

The Enterobacteriaceae are gram-negative, short rods. They are mesophilic, nonfastidious organisms that multiply in many foods and water sources. They are all non–spore-formers and susceptible to destruction by common physical and chemical agents. They are resistant to destruction by low temperatures and can therefore frequently survive in soil, sewage, water, and many foods for extended periods.

From a medical point of view, the pathogenic Enterobacteriaceae are salmonellae and shigellae. Salmonellae are responsible for enteric fevers, **typhoid,** the milder **paratyphoid**, and **gastroenteritis.** In typhoid, *Salmonella typhi* penetrates the intestinal mucosa and enters the bloodstream, thus infecting organs such as the gallbladder, intestines, liver, kidney, spleen, and heart. Ulceration of the intestinal wall, caused by the release of the lipopolysaccharide endotoxin into the blood over a long febrile period, and enteric symptoms are common. **Gastroenteritis** is caused by a number of *Salmonella* species. Symptoms associated with this type of food poisoning include abdominal pain, nausea, vomiting, and diarrhea, which develop within 24 hours of ingestion of contaminated food and last for several days.

Several shigellae are responsible for **shigellosis,** a bacillary dysentery that varies in severity. Ulceration of the large intestine, explosive diarrhea, fever, and dehydration occur in the more severe cases.

Isolation and identification of enteric bacteria from feces, urine, blood, and fecally contaminated materials are of major importance in the diagnosis of enteric infections. Although the Enterobacteriaceae are morphologically alike and in many ways metabolically similar, laboratory procedures for the identification of these bacteria are based on differences in biochemical activities (Figure 64.1).

In the past, several **multitest systems** have been developed for differentiation and identification of members of the Enterobacteriaceae. They use microtechniques that incorporate a number of media in a single unit. At least six multitest systems are commercially available. The obvious advantages of these units are the need for minimal storage space, the use of less media, the rapidity with which results may be obtained, and the applicability of the results to a computerized system for identification of organisms. There are also certain disadvantages with these systems, including difficulty in obtaining the proper inoculum size since some media require heavy inoculation while others need to be lightly inoculated, the possibility of media carryover from one compartment to another, and the possibility of using inoculum of improper age. Despite these difficulties, when properly correlated with other properties such as Gram stain and colonial morphology on specialized solid media, these systems are acceptable for the identification of Enterobacteriaceae. The most frequently used systems are discussed.

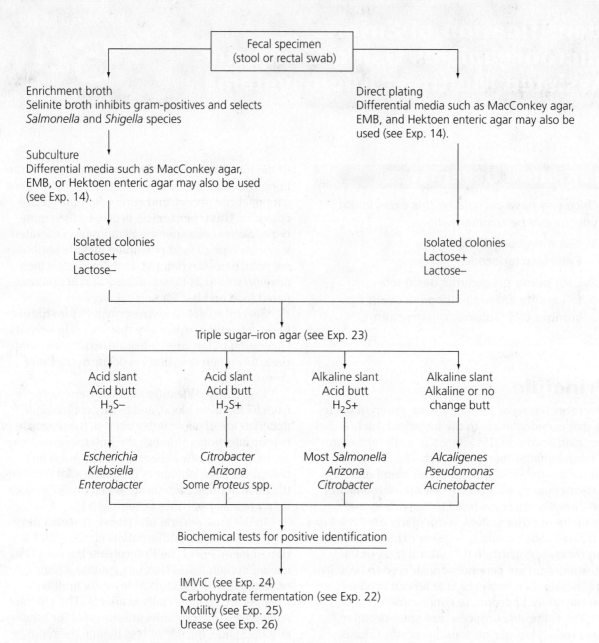

Figure 64.1 Conventional laboratory procedures for isolation and identification of enteric microorganisms

Enterotube II Multitest System and ENCISE II

The **Enterotube™ II Multitest System** (Roche Diagnostics, Division of Hoffmann-La Roche, Inc.) consists of a single tube containing 12 compartments (**Figure 64.2a**) and a self-enclosed inoculating needle. This needle can touch a single isolated colony and then in one operation be drawn through all 12 compartments, thereby inoculating all of the test media. In this manner,

15 standard biochemical tests can be performed in one inoculating procedure. Following incubation, the color changes that occur in each of the compartments are interpreted according to the manufacturer's instructions to identify the organisms (**Figure 64.2b**). This method has been further refined to permit identification of the enteric bacteria by means of a computer-assisted system called ENCISE (Enterobacteriaceae numerical coding and identification system for Enterotube).

Cap Handle Individual compartments Inoculating needle

(a) Diagram of Enterotube

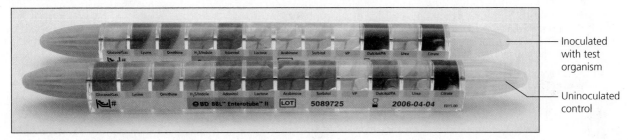

Inoculated with test organism

Uninoculated control

(b) Inoculated and uninoculated control

Figure 64.2 **Enterotube II multitest system**

Plastic strip
Cupule
Microtube
Dehydrated media

ONPG ADH LDC | ODC |CIT| H2S | URE TDA IND | |VP| |GEL| GLU | MAN INO SOR | RHA SAC MEL | AMY ARA

(a) Uninoculated control

ONPG ADH LDC | ODC |CIT| H2S | URE TDA IND | |VP| |GEL| GLU | MAN INO SOR | RHA SAC MEL | AMY ARA

(b) Inoculated with test organism

Figure 64.3 **The API 20-E system**

API (Analytical Profile Index) System

The API® 20-E employs a plastic strip composed of 20 individual microtubes, each containing a dehydrated medium in the bottom and an upper cupule as shown in **Figure 64.3**. The media become hydrated during inoculation of a suspension of the test organism, and the strip is then incubated in a plastic-covered tray to prevent evaporation. In this manner, 22 biochemical tests are performed. Following incubation, identification of the organism is made by using differential charts supplied by the manufacturer or by means of a

computer-assisted system called PRS (Profile Recognition System). PRS includes an API coder, profile register, and selector.

In the following experiment, you will inoculate an Enterotube and an API strip with an unknown enteric organism. Following incubation, you will make your identification by two methods: (1) the traditional method of noting the characteristic color changes and interpreting them according to manufacturers' instructions, and (2) the computer-assisted methods illustrated in **Figure 64.4**.

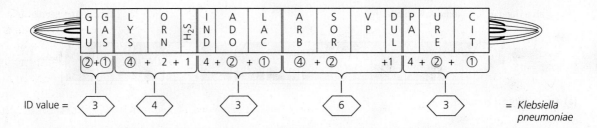

ID value = 3   4   3   6   3   = *Klebsiella pneumoniae*

1. Each positive reaction is indicated by circling the number directly below its compartment.
2. The circled numbers in each bracket are added together, and the sum is placed in the box below.
3. The resultant 5-digit number (ID value) is then located in the computer coding manual to identify the organism.

(a) The Enterotube® II

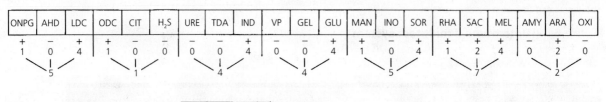

7-digit profile number = 5 1 4 4 5 7 2 = *E. coli*

1. The 21 tests are divided into seven groups of three each.
2. A value of 1 is assigned to the first positive test in each group.
3. A value of 2 is assigned to the second positive test in each group.
4. A value of 4 is assigned to the third positive test in each group.
5. A 7-digit number is obtained by totaling the positive values of each of the seven groups of three. This number is located in the analytical profile index to identify the organism.

(b) The API® strip

Figure 64.4   Computer-assisted techniques for the identification of Enterobacteriaceae

CLINICAL APPLICATION

Enterobacteriacae Infections

The Enterobacteriacae are a very diverse group of bacteria that commonly inhabit the human colon, but can cause a variety of infections throughout the body. In the hospital environment these often result from colonization of intravascular catheters, leading to bacteremia that can progress rapidly to sepsis and septic shock. Once identification of the infectious agent has been made, treatment with effective antimicrobials may be used.

AT THE BENCH

Materials

Cultures

Number-coded, 24-hour Trypticase soy agar streak plates of *Escherichia coli, Salmonella typhimurium* BSL-2 , *Klebsiella pneumoniae* BSL-2 , *Enterobacter aerogenes, Shigella dysenteriae* BSL-2 , and *Proteus vulgaris.*

Media

Per designated student group: one Enterotube II, one API 20-E strip, and one 5-ml tube of 0.85% sterile saline.

Reagents

Sterile mineral oil, 10% ferric chloride, Kovac's reagent, VP reagent for API system, nitrate reduction reagents, Barritt's reagent (VP test reagent for Enterotube II system), 1.5% hydrogen peroxide, and 1% p-aminodimethylaniline oxalate (oxidase reagent).

Equipment

Microincinerator or Bunsen burner, inoculating loop, 5-ml pipette, mechanical pipetting device, sterile Pasteur pipettes, glassware marking pencil, API profile recognition system and differential identification charts, and Enterotube II ENCISE pads and color reaction charts.

Procedure Lab One

Enterotube II System

1. Familiarize yourself with the components of the system: screw caps at both ends, medium-containing compartments, self-enclosed inoculating needle, plastic side bar, and blue-taped section.

2. Label the Enterotube II with your name and the number of the unknown culture supplied by the instructor.

3. Remove the screw caps from both ends of the Enterotube II. Using the inoculating needle contained in the Enterotube II, aseptically pick some inoculum from an isolated colony on the provided streak-plate culture.

4. Inoculate the Enterotube II as follows:

    a. Twist the needle in a rotary motion and withdraw it slowly through all 12 compartments.

    b. Replace the needle in the tube and with a rotary motion push the needle into the first three compartments (GLU/GAS, LYS, and ORN). The point of the needle should be visible in the H₂S/IND compartment.

    c. Break the needle at the exposed notch by bending, discard the needle remnant, and replace the caps at both ends. The presence of the needle in the three compartments maintains anaerobiosis, which is necessary for glucose fermentation, CO_2 production, and the decarboxylation of lysine and ornithine.

5. Remove the blue tape covering the ADO, LAC, ARB, SOR, VP, DUL/PA, URE, and CIT compartments. Beneath this tape are tiny air vents that provide aerobic conditions in these compartments.

6. Place the clear plastic slide band over the GLU/GAS compartment to contain the wax, which may be spilled by the excessive gas production of some organisms.

7. Incubate the tube on a flat surface for 24 hours at 37°C.

API 20-E System

1. Familiarize yourself with the components of the system: incubation tray, lid, and the strip with 20 microtubes.

2. Label the elongated flap on the incubation tray with your name and the number of the unknown culture supplied by the instructor.

3. With a pipette, add approximately 5 ml of tap water to the incubation tray.

4. Using a sterilized loop, touch an isolated colony on the provided streak-plate culture, transfer the inoculum to a 5-ml tube of sterile saline, and mix well to effect a uniform suspension.

5. Remove the API strip from its sterile envelope and place it in the incubation tray.

6. Tilt the incubation tray. Using a sterile Pasteur pipette containing the bacterial saline suspension, fill the tube section of each compartment by placing the tip of the pipette against the side of the cupule. Fill the cupules in the CIT, VP, and GEL microtubes with the bacterial suspension.

7. Using a sterile Pasteur pipette, fill the cupules of the AHD, LDC, ODC, and URE microtubes with sterile mineral oil to provide an anaerobic environment.

8. Cover the inoculated strip with the tray lid and incubate for 18 to 24 hours at 37°C.

Procedure Lab Two

Enterotube II System

1. Observe all reactions in the Enterotube II except IND and VP, and interpret your observations using the manufacturer's instructions. Record your observations and results in the Lab Report.

2. Perform the IND and VP tests as follows:

   a. Place the Enterotube II in a rack with the GLU and VP compartments facing downward.

   b. With a needle and a syringe, gently pierce the plastic film of the H_2S/IND compartment and add 2 or 3 drops of Kovac's reagent. Read the results after 1 minute.

   c. As in Step 2b, add 2 drops of Barritt's reagent to the VP compartment and read the results after 20 minutes.

   d. Record your IND and VP observations and results in the Lab Report.

3. Based on your results, identify your unknown organism using the manufacturer's color identification charts.

4. Determine and record in the Lab Report the five-digit ID value as described in Figure 64.4a on page 460. Identify your unknown organism by referring to the computer coding manual.

API 20-E System

1. Observe all reactions in the API strip that do not require addition of a test reagent, and interpret your observations using the manufacturer's instructions. Record your observations and results in the Lab Report.

2. Add the required test reagents in the following order: Kovac's reagent to IND, VP reagent to VP (read the result after 15 minutes), ferric chloride to TDA, nitrate reagents to GLU, and oxidase reagent to OXI. Note color changes and interpret your observations according to the manufacturer's instructions. Record your observations and results in the Lab Report.

3. Based on your results, identify your unknown organism using the differential identification chart.

4. Determine and record in the Lab Report the seven-digit profile number as described in Figure 64.4b on page 460. Identify your unknown organism by referring to the Profile Recognition System.

Name: _____

Date: _____ Section: _____

Observations and Results

Code	Name	API 20-E Appearance (color)	API 20-E Result (+) or (−)	Enterotube II Appearance (color)	Enterotube II Result (+) or (−)
ONPG	β-Galactosidase				
AHD	Arginine dihydrolase				
LDC/LYS	Lysine decarboxylase				
ODC/ORN	Ornithine decarboxylase				
CIT	Citrate				
H₂S	Hydrogen sulfide				
URE	Urease				
TDA	Tryptophan deaminase				
IND	Indole				
VP	Acetonin				
GEL	Gelatin				
GLU	Glucose				
MAN	Mannitol				
INO	Inositol				
SOR	Sorbitol				
RHA	Rhamnose				
SAC	Sucrose				
MEL	Melibiose				
AMY	Amygdalin				
ARA/ARB	Arabinose				
OXI	Oxidase				
ADO	Adonitol fermentation				
GAS	Gas production				
PHE/PA	Phenylalanine				
LAC	Lactose				
DUL	Dulcitol				
Organism					

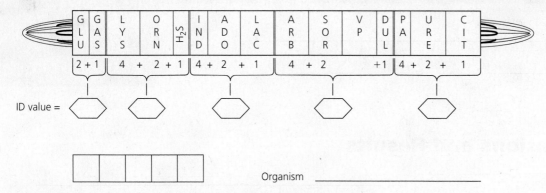

ID value =

Organism _____

Determination of Enterotube II five-digit identification number

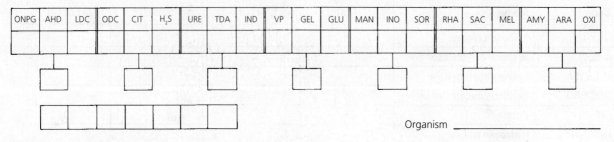

Organism _____

Determination of API 20-E seven-digit profile number

Review Questions

1. What are the advantages of multitest systems?

    Disadvantages?

2. What Enterobacteriaceae are of medical significance?

List and describe the infections caused by these organisms.

3. Would similar results be obtained by use of the computer-assisted method and the traditional color-change method?

4. What is the clinical justification for the use of a rapid test procedure such as the Enterotube II System for the identification of enteric microorganisms?

Isolation and Presumptive Identification of *Campylobacter*

LEARNING OBJECTIVE

Once you have completed this experiment, you should understand

1. The laboratory procedures required for the isolation, cultivation, and presumptive identification of the genus *Campylobacter*.

Principle

Clinicians are aware of the medical significance of *Campylobacter* strains as the etiological agents of enteric infections. The incidence of enteritis caused by *Campylobacter jejuni* equals or exceeds that of salmonellosis or shigellosis. The clinical syndrome, although varying in severity, is generally characterized by acute gastroenteritis accompanied by the rapid onset of fever, headache, muscular pain, malaise, nausea, and vomiting. Twenty-four hours following this acute phase, diarrhea develops that may be mucoid, bloody, bile-stained, and watery. The precise epidemiology of the infection is not clear; however, contact with animals, waterborne organisms, and fecal-oral transmission remain suspect.

The organisms (*campylo*, curved; *bacter*, rod) were formerly called vibrios because of their curved and spiral morphology. In the early 1980s they were reclassified in the genus *Campylobacter*. They are gram-negative and curved or spiral, with a single flagellum located at one or both poles of the cell. In pure culture, two types of colonies have been recognized and designated as Types I and II. The more commonly observed Type I colonies are large, flat, and spread with uneven margins. They are nonhemolytic, watery, and grayish. Type II colonies are also nonhemolytic, but they are smaller (1 to 2 mm), with unbroken edges. They are convex and glistening.

Initially, the isolation of *Campylobacter* organisms from fecal specimens was difficult because of their microaerophilic nature and their 42°C optimal growth temperature. Furthermore, in the absence of selective media, their growth was masked by the overgrowth of other enteric organisms, and they were often overlooked on primary isolation. This situation has been rectified with the development of selective media that are designed specifically for isolating *Campylobacter* species and that inhibit the growth of other enteric organisms. These media are nutritionally enriched and supplemented with 5% to 10% sheep or horse blood. In addition they contain three to five antimicrobial agents, depending on the medium. For example, cephalosporins, one of the antimicrobial agents present in the Campy-BAP medium, is selective for *C. jejuni* and inhibits the species *C. intestinalis*, which is rarely responsible for enteric infections.

The most essential requirement for cultivating *Campylobacter* is a microaerophilic incubation atmosphere. High concentrations of oxygen are toxic to these organisms, and an atmosphere of 3% to 10% carbon dioxide and 5% to 10% oxygen is optimal for their growth. The incubation temperature for *C. jejuni* is 42°C. At this temperature the organism grows optimally, while growth of *C. intestinalis* is inhibited.

In the experiment to follow, a simulated fecal specimen (a culture containing an attenuated strain of *C. jejuni* and other enteric organisms) is used. You will attempt to isolate the *Campylobacter* organisms by using the following two procedures:

1. A conventional method uses MacConkey agar directly, circumventing enrichment procedures, using a mixed simulated fecal population as the test culture.

2. A special method employs Campy-BAP agar and the CampyPak® and GasPak® jar, which are illustrated in Figure 65.1.

Presumptive identification is made on the basis of colonial morphology and the microscopic appearance of the organisms obtained from a typical isolated colony. You may perform the catalase and oxidase tests as described in Experiments 28 and 29 for further presumptive identification. In the case of *C. jejuni*, both tests should be positive.

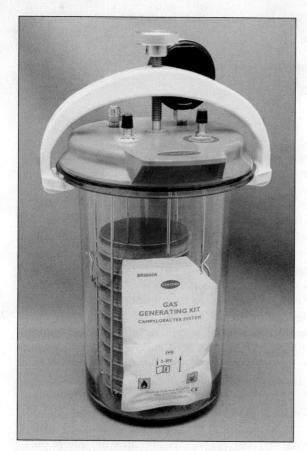

Figure 65.1 CampyPak and GasPak jar

CLINICAL APPLICATION

Traveler's Diarrhea

Campylobacter is the most common cause of bacterial diarrheal diseases worldwide. It is also referred to as "Traveler's Diarrhea." Poultry and poultry products have been associated with *Campylobacter* infections. Other foods have also been implicated in its transmission. *Campylobacter jejuni* and *Campylobacter coli* are the two most clinically significant and may be isolated from the intestinal tract of poultry. They are slow-growing organisms and are identified by biochemical, immunological, and molecular techniques.

AT THE BENCH

Materials

Cultures

Mixed saline suspensions of *Campylobacter jejuni* BSL-2 cultured on a sheep blood–enriched medium, *Salmonella typhimurium* BSL-2 , and *Escherichia coli*.

Media

Per designated student group: one Campy-BAP agar plate and one MacConkey agar plate.

Reagents

Crystal violet, Gram's iodine, 95% ethyl alcohol, and 0.8% carbol fuchsin.

Equipment

Microincinerator or Bunsen burner, inoculating loop, glassware marking pencil, CampyPak and GasPak jars, and 10-ml pipettes.

Procedure Lab One

1. Aseptically perform a four-way streak inoculation as described in Experiment 2 for the isolation of discrete colonies on both appropriately labeled agar plates.
2. Place the inoculated Campy-BAP agar plate in the GasPak jar in an inverted position. Following the manufacturer's instructions, open the CampyPak envelope and place it in the jar. With a pipette, add 10 ml of water to each envelope and immediately seal the jar to establish a microaerophilic environment.
3. Incubate the jar for 48 hours at 42°C.
4. Incubate the MacConkey agar plate culture in an inverted position for 48 hours at 37°C.

Procedure Lab Two

1. Observe both plate cultures for the presence of discrete colonies. Record your observations in the chart provided in the Lab Report.
2. Prepare a Gram stain, using 0.8% carbol fuchsin as the counterstain, of a representative colony agar plate culture. Observe microscopically and record in the Lab Report the microscopic morphology and Gram reaction of each preparation.
3. Based on your observations, identify your isolates and record in the Lab Report.
4. Optional: Perform the catalase and oxidase tests on the representative isolates.

Name: _____

Date: _____   Section: _____

Observations and Results

1. In the chart below, diagram the appearance of representative colonies on both plates and describe their colonial characteristics. Also, note and record the color of the medium surrounding the representative colonies on the MacConkey plate. (Refer to Experiment 13 for an explanation of the selective and differential nature of MacConkey agar.)

Plate Culture	Diagram of Colonies	Colonial Characteristics	Color of Medium
Campy-BAP agar			
MacConkey agar			

2. Record your observations of the Gram reactions in the chart below.

| Gram Stain Preparation | Campy-BAP Plate Isolate | MACCONKEY AGAR PLATE | |
		Isolate 1	Isolate 2
Draw a representative field.			
Microscopic morphology			
Gram reaction			

3. Based on your observations, identify your isolates:

Campy-BAP agar culture isolate: _____

MacConkey agar culture Isolate 1: _____

MacConkey agar culture Isolate 2: _____

Review Questions

1. How would you describe the clinical syndrome induced by *C. jejuni*?

2. What are the purposes of the antimicrobial agents present in the selective media used for the isolation of *Campylobacter*?

3. How may *C. jejuni* be separated from *C. intestinalis*?

4. Why might members of *Campylobacter* not be isolated from a stool specimen in a diagnostic laboratory?

Microbiological Analysis of Urine Specimens

LEARNING OBJECTIVES

Once you have completed this experiment, you should be familiar with

1. The organisms responsible for infections of the genitourinary tract.

2. Laboratory methods for detection of bacteriuria and identification of microorganisms associated with the urinary tract.

Principle

The anatomical structure of the mammalian urinary system is such that the external genitalia and the lower aspects of the urethra are normally contaminated with a diverse population of microorganisms. The tissues and organs that compose the remainder of the urinary system, the bladder, ureters, and kidneys, are sterile, and therefore urine that passes through these structures is also sterile. When pathogens gain access to this system, they can establish infection. Some etiological agents of urinary tract diseases are illustrated on this page.

Urinary tract infections may be limited to a single tissue or organ, or they may spread upward and involve the entire system. Infections, such as **cystitis,** involve the bladder but may spread through the ureters to the kidneys. Infections limited to the ureters and kidneys are called **pyelitis. Glomerulonephritis** is an inflammation that results in the destruction of renal corpuscles; **pyelonephritis** results in the destruction of renal tubules. Organisms other than bacteria may also act as etiological agents of urogenital infections. *Trichomonas vaginalis*, a pathogenic flagellated protozoan, is commonly found in the vagina, and under appropriate conditions, it is responsible for a severe inflammatory **vaginitis.** *Candida albicans*, a pathogenic yeast, is normally found in low numbers in the intestines. Under suitable conditions, such as the use of antibacterial antibiotics, which disrupt the normal intestinal flora and allow

Candida to proliferate, it can enter the urogenital systems, where it gives rise to vaginal infections. *Schistosoma haematobium* is a pathogenic fluke, a helminth, responsible for severe bladder infections.

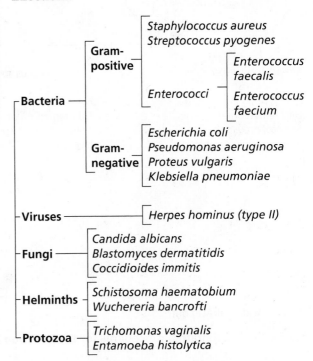

The initial step in diagnosis of a possible urinary tract infection is laboratory examination of a urine specimen. The sample must be collected midstream in a sterile container following adequate cleansing of the external genitalia. It is imperative to culture the freshly voided, unrefrigerated urine sample immediately to avoid growth of normal indigenous organisms, which may overtake the growth of the more slowly growing pathogens. In this event the infectious organism might be overlooked, resulting in an erroneous diagnosis.

Clinical evaluation of the specimen requires a quantitative determination of the microorganisms per ml of urine. Urine in which the bacterial count per ml exceeds 100,000 (10^5) denotes significant **bacteriuria** and is indicative of a urinary tract infection. Urine in which counts range from 0 to 1000 per ml are generally normal.

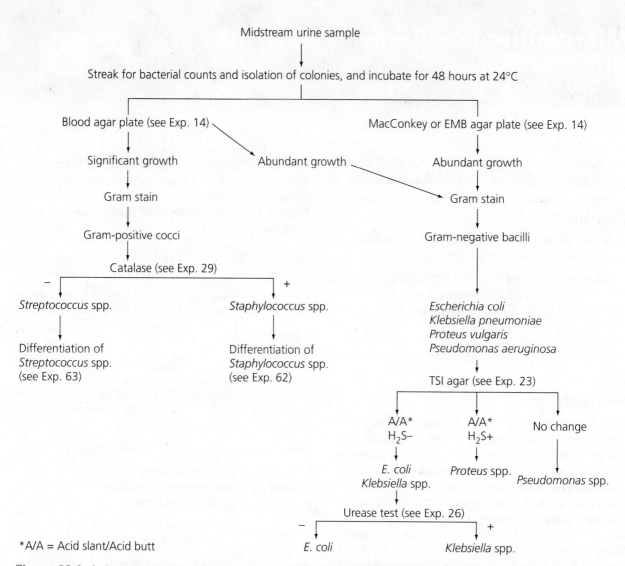

Figure 66.1 **Laboratory procedures for the isolation and identification of urinary tract pathogens**

In the conventional method, a urine sample is streaked over the surface of an agar medium with a special loop calibrated to deliver a known volume. Following incubation, the number of isolated colonies present on the plate is determined and multiplied by a factor that converts the volume of urine to 1 ml. The final calculation is then equal to the number of organisms per ml of sample. Example: Twenty-five colonies were present on a plate inoculated with a loop calibrated to deliver 0.01 ml of a urine specimen.

$$\begin{array}{c} \text{number of} \\ \text{colonies} \end{array} \times \begin{array}{c} \text{factor that} \\ \text{converts} \\ \text{0.01 ml to 1 ml} \end{array} = \begin{array}{c} \text{organisms} \\ \text{per ml} \end{array}$$

$$25 \times 100 = \begin{array}{c} 2500 \text{ organisms} \\ \text{per ml} \end{array}$$

If the specimen is turbid, dilution is necessary prior to culturing. In this case, conventional 10-fold dilutions are prepared in physiological saline to effect a final dilution of 1:1000 (see Experiment 20). Each of the dilutions (10^{-1}, 10^{-2}, and 10^{-3}) is then streaked on the surface of a suitable agar plate medium for isolation of colonies. Following incubation, the number of microorganisms per ml of sample is determined by the following formula:

organisms per ml = number of colonies
$\times$ factor that converts the volume of urine to 1 ml
$\times$ dilution factor

Example: Twenty-five colonies were counted on a 10^{-2} dilution plate inoculated with a loop calibrated to deliver 0.01 ml of urine.

Calculation:

$$25 \times 100 \times 100 = 250,000 \text{ organisms per ml}$$

On determination of bacteriuria, identification of the infectious organism can be accomplished by the laboratory procedures outlined in Figure 66.1.

A newer, less conventional, and less time-consuming method uses a diagnostic urine-culture tube, Bacturcult®, devised by Wampole Laboratories (Figure 66.2). Bacturcult is a sterile, disposable plastic tube coated on the interior with a special medium that allows detection of the bacteriuria and a presumptive class identification of urinary bacteria.

Following incubation of the Bacturcult urine culture, bacteriuria can be detected with a bacterial count. This is performed by placing the counting strip around the Bacturcult tube over an area of even colony distribution and counting the number of colonies within the circle. The average number of colonies counted is interpreted inTable 66.1.

For the presumptive identification of bacteria, the medium contains two substrates, lactose and urea, and the pH indicator phenol red. Depending on the organism's enzymatic action on these substrates, differentiation of urinary bacteria into three groups following incubation is possible based on observable color changes that occur in the culture:

Group I: *E. coli* and *Enterococcus*—yellow.
Group II: *Klebsiella*, *Staphylococcus*, and *Streptococcus*—rose to orange.
Group III: *Proteus* and *Pseudomonas*—purplish-red.

Mixed cultures do not always produce clear-cut color changes, however. Therefore, if additional testing is required, the discrete colonies that develop on the medium can be used as the source for subculturing into other media.

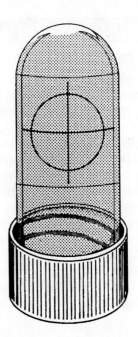

Figure 66.2 Bacturcult culture tube

In this experiment, seeded saline cultures will be used to simulate urine specimens. This is done to minimize the risk of using a potentially infectious body fluid, urine, as the test sample. The conventional procedure performed with the calibrated loop will be used to determine the number of cells in the specimens. The Bacturcult tube will be used for enumeration and presumptive group identification. If your instructor desires to emulate more closely a clinical evaluation of urine, then a mixed seeded culture must be used. Representative colonies isolated from the blood agar streak-plate culture for detection of bacteriuria can then be identified following the schema in Figure 66.1.

TABLE 66.1 Bacturcult: Interpretation of Colony Counts		
AVERAGE NUMBER OF COLONIES WITHIN CIRCLE	**APPROXIMATE NUMBER OF BACTERIA PER ML**	**DIAGNOSTIC SIGNIFICANCE**
< 25	< 25,000	Negative bacteriuria
25 to 50	25,000 to 100,000	Suspicious*
< 50	< 100,000	Positive bacteriuria

Source: Wampole Laboratories Division, Carter-Wallace, Inc., Cranbury, NJ 08512. Reprinted with permission.

*Additional testing recommended.

CLINICAL APPLICATION

The Oldest Clinical Test

Urinary tract infections are among the most frequently occurring problems in clinical medicine. Urine is composed of 95% water with the remainder consisting mainly of urea, uric acid, ammonia, hormones, sloughed squamous cells, proteins, salts, and minerals. Urinalysis is performed for the diagnosis of metabolic or systemic diseases that affect kidney function, for disorders of the kidney and urinary tract, screening for drug abuse, and monitoring patients with diabetes. Urinalysis is considered to be the oldest clinical test, with physical examination of urine for diagnosis having been performed as long as 6000 years ago. Hippocrates, in the 4th century BCE, first realized that urine was a filtrate from the kidneys.

AT THE BENCH

Materials

Cultures

Six saline cultures, each seeded with one of the following 24-hour cultures: *Enterococcus faecalis* BSL-2 , *Staphylococcus aureus* BSL-2 , *Proteus vulgaris*, *Escherichia coli*, *Pseudomonas aeruginosa* BSL-2 , and *Klebsiella pneumoniae* BSL-2 . Optional: Saline culture seeded with a gram-positive and a gram-negative organism.

Media

Per designated student group: three blood agar plates, three sterile 9-ml tubes of saline, and six Bacturcult culture tubes.

Equipment

Bunsen burner, calibrated 0.01-ml platinum loop, glassware marking pencil, sterile 1-ml pipettes, and mechanical pipetting device.

Procedure Lab One

Bacturcult

1. Label each Bacturcult tube with the name of the bacterial organism present in the urine sample.

2. Fill each tube almost to the top with urine.

3. Immediately pour the urine out of each tube, allowing all the fluid to drain for several seconds. Replace the screw cap securely.

4. Immediately prior to incubation, loosen the cap on each tube by turning the screw cap counterclockwise for one-half turn.

5. Incubate the tubes with the caps down for 24 hours at 37°C.

Calibrated Loop for Bacterial Counts

1. Label the three 9-ml sterile saline tubes and the three blood agar plates 10^{-1}, 10^{-2}, and 10^{-3}, respectively.

2. Using the three 9-ml saline blanks, aseptically prepare a 10-fold dilution of the urine sample to effect 10^{-1}, 10^{-2}, and 10^{-3} dilutions.

3. With a calibrated loop, aseptically add 0.01 ml of the 10^{-1} urine dilution to the appropriately labeled blood agar plate and streak for isolation of colonies as illustrated.

4. Repeat Step 3 to inoculate the remaining urine sample dilutions.

5. Incubate all plates in an inverted position for 24 hours at 37°C.

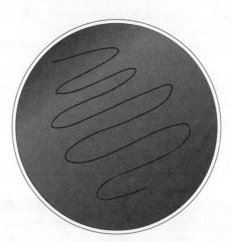

Procedure Lab Two

1. Determine the number of colonies in each of the Bacturcult urine cultures (refer to Lab Report for further instructions).

2. Record your results in the Lab Report.

Name: _____

Date: _____ Section: _____

Observations and Results

Bacturcult Procedure

1. Determine the number of colonies in each of the Bacturcult urine cultures as follows:

    a. Place the counting strip around the tube over an area of even colony distribution and count the number of colonies within the circle.

    b. Repeat the count in another area of the tube.

    c. Average the two counts.

    d. Record in the Lab Report the average number of colonies counted within the circle.

2. Based on your colony count, determine and record in the Lab Report the approximate number of bacteria per ml of each sample and its diagnostic significance as negative bacteriuria, suspicious, or positive bacteriuria.

3. Observe and record in the Lab Report the color of the medium in each of the urine cultures and the presumptive bacterial group.

Urine Culture	Number of Colonies	Number of Bacteria per ml	Diagnostic Significance	Color of Medium	Presumptive Group
E. faecalis					
S. aureus					
K. pneumoniae					
P. vulgaris					
P. aeruginosa					
E. coli					

Calibrated Loop Procedure

Determine the number of colonies on each blood agar culture plate and calculate the number of organisms per ml of the urine. Record your results in the Lab Report.

Urine Sample Dilution	Number of Colonies	Organisms per ml of Sample	Bacteriuria (+) or (−)
10^{-1}			
10^{-2}			
10^{-3}			

Review Questions

1. What types of urinary infections may be caused by different microorganisms?

2. How is a clinical diagnosis of a bacteriuria established?

3. If five colonies were counted on a 10^{-3} dilution plate streaked with 0.01 ml of urine, what was the number of organisms per ml of the original specimen, and is this count indicative of bacteriuria? Explain.

4. How accurate is a laboratory analysis of a 24-hour, unrefrigerated, non-midstream urine sample? Explain.

5. A male patient is diagnosed as having a urinary tract infection. A urine culture is ordered by his physician. She requests that a voided specimen be used rather than a catheterized sample. Why does she make this request?

Microbiological Analysis of Blood Specimens

LEARNING OBJECTIVES

Once you have completed this experiment, you should be familiar with

1. The microorganisms most frequently associated with septicemia.

2. Laboratory methods for the isolation and presumptive identification of the etiological agents of septicemia.

Principle

Blood is normally a sterile body fluid. This sterility may be breached, however, when microorganisms gain entry into the bloodstream during the course of an infectious process. The transient occurrence of bacteria in the blood is designated as **bacteremia** and implies the presence of nonmultiplying organisms in this body fluid.

Bacteremias may be encountered in the course of some bacterial infections such as pneumonia, meningitis, typhoid fever, and urinary tract infections. A bacteremia of this nature does not present a life-threatening situation because the bacteria are present in low numbers and the activity of the host's innate (nonspecific) immune system is generally capable of preventing further systemic invasion of tissues. A more dangerous and clinically alarming syndrome is **septicemia,** a condition characterized by the rapid multiplication of microorganisms, with the possible elaboration of their toxins into the bloodstream. The clinical picture frequently present in septicemia is that of septic shock, which is recognized by a severe febrile episode with chills, prostration, and a drop in blood pressure.

A large and diverse microbial population has been implicated in septicemia. The major offenders include the following:

1. Gram-negative bacteria, because of their endotoxic properties, are the most frequently encountered etiological agents of the serious complications of septicemia. Among these agents are *Haemophilus influenzae, Neisseria meningitidis, Serratia marcescens, Es-cherichia coli, Pseudomonas aeruginosa,* and *Salmonella* spp. Less frequently implicated are *Francisella tularensis* and members of the genera *Campylobacter* and *Brucella.*

2. Gram-positive bacteria that generally do not produce the presenting signs of septic shock include primarily members of the genera *Streptococcus* and *Staphylococcus.*

3. *Candida albicans* is the major fungal invader of the bloodstream.

In the clinical setting, to facilitate the rapid initiation of effective chemotherapy, a culture of the suspect blood sample is required for the isolation and identification of the offending organisms. A blood sample is drawn and cultured in an appropriate medium under both aerobic and anaerobic conditions. Over a period of 3 to 7 days, the cultures are observed for turbidity and Gram-stained smears are prepared to ascertain the presence of microorganisms in the blood. On detection of microbial growth in the cultures, transfers onto a variety of specialized agar media are made for the identification of the infectious agent. The schema for this protocol is shown in Figure 67.1.

Two methods are outlined in this exercise. Either method or both methods may be used for the isolation and presumptive identification of the microorganisms in the experimental culture. Both procedures use a simulated blood specimen: a prepared culture containing blood previously seeded with selected microorganisms. The traditional method is a modification of the schema shown in Figure 67.1. This procedure requires the preparation of Gram-stained smears for the morphological study of the organisms and the inoculation of selected agar media for their isolation and preliminary identification. The alternative method uses the commercially available **BBL Septi-Chek™ System,** a single unit composed of the Septi-Chek culture bottle and the Septi-Chek slide

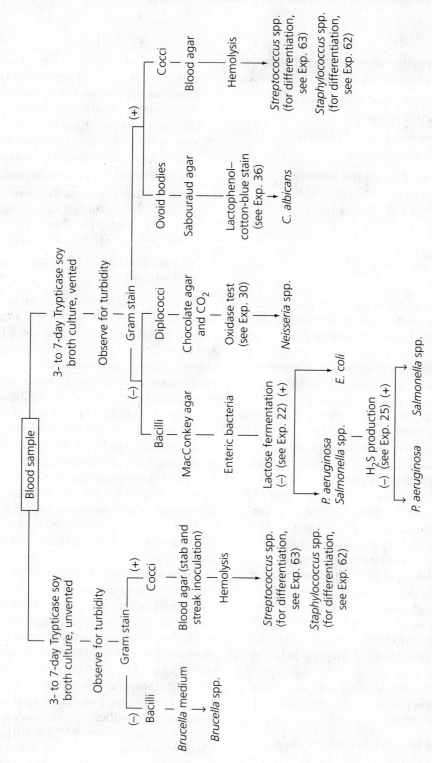

Figure 67.1 Schema for the isolation and identification of the etiological agents of septicemia

as illustrated in Figure 67.2. The culture-bottle component permits the qualitative determination of the presence of microorganisms in the blood sample, and the slide component is designed for the simultaneous subculturing of the organisms onto a plastic slide containing three differential media (chocolate, MacConkey, and malt agar). Differential growth on these media provides preliminary information as to the nature of the infectious agent and isolated colonies for further study.

Figure 67.2 Septi-Chek System

CLINICAL APPLICATION

Drawing Blood for Cultures

Normally, drawing blood for hematological analysis simply requires cleansing of the skin with an alcohol pad, but those draws intended for microbiological testing require a different protocol. The area of the draw is cleaned thoroughly with alcohol followed by a disinfectant such as Chloraprep® One-Step. All palpation after cleansing of the skin must be done with sterile gloves, and the phlebotomist wears a face shield. While it is nearly impossible to eliminate all bacteria from the skin, these techniques attempt to minimize contamination of drawn blood from contact with the skin flora that could produce false positive blood cultures.

AT THE BENCH

Materials

Culture

48- to 72-hour simulated blood culture prepared as follows: 10 ml of citrated blood, obtained from a blood bank, or 10 ml of saline seeded with 2 drops each of *Escherichia coli*, *Neisseria perflava*, and *Saccharomyces cerevisiae*, each adjusted to an absorbance of 0.1 at 600 nm, in 90 ml of Trypticase soy broth containing 0.05% of sodium polyanetholesulfonate (SPS) used to prevent clotting of the blood sample.

Media

Per designated student group: one each of blood agar plate, MacConkey agar plate, chocolate agar plate, Sabouraud agar plate, and Septi-Chek System.

Reagents

Crystal violet, Gram's iodine, 95% ethyl alcohol, safranin, lactophenol–cotton-blue stain, 70% isopropyl alcohol, and 1% p-aminodimethylaniline oxalate.

Equipment

Sterile 20-gauge, 1½-inch needles; sterile 1-ml and 10-ml syringes; microincinerator or Bunsen burner; staining tray; inoculating loop; glass microscope slide; lens paper; bibulous paper; microscope; glassware marking pencil; and disposable gloves.

Procedure Lab One

⚠ **Use gloves throughout the procedure.**

1. Swab the rubber stopper of the blood-culture bottle with 70% isopropyl alcohol and allow to air-dry.
2. Using a sterile needle and 1-ml syringe, aseptically remove 0.5 ml of the blood culture by penetrating the rubber stopper.

⚠ **Dispose of the needle and syringe, as a single unit, into the provided puncture-proof receptacle.**

4. To prepare a smear, place a small drop of the culture on a clean glass slide and spread evenly with an inoculating loop.

5. Place 1 drop of culture in one corner of the appropriately labeled blood agar plate and prepare a four-way streak inoculation as described in Experiment 2.

6. Repeat Step 4 to inoculate the MacConkey, chocolate, and Sabouraud agar plates.

7. Incubate the agar plate cultures in an inverted position for 24 to 48 hours as follows: Sabouraud agar culture at 25°C, chocolate agar culture in a 10% CO_2 atmosphere at 37°C, and the remaining cultures at 37°C.

8. Follow the **Septi-Chek System procedure** as follows:

   a. Remove the protective top of the screw cap of the culture bottle, disinfect the rubber stopper with 70% isopropyl alcohol, and allow to air-dry.

   b. Using the 10-ml syringe, aseptically transfer 10 ml of the experimental culture to the appropriately labeled Septi-Chek culture bottle.

   c. Aseptically vent the bottle for aerobic incubation.

   d. Replace the protective top of the screw cap on the bottle.

   e. Gently invert the bottle two or three times to disperse the blood evenly throughout the medium.

   f. Incubate the culture for 4 to 6 hours at 37°C.

   g. Attach the Septi-Chek slide according to the manufacturer's instructions.

   h. Tilt the combined system to a horizontal position and hold until the liquid medium enters the slide chamber and floods the agar surfaces. While maintaining this position, rotate the entire system one complete turn to ensure that all agar surfaces have come in contact with the liquid medium. Return the system to an upright position.

   i. Incubate the system in an upright position at 37°C.

   j. Check the culture bottle daily for turbidity and the slide for visible colony formation.

Procedure Lab Two

1. Examine the blood agar plate culture for the presence (+) or absence (–) of hemolytic activity. (Refer to Figure 60.1.) If hemolysis is present, determine the type observed. Record your observations in the Lab Report.

2. For the performance of the oxidase test, add p-aminodimethylaniline oxalate to the surface of the growth on the chocolate agar plate. The presence of pink-to-purple colonies is indicative of *Neisseria* spp. (Refer to Figure 60.2.) Record your observations and the oxidase test results in the Lab Report.

3. Examine the MacConkey agar plate culture for determination of lactose fermentation. Lactose fermenters exhibit a pink-to-red halo in the medium, a red coloration on the surface of their growth, or both a halo and red coloration. (Refer to Figure 13.2b.) Record your observations and indicate the presence or absence of lactose fermenters in the Lab Report.

4. Examine the Sabouraud agar plate culture for the presence of growth. Prepare a lacto-phenol–cotton-blue–stained smear from an isolated colony (see Experiment 35). Examine the smear microscopically for the presence of large ovoid bodies indicative of the yeast cells. Record your morphological observations in the Lab Report.

5. Observe the Septi-Chek slide system for the presence of growth on the three agar surfaces. If growth is present on:

   a. Medium 1 (MacConkey agar), examine for fermentative patterns as described in Step 3 and record your observations in the Lab Report.

   b. Medium 2 (chocolate agar), perform the oxidase test as described in Step 2 and record your observations in the Lab Report.

   c. Medium 3 (malt agar), prepare and examine microscopically a lactophenol–cotton-blue–stained smear as described in Step 4. Record your observations in the Lab Report.

Name: _____

Date: _____ Section: _____

Observations and Results

Culture	Traditional Procedure	Septi-Chek System
Blood agar		
Hemolysis: (+) or (−)	_____	_____
Type of hemolysis	_____	_____
Chocolate agar		
Color of colonies	_____	_____
Oxidase test: (+) or (−)	_____	_____
MacConkey agar		
Color of colonies	_____	_____
Color of medium	_____	_____
Lactose fermentation: (+) or (−)	_____	_____
Sabouraud or malt agar		
Cell morphology	_____	_____
Presumptive identification of organisms present	_____	_____

Review Questions

1. Differentiate between septicemia and bacteremia, and explain the medical significance of each.

2. Why are blood samples cultured in both vented and unvented systems?

3. A 15-year-old boy is admitted to the hospital and presents the following symptoms: chills, fever, increased pulse rate, and a drop in blood pressure. The patient indicates that these symptoms have occurred intermittently. The physician suspects a bacteremia and orders a series of three blood cultures over a 24-hour period. Explain the following:
a. Why did the physician order more than one blood culture?

b. Why does blood culture medium contain an anticoagulant?

4. Prior to the introduction of antibiotic therapy, what was the prognosis for patients with septicemia? What significant factors played roles in recovery in the absence of antibiotics?

Species Identification of Unknown Bacterial Cultures

EXPERIMENT
68

LEARNING OBJECTIVES

Once you have completed this experiment, you should be able to

1. Identify an unknown bacterial species by the use of dichotomous keys and *Bergey's Manual of Systematic Bacteriology.*

Principle

At this point in the course, you have developed the manipulative skills and the cognitive microbiological knowledge to identify microorganisms beyond their genus classification to the level of their species identification. Therefore, in this experiment, you will use dichotomous keys, *Bergey's Manual of Systematic Bacteriology,* and information accrued from previously performed laboratory procedures to help identify the species of an unknown culture.

In Experiment 31, "Genus Identification of Unknown Bacterial Cultures," you were required to use a variety of biochemical tests to successfully accomplish the experimental purpose. Your review of the required procedures and ensuing results should indicate that only a few of these tests were actually necessary, in most instances, for the identification of the unknown culture. Similarly, species identification can be accomplished by using a limited number of carefully selected laboratory procedures. Notice that what appears to be a spurious result in some cases, one that departs from the expected norm for a particular species, may be attributable to strain differences within the given species. These nonconforming results may be verified by the use of *Bergey's Manual* to ascertain the existence of variable biochemical test results for the particular species being studied.

In this experimental procedure, you will receive a mixed culture containing a gram-positive and a gram-negative organism. The protocol will require (1) Gram staining, (2) streak plating for observation of colonial characteristics, (3) use

of selective media for the preparation of pure cultures, (4) the performance of appropriate biochemical tests as indicated in the dichotomous keys outlined in Figure 68.1 and Figure 68.2, and (5) information in *Bergey's Manual.*

CLINICAL APPLICATION

New Molecular Techniques for Rapid Species Identification

Once bacteria from blood or other tissues has been cultured, the organisms must be positively identified. While biochemical and serological tests are the norm for such identification, a recently developed technique of mass spectrometry using matrix-assisted laser desorption/ionization (MALDI) offers a quicker (less than 1 hour after detection in blood) way to identify organisms. This technique releases key molecules from the organisms in question and, using analysis of the size-to-charge ratios of the molecules and specialized computer software, provides accurate identification of infectious organisms and may provide a future alternative or addition to both biochemical and genomic identification schemes.

AT THE BENCH

Materials

Cultures

Per student: number-coded, 24- to 48-hour mixed Trypticase soy broth cultures each containing a gram-positive and a gram-negative organism selected from the species listed in Figures 68.1 and 68.2.

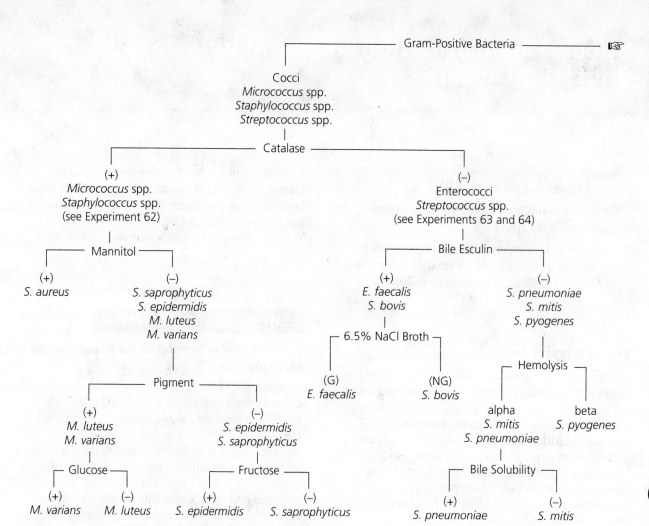

NG=No growth; G=Growth; A/G=Acid and gas; A=Acid only

Figure 68.1 Schema for the identification of gram-positive bacteria

Media

Per student: one Trypticase soy agar plate, two Trypticase soy agar slants, one Trypticase soy broth, one phenylethyl alcohol agar plate, and one MacConkey agar plate.

Required media for the biochemical tests listed in Figures 68.1 and 68.2 should be available on your request.

Reagents

Crystal violet, Gram's iodine, 95% ethyl alcohol, safranin, and required reagents for the interpretation of the biochemical reactions listed in Figures 68.1 and 68.2.

Equipment

Microincinerator or Bunsen burner, inoculating loop and needle, staining tray, immersion oil, lens paper, bibulous paper, microscope, and glassware marking pencil.

Procedure Lab One

Separation of the Bacteria in Mixed Unknown Culture

1. Prepare a Trypticase soy agar broth subculture of the unknown and refrigerate following incubation. You will use this culture if contamination of the test culture is suspected during the identification procedure.

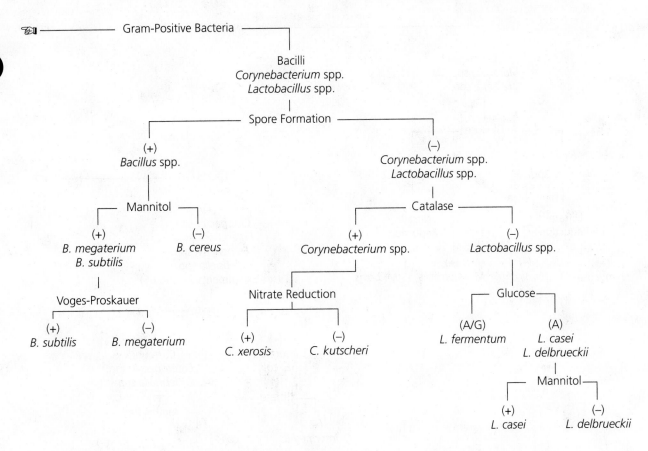

Gram-Positive Bacteria

Bacilli
Corynebacterium spp.
Lactobacillus spp.

Spore Formation

(+)
Bacillus spp.

(−)
Corynebacterium spp.
Lactobacillus spp.

Mannitol

(+)
B. megaterium
B. subtilis

(−)
B. cereus

Catalase

(+)
Corynebacterium spp.

(−)
Lactobacillus spp.

Voges-Proskauer

(+)
B. subtilis

(−)
B. megaterium

Nitrate Reduction

(+)
C. xerosis

(−)
C. kutscheri

Glucose

(A/G)
L. fermentum

(A)
L. casei
L. delbrueckii

Mannitol

(+)
L. casei

(−)
L. delbrueckii

NG= No growth; G= Growth; A/G= Acid and gas; A= Acid only

Figure 68.1 (continued) Schema for the identification of gram-positive bacteria

2. Prepare a Gram-stained smear of the original unknown culture. Examine the smear and record your observations in the Lab Report.

3. Prepare four-way streak inoculations (see Experiment 2) on the following media for the separation of the microorganisms in the mixed cultures:

4. Trypticase soy agar for observation of colonial characteristics.

5. Phenylethyl alcohol agar for isolation of gram-positive bacteria.

6. MacConkey agar for isolation of gram-negative bacteria.

7. Incubate all the plates in an inverted position and then subculture for 24 to 48 hours at 37°C.

Procedure Lab Two
Preparation of Pure Cultures

1. Isolate a discrete colony on both the phenylethyl alcohol agar plate and the MacConkey agar plate and aseptically transfer each onto a Trypticase soy agar slant (see Experiment 2).

2. Incubate the Trypticase soy agar slants for 24 to 48 hours at 37°C.

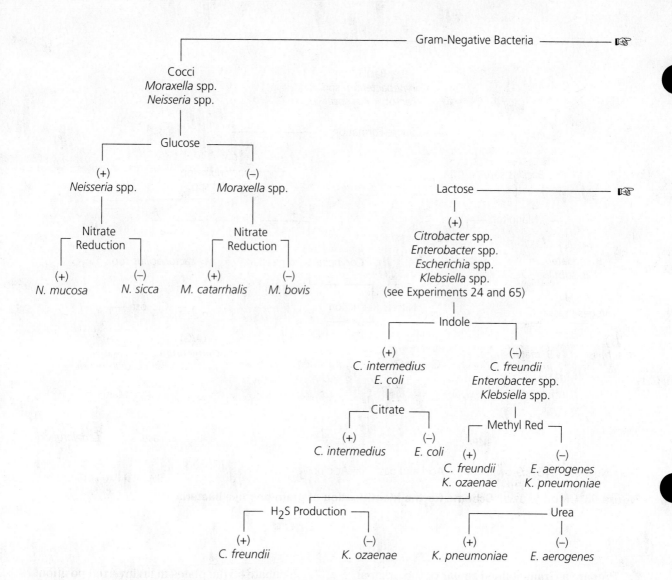

NG=No growth; G=Growth; A/G=Acid and gas; A=Acid only

Figure 68.2 Schema for the identification of gram-negative bacteria

Procedure Lab Three
Preparation of Pure Cultures

1. Examine the Trypticase soy agar plate for the appearance of discrete colonies. Select two colonies that differ in appearance and record their colonial morphologies in the Lab Report.

2. Examine the phenylethyl alcohol and MacConkey agar plates. Record your observations in the Lab Report.

Identification of Unknown Bacterial Species

1. Prepare a Gram-stained smear from each of the Trypticase soy agar slant cultures to verify their purity by means of the Gram reaction and cellular morphology. Examine the smears and record your observations in the Lab Report.

2. If each Gram-stained preparation is not solely gram-positive or gram-negative, repeat the steps in Labs One and Two, using the

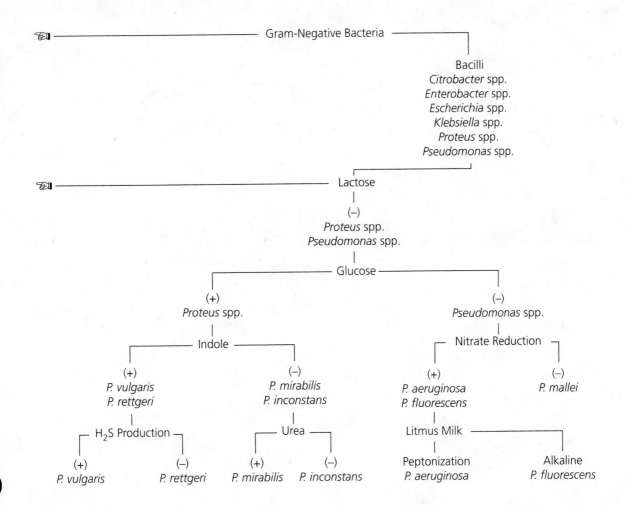

NG = No growth; G = Growth; A/G = Acid and gas; A = Acid only

Figure 68.2 (continued) **Schema for the identification of gram-negative bacteria**

refrigerated Trypticase soy agar subculture as the test culture.

3. If the isolates are deemed to be pure on the basis of their cultural and cellular morphologies, continue with the identification procedure. During this period and in subsequent sessions, use the dichotomous keys in Figures 68.1 and 68.2 to select and perform the necessary biochemical tests on each of your isolates for identification of their species. Incubate all cultures for 24 to 48 hours at 37°C.

Procedure Lab Four
Identification of Unknown Bacterial Species

1. Examine all the biochemical test cultures and record your observation and results in the Lab Report.

Name: _____

Date: _____   Section: _____

Observations and Results

Separation of the Bacteria in Mixed Unknown Culture

Record your observations of the Gram-stained smear of your mixed unknown culture in the chart below.

Organism	Cellular Morphology	Gram Reaction
1		
2		

Preparation of Pure Cultures

1. Select from the Trypticase soy agar plates two colonies that differ in appearance, and record their colonial morphologies.

   Isolate 1:

   Isolate 2:

2. Record your observations of the phenylethyl alcohol and MacConkey agar plates in the chart below.

Medium	Growth (+) or (−)	Colonial Morphology	Coloration of Medium
Phenylethyl alcohol agar			
MacConkey agar			

Identification of Unknown Bacterial Species

1. Record your observations of the Gram-stained smears of the Trypticase soy agar cultures obtained from the phenylethyl alcohol and MacConkey agar plates in the chart below.

Agar Slant From	Cellular Morphology	Gram Reaction
Phenylethyl alcohol agar plate culture		
MacConkey agar plate culture		

2. Record your observations and results of all the biochemical tests in the charts below.

Gram-Positive Isolate		
Biochemical Test	**Observation**	**Result**

Unknown gram-negative organism: _____

PART 16

Immunology

Once you have completed the experiments in this section, you should be familiar with

1. The basic principles of nonspecific (innate) and specific (acquired) immunity.
2. Serological procedures that demonstrate immunological reactions of agglutination and precipitin formation.
3. Rapid immunodiagnostic screening procedures.

Introduction

Immunity, or resistance, is a state in which a person, either naturally or by some acquired mechanism, is protected from contracting certain diseases or infections. The ability to resist disease may be innate (nonspecific), or it may be adaptive (also called acquired, or specific), when the disease state is stimulated in the host.

Innate immunity is native or natural. It is inborn and provides the basic mechanisms that defend the host against intrusion of foreign substances or agents of disease. This defense is not restricted to a single or specific foreign agent, but it provides the body with the ability to resist many pathological conditions. The mechanisms responsible for this native immunity include the **mechanical barriers**, such as the skin and mucous membranes; **biochemical factors**, such as antimicrobial substances present in the body fluids; and the more sophisticated process of **phagocytosis** and action of the **reticuloendothelial** system.

Adaptive immunity, **cell-mediated** and **humoral**, is acquired by the host in response to the presence of a single or particular foreign substance, usually protein, called an **antigen** (immunogen). In humoral immunity, antigens that penetrate the mechanical barriers of the host, namely the skin and mucous membranes, stimulate formation of **antibodies**. The function of the antibodies is to bind to the specific antigens that are responsible for their production and to inactivate or destroy them. Antibodies are a group of homologous proteins called **immunoglobulins**, which are found in serum and represent five distinct classes: immunoglobulin G (IgG), immunoglobulin A (IgA), immunoglobulin M (IgM), immunoglobulin D (IgD), and immunoglobulin E (IgE).

The primary immunological complexes (antigen + antibody) are as follows:

1. **Agglutination**: This type of reaction uses specific antibodies, **agglutinins**, that are formed in response to the introduction of particulate antigens into host tissues. When these particulate antigens combine with a homologous antiserum, a three-dimensional mosaic complex occurs. This is called an agglutination reaction and can be visualized microscopically and in some cases macroscopically.

2. **Precipitin formation**: This reaction requires specific antibodies, **precipitins**, that are formed in response to the introduction of soluble, nonparticulate antigens into host tissues. These antibodies, when present in serum, form a complex with the specific homologous nonparticulate antigen and result in a visible precipitate.

Advances in chemistry, especially immuno-chemistry, have enabled us to study the interaction of antigens and immunoglobulins outside the body, in a laboratory setting. These advances have provided an immunological discipline known as **serology**, which studies these *in vitro* reactions that have diagnostic, therapeutic, and epidemiological implications.

In the experiments to follow, you will study several serological procedures based on the principles of agglutination and precipitin formation for the detection of serum antibodies or antigens. The techniques presented in these experiments span a spectrum of methods, ranging from basic reactions to more sophisticated forms of antigen and antibody interactions.

⚠️ Note that some of the experimental protocols use positive and negative controls provided in the test kits to demonstrate the desired immunological reactions. These controls do not represent the source of potential pathogens capable of inducing infection in students and instructional staff.
The rationale for this design is that body fluids, particularly blood of unknown origin, may serve as a major vehicle for the transmission of infectious viral agents. Thus, our concern with the spread of AIDS and hepatitis precludes the use of blood as a test specimen in a college laboratory.

It is further suggested that your instructor present experiments that use positive and negative control test kits as laboratory demonstrations. This will reduce the cost of the required materials, which may otherwise be prohibitive at many academic institutions, but will still allow you and your fellow students to observe the advances in immunological serology.

Precipitin Reaction: The Ring Test

LEARNING OBJECTIVE

Once you have completed this experiment, you should be able to

1. Demonstrate a precipitin reaction by means of the ring test.

Principle

The **ring** or **interfacial test** is a simple serological technique that illustrates the precipitin reaction in solution. This antigen-antibody reaction can be demonstrated by the formation of a visible precipitate, a flocculent or granular turbidity, in the test fluid. Antiserum is introduced into a small-diameter test tube, and the antigen is then carefully added to form a distinct upper layer. Following a period of incubation of up to 4 hours, a ring of precipitate forms at the point of contact (interface) in the presence of the antigen-antibody reaction. The rate at which the visible ring forms depends on the concentration of antibodies in the serum and the concentration of the antigen.

To detect the precipitin reaction, a series of dilutions of the antigen is used because both insufficient (zone of antibody excess) and excessive (zone of antigen excess) amounts of antigen will prevent the formation of a visible precipitate (zone of equivalence), as shown in Figure 69.1. In addition, you will be able to determine the optimal antibody:antigen ratio by the presence of a pronounced layer of granulation at the interface of the antiserum and antigen solution. This immunological reaction is illustrated in Figure 69.2.

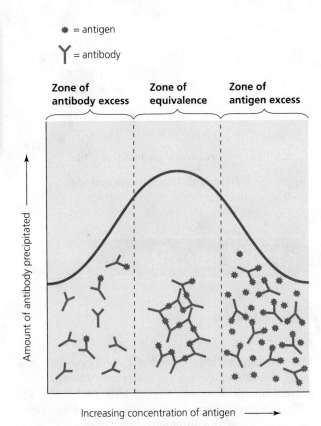

Figure 69.1 **The precipitin reaction**

CLINICAL APPLICATION

Criminology

The precipitin reaction is a serological test in which an antibody reacts with a specific soluble antigen to form a visible precipitate ring in the tube. This test is mainly used today in criminology for the identification of human blood or other bloodstains, in cases of disputed parentage, and for the determination of the cause of death.

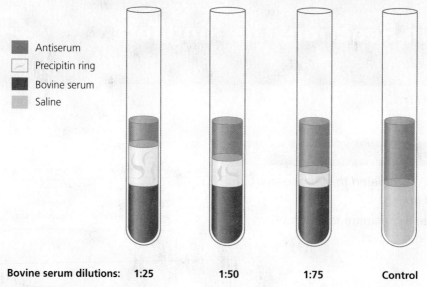

Antiserum
Precipitin ring
Bovine serum
Saline

Bovine serum dilutions: 1:25   1:50   1:75   Control

Figure 69.2 Ring test: Precipitin reactions

Materials

Reagents

Physiological saline (0.85% NaCl), and commercially available bovine globulin antiserum and normal bovine serum diluted to 1:25, 1:50, and 1:75 with physiological saline. The normal bovine serum contains the antigen (bovine globulin), to which antibodies were made commercially in another animal species and provided as antiserum to bovine globulin.

Equipment

Serological test tubes (8 × 75 mm), 0.5-ml pipettes, serological test tube rack, mechanical pipetting device, glassware marking pencil, and 37°C incubator.

Procedure

1. Label three serological test tubes according to the antigen dilution to be used (1:25, 1:50, and 1:75) and the fourth test tube as a saline control.

2. Using a different 0.5-ml pipette, transfer 0.3 ml of each of the normal bovine serum dilutions into its appropriately labeled test tube.

3. Using a clean 0.5-ml pipette, transfer 0.3 ml of saline into the test tube labeled as control.

4. Carefully overlay all four test tubes with 0.3 ml of bovine globulin antiserum. To prevent mixing of the sera, tilt the test tube and allow the antiserum to run down the side of the test tube.

5. Incubate all test tubes for 30 minutes at 37°C.

6. Examine all test tubes for the development of a ring of precipitation at the interface. Indicate the presence or absence of a ring in the Lab Report.

7. Determine and record the antigen dilution that produced the greatest degree of precipitation; this is indicative of the optimal antibody:antigen ratio.

Name: _____

Date: _____   Section: _____

Observations and Results

	ANTIGEN DILUTIONS			
	1:25	**1:50**	**1:75**	**Saline Control**
Presence of interfacial ring: (+) or (−)				

Dilution showing optimal antibody: antigen ratio is _____.

Review Questions

1. How do precipitin and agglutination reactions differ?

2. How would you determine the optimal antigen: antibody ratio by means of the ring test?

3. Why is it essential to use a series of antigen dilutions in this procedure?

4. How would you explain the absence of visible precipitate?

Agglutination Reaction: The Febrile Antibody Test

LEARNING OBJECTIVE

Once you have completed this experiment, you should be able to

1. Demonstrate the agglutination reaction by means of the febrile antibody test and an antibody titer determination.

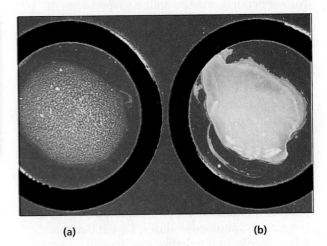

(a)   (b)

Figure 70.1 Agglutination reaction. (a) Visible clumping indicates the presence of homologous antibodies in the serum, and a positive reaction. **(b)** The lack of visible clumping indicates the absence of homologous antibodies, and a negative reaction.

Principle

The **febrile antibody test** is used in the diagnosis of diseases that produce febrile (fever) symptoms. Some of the microorganisms responsible for febrile conditions are salmonellae, brucellae, and rickettsiae. **Febrile antigens,** such as endotoxins, enzymes, and other toxic end products, elaborated by these organisms are used specifically to detect or exclude the homologous antibodies that develop in response to these antigens during infection.

In this procedure, the antigen is mixed on a slide with the serum being observed. Cellular clumping is indicative of the presence of homologous antibodies in the serum; the absence of homologous antibodies is indicated when there is no visible clumping. Only the febrile antigens and antibodies of *Salmonella* spp. will be used. Figure 70.1 shows a positive and a negative agglutination reaction.

The second part of this experiment is designed to illustrate that agglutination reactions such as the febrile antibody test can be used to identify an unknown microorganism through serotyping. A specific antiserum prepared in a susceptible, immunologically competent laboratory animal is mixed with a variety of unknown bacterial antigen preparations on slides. The bacterial antigen that is agglutinated by the antiserum is identified and confirmed to be the agent of infection.

These tests are strictly qualitative. A quantitative result can be obtained by performing the **antibody titer test,** which measures the concentration of an antibody in the serum and allows the physician to follow the course of an infection. The patient's serum is titrated (diluted), and the decreasing concentrations of the antiserum are mixed with a constant concentration of homologous antigen. The end point of the test will occur in the test tube containing the serum having the highest dilution showing agglutination.

CLINICAL APPLICATION

Febrile Disease Diagnosis

Febrile antigens are standardized suspensions of bacteria or bacterial antigens used to qualify or quantify specific serum antibodies that develop during some febrile infections. Based on the Widal Agglutination test for the diagnosis of typhoid fever, serum from patients can be tested for the presence of antibodies correlating to infectious diseases such as brucellosis, salmonellosis, and some rickettsial infections.

AT THE BENCH

Materials

Cultures

Number-coded, washed saline suspensions of *Escherichia coli, Proteus vulgaris, Salmonella typhimurium* **BSL-2** , and *Shigella dysenteriae* **BSL-2** .

Reagents

Physiological saline (0.85% NaCl), commercial preparations of *Salmonella typhimurium* H antigen, and *Salmonella typhimurium* H antiserum (Abcam, Inc.).

Equipment

Microincinerator or Bunsen burner, inoculating loop, glass microscope slides, 13 × 100-mm test tubes, sterile 1-ml pipettes, mechanical pipetting device, applicator sticks, glassware marking pencil, microscope, and waterbath.

Procedure

Febrile Antibody Test

1. With a glassware marking pencil, make two circular areas about ½ inch in diameter on a microscope slide. Label the circles A and B.

2. To Area A, add 1 drop of *S. typhimurium* H antigen and 1 drop of 0.85% saline. Mix the two with an applicator stick.

3. To Area B, add 1 drop of *S. typhimurium* H antigen and 1 drop of *S. typhimurium* H antiserum. Mix the two with a clean applicator stick.

4. Pick up the slide, and with two fingers of one hand, rock the slide back and forth.

5. Observe the slide both macroscopically and microscopically, under low power, for cellular clumping (agglutination).

6. Indicate the presence or absence of macroscopic and microscopic agglutination, and draw a representative field of Areas A and B in the Lab Report.

Serological Identification of an Unknown Organism

1. Prepare two microscope slides as in the previous procedure. Label the four areas on the slides with the numbers of your four unknown cultures.

2. Into each area on both slides, place 1 drop of *S. typhimurium* H antiserum.

3. With a sterile inoculating loop, suspend a loopfull of each number-coded unknown culture in the drop of antiserum in its appropriately labeled area on the slides.

4. Pick up the slides and slowly rock them back and forth.

5. Observe both slides macroscopically and microscopically, under low power, for agglutination.

6. In the Lab Report, indicate the presence or absence of macroscopic and microscopic agglutination in each of the suspensions. Also, indicate the suspension that is indicative of a homologous antigen-antibody reaction.

Determination of Antibody Titer

Refer to Figure 70.2 when reading the following instructions.

1. Place a row of 10 test tubes (13 × 100 mm) in a rack and number the tubes 1 through 10.

2. Pipette 1.8 ml of 0.85% saline into the first tube and 1 ml into each of the remaining nine tubes.

3. Into Tube 1, pipette 0.2 ml of *Salmonella typhimurium* H antiserum. Mix thoroughly by pulling the fluid up and down in the pipette. *Note: Avoid vigorous washing.* The antiserum has now been diluted 10 times (1:10).

4. Using a clean pipette, transfer 1 ml from Tube 1 to Tube 2 and mix thoroughly as described. Using the same pipette, transfer 1 ml from Tube 2 to Tube 3. Continue this procedure through Tube 9.

5. Discard 1 ml from Tube 9. Tube 10 will serve as the antigen control and therefore will not contain antiserum.

PROCEDURE

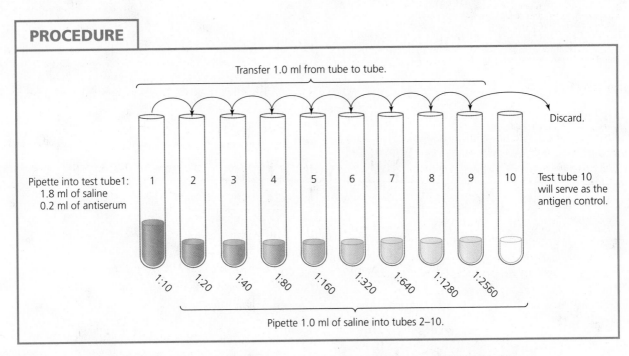

Transfer 1.0 ml from tube to tube.

Discard.

Pipette into test tube1:
 1.8 ml of saline
 0.2 ml of antiserum

Test tube 10
will serve as the
antigen control.

1:10 1:20 1:40 1:80 1:160 1:320 1:640 1:1280 1:2560

Pipette 1.0 ml of saline into tubes 2–10.

Figure 70.2 **Antibody titer test.** Serial dilution of *Salmonella typhimurium* H antibody

6. The antiserum has been diluted during this twofold dilution to give final dilutions of 1:10, 1:20, 1:40, 1:80, 1:160, 1:320, 1:640, 1:1280, and 1:2560.

7. Add 1 ml of the *Salmonella typhimurium* H antigen suspension adjusted to an absorbance of 0.5 at 600 nm to all tubes.

8. Mix the contents of the test tubes by shaking the rack vigorously.

9. Incubate the test tubes in a 55°C waterbath for 2 to 3 hours.

10. In the Lab Report, indicate the presence or absence of agglutination in each of the antiserum dilutions. Also, indicate the end point of the reaction.

Name: _____

Date: _____ Section: _____

Observations and Results

Febrile Antibody Test

Draw the appearance of the mixture and of the control.	**Area A** Saline *S. typhimurium* H antigen	**Area B** *S. typhimurium* H. antiserum *S. typhimurium* H antigen
Macroscopic agglutination (+) or (−)		
Microscopic agglutination (+) or (−)		

Serological Identification of an Unknown Organism

		AGGLUTINATION		
Cell Antigen	**Antiserum**	**Macroscopic (+) or (−)**	**Microscopic (+) or (−)**	**Homologous Antigen-Antibody Reaction**
Unknown No: ____	*S. typhimurium* H			
Unknown No: ____	*S. typhimurium* H			
Unknown No: ____	*S. typhimurium* H			
Unknown No: ____	*S. typhimurium* H			

Determination of Antibody Titer

Tube	Dilution	Agglutination	Titer
1	1:10		
2	1:20		
3	1:40		
4	1:80		
5	1:160		
6	1:320		
7	1:640		
8	1:1280		
9	1:2560		
10	Antigen control		

Review Questions

1. What are febrile antibodies?

   What is their clinical significance?

2. What is the purpose of determining an antibody titer?

3. Why does the antibody titer determination use twofold dilutions of the antiserum rather than 10-fold dilutions?

LEARNING OBJECTIVE

Once you have completed this experiment, you should be able to

1. Demonstrate a method for the identification of either an antigen or an antibody by use of an enzyme-labeled antibody test procedure.

Principle

The **enzyme-linked immunosorbent assay (ELISA)** procedure is a widely accepted method that is used for the detection of specific antigens or antibodies. The procedure is predicated on the use of an enzyme-linked (labeled) specific antibody to demonstrate the agglutination reaction for the interpretation of the test result. This test can be performed as a double-antibody technique or as an indirect immunosorbent assay. The former method is used for the detection of test antigens; the latter is used for the detection of the test antibodies. In both methods the reactions are carried out in a well of a plastic microtiter plate.

The double-antibody system requires that the unlabeled antibody be allowed to adsorb to the inner surface of the plastic well in the microtiter plate. Any unbound antibody is then washed away, and a specific test antigen is added to the well. If the antigen binds with the antibody adhering to the walls of the well, this immunocomplex will not be removed by the subsequent washing for the removal of any unbound antigen. An enzyme-linked antibody, specific for the antigen, is now added. If the antigen is present in the well, this labeled antibody binds to the antigen, forming an antibody-antigen-antibody complex. Any unbound enzyme-linked antibody is again removed by

washing. This is followed by the addition of a substrate that is capable of producing a colored end product upon its reaction with the enzyme. The resultant enzymatically produced color change may be observed by eye or spectrophotometrically.

The indirect immunosorbent test procedure is similar to the double-antibody technique in that it requires the use of an enzyme-linked antibody. However, an antigen, rather than an antibody, is adsorbed onto the inner surface of the well.

Enzyme-linked immunosorbent assays are used extensively for the diagnosis of human infectious diseases. Included among these are viral infections, such as AIDS, influenza, respiratory syncytial viral infection, and rubella. Bacterial infections such as syphilis, brucellosis, salmonellosis, and cholera can also be ascertained by means of this technique. This procedure also can be used for the detection of drugs in blood or tissues.

In this experiment you will use the **Directigen™ Flu A Test** (Becton, Dickinson and Company) to demonstrate the application of an *in vitro* enzyme immunoassay. This rapid, qualitative test employs an enzyme immunomembrane filter assay to detect influenza A antigen extracted from nasopharyngeal or pharyngeal specimens of symptomatic patients. These specimens are added to a ColorPAC™ test device, and any influenza A antigen present is nonspecifically bound to the membrane surface. Detector enzyme conjugated to monoclonal antibodies specific for the influenza A nucleoprotein antigen is bound to the trapped antigen following its addition to the ColorPAC membrane. Two substrates are then added sequentially and allowed to incubate for 5 to 30 minutes prior to determination of the result.

In the experimental procedure to be followed, the positive control will simulate the nasopharyngeal specimen of a symptomatic patient and will be indicative of a positive result. A pharyngeal swab sample of an asymptomatic individual will be used to illustrate a negative result.

CLINICAL APPLICATION

Lyme Disease

The ELISA test is commonly used in the diagnosis of Lyme disease for the detection of antibodies to *Borrelia burgdorferi*. Because of the test's sensitivity, it can sometimes produce false positive results and it is not used as the sole basis for diagnosis of Lyme disease. It is generally followed up by a Western Blot test to confirm the diagnosis.

AT THE BENCH

Materials

Cultures

Directigen Flu A positive control and pharyngeal swab specimen from an asymptomatic individual.

Media

Per designated student group or demonstration: one test tube with 2 ml of sterile saline.

Equipment

Directigen Flu A Test kit, sterile cotton swabs, sterile 0.2-ml (200-µl) pipette, mechanical pipetting device, and disposable gloves.

Procedure

Note: This test may be performed as a demonstration for economic reasons or conservation of laboratory time.

> ⚠ **Wear disposable gloves during the procedure.**

Preparation of Negative Result by Use of a Pharyngeal Specimen

1. Using a sterile cotton swab, obtain a specimen from the palatine tonsil by rotating the swab vigorously over its surface.

2. Immerse the cotton swab into a test tube containing 2 ml of sterile saline. Mix well. Remove as much liquid from the swab as possible by pressing the swab against the inner surface of the tube. Discard the swab into a container of disinfectant.

3. Using a 0.2-ml (200-µl) pipette and a mechanical pipetting device, transfer 124 µl of the pharyngeal specimen into a DispensTube™ provided in the kit.

4. Gently mix and add 8 drops of Reagent A into the DispensTube. Mix well.

5. Insert a tip, provided in the kit, into the DispensTube. Dispense all of the extracted specimen into the ColorPAC test well in drops with the sealed flow controller in position. Allow for complete adsorption.

6. Gently mix and rapidly add drops of Reagent 1 until the test well is filled. Allow sufficient time for complete adsorption.

7. Remove the flow controller from the ColorPAC well and discard it into a container of disinfectant.

8. Gently mix and add 4 drops of Reagent 2 onto the ColorPAC membrane. Allow sufficient time for complete adsorption.

9. Gently mix and add 4 drops of Reagent 3 onto the ColorPAC membrane. Allow sufficient time for complete adsorption. Let stand for 2 minutes.

10. Rapidly add enough drops of Reagent 4 to fill the ColorPAC well. Allow sufficient time for complete adsorption.

11. Gently mix and add 4 drops of Reagent 5 onto the ColorPAC membrane. Allow sufficient time for complete adsorption.

12. Gently mix and add 4 drops of Reagent 6 onto the ColorPAC membrane. Allow sufficient time for complete adsorption. *Note: The membrane will turn yellow.*

13. Gently mix and add 4 drops of Reagent 7 onto the ColorPAC membrane. Allow sufficient time for complete adsorption.

14. Wait at least 5 minutes, but no longer than 30 minutes, and then read the results in a well-lighted area.

15. Observe the appearance of the inner surface of the ColorPAC test wells and record your results in the Lab Report.

Preparation of Positive Result by Use of Positive Control

1. Dispense 4 drops of the positive control, provided in the test kit, into a DispensTube.

2. Repeat Steps 4 through 15 as outlined previously for the preparation of the negative pharyngeal specimen.

Name: _____

Date: _____ Section: _____

Observations and Results

Record your results below based on the following interpretations of your observations:

 Positive test (antigen present): The appearance of a purple triangle (of any intensity) on the ColorPAC membrane indicates the presence of the influenza A antigen in the specimen. A purple dot may be evident in the center of the triangle. The background area should be grayish white.

 Negative test (no antigen detected): The appearance of a purple dot on the ColorPAC membrane indicates the absence of the influenza A antigen in the specimen. The background area should be grayish white.

 Uninterpretable test: The absence of a purple dot, a purple triangle, or an incomplete purple triangle indicates an uninterpretable test.

Negative pharyngeal specimen:

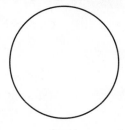

Result

Positive control specimen:

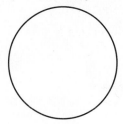

Result

Review Question

1. Why is the ELISA test used to screen human serum for the AIDS virus, while the Western blot procedure is used only as the confirmation test?

Sexually Transmitted Diseases: Rapid Immunodiagnostic Procedures

Sexually transmitted diseases (STDs) represent a diverse group of infectious syndromes that share the same mode of transmission, direct sexual contact. Their etiological agents represent a broad spectrum of pathogenic microorganisms that include bacteria, viruses, yeasts, and protozoa. The bacterial STDs include **gonorrhea, syphilis, nongonococcal urethritis,** and **lymphogranuloma venereum.** The representative viral infections are **genital herpes, genital warts, hepatitis B,** and the latest member of this group, AIDS. The protozoal and fungal infections, namely **trichomoniasis** and **candidiasis,** are diseases of lesser magnitude in the spectrum of STDs.

The experimental procedures that follow were chosen to demonstrate some of the rapid tests that are currently available for the diagnosis of selected STDs, specifically syphilis, genital herpes, and the chlamydial infections. In the methods that follow, you will perform modified procedures in the absence of clinical specimens. Commercially available positive and negative controls will be used to simulate clinical materials. *It is suggested that any of these tests, if performed, should be done as demonstrations.*

PART A Rapid Plasma Reagin Test for Syphilis

LEARNING OBJECTIVE

Once you have completed this experiment, you should be able to

1. Perform a rapid screening procedure for diagnosis of syphilis.

Principle

Treponema pallidum, the causative agent of **syphilis,** is a tightly coiled, highly motile, delicate spirochete that can be cultivated only in rabbit tissue cultures or rabbit testes. The organisms are resistant to common staining procedures and are best observed under darkfield microscopy.

Syphilis is a systemic infection that, if untreated, progresses through three clinical stages. The first stage, primary syphilis, is characterized by the formation of a painless papule, called a **chancre,** at the site of infection. Secondary syphilis represents the systemic extension of the infection and presents itself in the form of a **maculopapular rash,** malaise, and lymphadenopathy. Following this stage, the disease becomes self-limiting, and the patient appears asymptomatic until the development of tertiary syphilis. In this final stage, life-threatening complications may develop as a result of the extensive cardiovascular and nervous tissue damage that has ensued.

The **rapid plasma reagin (RPR) test,** which has to a large extent replaced the VDRL (Venereal Disease Research Laboratory) agglutination test, determines the presence of **reagin,** the nonspecific antibody present in the plasma of individuals with a syphilitic infection. The reagin appears in the plasma within 2 weeks of infection and will remain at high concentrations until the disease is eradicated. In the RPR test, if the reagin is present in the blood, it will react with a soluble antigen bound to carbon particles to produce a macroscopically visible antigen, or carbon-antibody complex. This procedure has several advantages over the VDRL test:

1. The serum does not require inactivation by heat for 30 minutes.

2. The serum may be obtained from a finger puncture, unlike the VDRL test, which requires a venous blood sample.

3. The required materials, which include the antigen suspension with a dispensing bottle, diagnostic cards, and capillary pipettes, are all contained in individual kits that do not require additional equipment and are disposable.

In the qualitative form of the RPR test, the patient's blood serum and the carbon-bound antigen suspension are mixed within a circle on the diagnostic card. In the presence of a positive

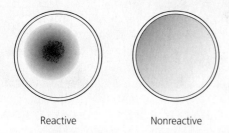

Reactive   Nonreactive

Figure 72.1 Test card showing results of the rapid plasma reagin test

(reactive) serum, the antigen-antibody complex will produce a macroscopically visible black agglutination reaction. The macroscopic appearance of a light-gray suspension, devoid of any form of agglutination, is indicative of a negative (nonreactive) serum (Figure 72.1).

Since this is a nonspecific test, false-positive results may be obtained. It is believed that the reagin is an antibody against tissue lipids in general. Therefore, it may be present in uninfected individuals due to the release of lipids resulting from normally occurring wear and tear of body tissues. It has also been found that serum levels of reagin are elevated during the course of other infectious diseases such as viral pneumonia, lupus erythematosus, infectious mononucleosis, yaws, and pinta. The serum of patients with a reactive RPR result is subjected to additional serological testing, such as the FTA-ABS (fluorescent treponemal antibody-absorption) test, or the TPI (*Treponema pallidum* immobilization) test, using the Treponema pallidum bacterium as an antigen to detect specific antibodies that are also present in the serum during syphilitic infection.

CLINICAL APPLICATION

Spirochetes

The rapid plasma reagin (RPR) test detects non-specific antibodies in the blood of patients that may indicate the spirochete *Treponema pallidum* that causes syphilis. This test is used to screen asymptomatic patients, diagnose symptomatic infection, and track the progress of disease over the treatment period. High incidence of false positives due to cross reactivity and false negatives due to low antibody titers requires further testing using the Venereal Disease Research Lab (VDRL) test in many clinical labs.

AT THE BENCH

Materials

Reagents

Commercially prepared syphilitic serum 4+ and nonsyphilitic serum.

Equipment

RPR test kit (Inverness Medical Professional Diagnostics), disposable gloves, and rotating machine (optional).

Procedure

⚠ **Wear disposable gloves.**

1. Label circles on the diagnostic plastic card as reactive and nonreactive.

2. Use a capillary pipette with an attached rubber bulb; draw the reactive serum up to the indicated mark (0.05 ml).

3. Expel the serum directly onto the card in the circle labeled reactive serum. With a clean applicator stick, spread the serum to fill the entire circle.

4. Repeat Steps 2 and 3 for the nonreactive serum.

5. Shake the dispensing bottle to mix the suspension. Hold the bottle with attached 20-gauge needle in a vertical position and dispense 1 drop onto each circle containing the test serum.

6. If a mechanical rotator is available, place the card on the rotator set at 100 rpm or rotate the card back and forth manually for 8 minutes.

7. In the presence of direct light, while tilting the card back and forth, determine the presence or absence of black clumping in each of the serum-antigen mixtures. Record your observations and the reaction as (+) or (−) in the Lab Report.

Genital Herpes:
Isolation and Identification of Herpes Simplex Virus

LEARNING OBJECTIVE

Once you have completed this experiment, you should be able to

1. Perform a tissue culture procedure for the growth and identification of the herpes simplex virus.

Principle

The double-stranded DNA herpes simplex virus (HSV) is the etiological agent of a variety of human infections. Included among these are **herpes labialis,** fever blisters around the lips; **keratoconjunctivitis,** infection of the eyes; **herpes genitalis,** eruptions on the genitalia; **herpes encephalitis,** a severe infection of the brain; and **neonatal herpes.** The herpes simplex virus is divided into two antigenically distinct groups, HSV-1 and HSV-2. The former is most frequently implicated with infections above the waist, whereas the latter is predominantly responsible for genital infections.

Primary infection with HSV-2 manifests itself with the appearance of vesicular lesions, characterized by itching, tingling, or burning sensations on or within the male and female genitalia. These vesicles regress spontaneously within 2 weeks. Following this symptomatic phase, the virus reverts to a latent state in the sacral nerves and remains quiescent until exacerbated by some environmental factor. With no chemotherapeutic cure presently available, recurrent genital herpes with subclinical symptoms is common.

Detection of the herpes simplex virus requires the use of tissue culture techniques. The presence of the virus is then determined by the development of cytopathogenic effects in these cultures, such as the detection of intranuclear inclusion bodies. In recent years, these time-consuming, specialized procedures have been greatly facilitated by the availability of immunoenzymatic reagents for the identification of this clinically significant virus.

The **Cellmatics™ HSV Detection System** is a self-contained system providing for both the growth and identification of the virus from clinical specimens. In this procedure, the provided tissue culture tubes are inoculated with the clinical sample. Following a 24-hour incubation period and fixation, the presence of HSV antigens is determined by the addition of anti-HSV antibodies, which specifically bind to the HSV antigens. To demonstrate this antigen-antibody complex, a secondary antibody, substrate, and chromogen are added. Following this staining process, HSV-positive cultures viewed microscopically will exhibit brown-black areas of viral infection on a clear background of unstained cells.

In this exercise, you will perform a modified procedure. In the absence of a clinical specimen, the actual culturing and fixation process will not be performed. Instead, the positive and negative commercially available controls will be used to simulate the clinical samples.

CLINICAL APPLICATION

Culturing HSV

Genital herpes is caused by herpes simplex-2 virus (HSV-2). It is spread from person to person during sexual contact. The infection is transmitted by means of viral shedding, which may occur even when no signs or symptoms appear. A swab sample of a vesicular lesion or from the site of a previous lesion is taken from the patient and sent to the clinical or infectious disease lab for identification. This virus can only be grown in tissue culture and not cultivated on or in other laboratory media.

AT THE BENCH

Materials

Cultures

Cellmatics HSV Positive and Negative Control Tubes.

Reagents

Cellmatics Immunodiagnostic Reagents Kit, distilled water (Difco Labs).

Equipment

5-ml pipettes, mechanical pipetting device, and microscope.

Procedure

1. Warm immunodiagnostic reagents to room temperature.

2. Drain all fluid from the positive and negative control tubes.

3. Using a 5-ml pipette, wash the culture tubes twice with 5 ml of distilled water and drain. *Note: When washing, exercise care to prevent disruption of the monolayer.*

4. Add 10 drops of primary antiserum (Vial 1). *Note: When adding reagents, hold the vials vertically to ensure proper delivery.*

5. Incubate the *tightly capped* tubes in a *horizontal* position for 15 minutes at 37°C. To ensure complete coverage of the monolayer, occasionally rock the tubes gently during incubation.

6. Wash three times with 5 ml of distilled water and drain.

7. Add 10 drops of secondary antibody (Vial 2).

8. Incubate for 15 minutes at 37°C as described in Step 5.

9. Wash three times with 5 ml of distilled water and drain.

10. Add 10 drops of substrate (Vial 3) and 2 drops of chromogen (Vial 4). Mix gently.

11. Incubate for 15 minutes at 37°C as described in Step 5.

12. Wash three times with 5 ml of distilled water and drain.

13. Examine microscopically for the presence of stained cells at 40× and 100× magnifications. Scan the entire stained monolayer of both culture tubes for the presence of brown to blackish-brown stained cells. HSV infection is indicated by the presence of dark-colored cells when viewed against an unstained background of normal cells.

14. Record your observations in the Lab Report.

LEARNING OBJECTIVE

Once you have completed this experiment, you should be able to

1. Perform an immunofluorescent procedure for diagnosis of *Chlamydia* infections.

Principle

Members of the genus Chlamydia are a group of obligate intracellular parasites. Although they were once believed to be viruses, their morphological and physiological characteristics more closely resemble bacteria, so they are now considered small bacteria. Chlamydiae are gram-negative, nonmotile, thick-walled, spherical organisms possessing both DNA and RNA that reproduce by means of binary fission. Their dependence on living tissues for cultivation and their lack of an ATP-generating system emulate the characteristics of viruses, but their bacterial nature is further affirmed by their sensitivity to antibiotic therapy. *Chlamydia trachomatis*, the human pathogen, is now recognized to be responsible for two sexually transmitted diseases, **nongonococcal urethritis** (NGU) and **lymphogranuloma venereum** (LGV). The incidence of both diseases in contemporary society is increasing dramatically.

NGU is a urethritis (inflammation of the urethra) with symptoms similar to, but less severe than, those of gonorrhea. Undiagnosed and untreated infections may lead to **epididymitis** and **proctitis** in men and **cervicitis, salpingitis,** and **pelvic inflammatory disease** in women. Nongonococcal urethritis is also caused by other bacteria, such as *Ureaplasma urealyticum*, and

Mycoplasma hominis, as well as the protozoan *Trichomonas vaginalis*. LGV, the most severe of the genital chlamydial infections, initially develops with a painless lesion at the portal of entry, the genitalia. Systemic involvement is evidenced by swelling of the regional lymph nodes, which become tender and suppurative before disseminating the organisms to other tissues. In the absence of chemotherapeutic intervention, scarring of the lymphatic vessels can cause their obstruction, leading to **elephantiasis,** enlargement of the external genitalia in men, and narrowing of the rectum in women.

The MicroTrak® Direct Specimen Test is a rapid, immunofluorescent procedure for the detection of *C. trachomatis*. The procedure circumvents the need to culture these organisms in susceptible tissues prior to their identification. This slide test is designed to detect elementary bodies, the infectious particles produced during the life cycle of this organism, by the use of a staining reagent, a fluorescein-labeled monoclonal antibody specific for the principal protein of the *C. trachomatis* outer-membrane. In this procedure, a slide smear is prepared from the clinical specimen. Following fixation, when the slide is exposed to the Direct Specimen Reagent, the antibody binds to the organisms. Their presence is then determined by the appearance of apple-green chlamydiae against a red background of counterstained cells when viewed under a fluorescent microscope.

CLINICAL APPLICATION

Treating Chlamydia

Chlamydial infections are the most commonly reported sexually transmitted diseases. More than 50 million infections occur worldwide with 3 million cases occurring in the United States annually. Any sexually active person can contract Chlamydia, but it most frequently occurs in teenagers and young adults. The incidence appears higher in females than in males. Chlamydia may be transmitted by an infected mother to her newborn during birth. If a mother's medical or sexual history indicates possible exposure, a urogenital swab will be used to collect a sample for testing.

AT THE BENCH

Materials

Cultures

Commercially prepared positive and negative control slides.

Reagents

MicroTrak Direct Specimen Test for *Chlamydia trachomatis* (VWR Scientific).

Equipment

Fluorescent microscope.

Procedure

1. Stain the positive and negative control slides with the MicroTrak reagent for 15 minutes.
2. Incubate slides for 15 minutes.
3. Rinse the slides with distilled water.
4. Air-dry the slides.
5. Examine the slides under a fluorescent microscope for the presence of apple-green particles indicative of chlamydiae. The particles are evident against a reddish background of counterstained cells.
6. Record your results in the Lab Report.

Name: _____

Date: _____ Section: _____

Observations and Results

PART A: Rapid Plasma Reagin Test for Syphilis

	Reactive Serum	Nonreactive Serum
Appearance of serum-antigen mixture		
Reaction (+) or (−)		
Draw the observed reaction.		

PART B: Genital Herpes: Isolation and Identification of Herpes Simplex Virus

Indicate in the chart below the presence (+) or absence (−) of dark-stained patches.

CONTROL CULTURES			
Negative		Positive	
40×	100×	40×	100×

PART C: Detection of Sexually Transmitted Chlamydial Diseases

Record your results below, indicating the presence (+) or absence (−) of the apple-green chlamydiae on each of the slides.

Positive control slide: _____

Negative control slide: _____

Review Questions

1. Why is an adult who has a high antibody titer to herpes simplex virus 2 (HSV-2) subject to recurrent genital herpes infections?

2. A 20-year-old college student was informed following a physical examination that her blood test for syphilis was reactive. She indicated that she was a virgin and had never received a blood transfusion. A repeat RPR test was positive, but the TPI test was negative. How would you explain these bizarre results, and what is the clinical status of this patient?

Scientific Notation

Microbiologists are required to perform a variety of laboratory techniques, including preparing and diluting solutions; expressing concentrations of chemicals, antibiotics, and antiseptics in solution; making quantitative determinations of cell populations based on the standard method for plate counting; and making serial dilutions to accommodate the latter procedure. These techniques commonly involve the use of very large or very small numbers (e.g., 9,000,000,000 or 0.0000000009), which can be so cumbersome to manipulate that errors may result. Therefore, it is essential for microbiologists to have a good command of scientific exponential notation known as **scientific notation**.

The basis for this system is predicated on the fact that all numbers can be expressed as the product of two numbers, one of which is the power of the number 10. In scientific notation, the small superscript number next to the 10 is called the **exponent**. Positive exponents tell us how many times the number must be multiplied by 10, while negative exponents indicate how many times a number must be divided by 10 (that is, multiplied by one-tenth).

For example, a number written using the exponential form designated as scientific notation would appear as 7.5×10^3, meaning that $7.5 \times 10 \times 10 \times 10 = 7500$. Appendix Table 1.1 shows both large and small numbers written in the exponential form.

Appendix Table 1.1	Scientific (Exponential) Notation
NUMBERS GREATER THAN ONE	**NUMBERS LESS THAN ONE**
$1,000,000,000 = 1 \times 10^9$	$0.000\ 000\ 001 = 1 \times 10^{-9}$
$100,000,000 = 1 \times 10^8$	$0.000\ 000\ 01 = 1 \times 10^{-8}$
$10,000,000 = 1 \times 10^7$	$0.000\ 000\ 1 = 1 \times 10^{-7}$
$1,000,000 = 1 \times 10^6$	$0.000\ 001 = 1 \times 10^{-6}$
$100,000 = 1 \times 10^5$	$0.000\ 01 = 1 \times 10^{-5}$
$10,000 = 1 \times 10^4$	$0.000\ 1 = 1 \times 10^{-4}$
$1000 = 1 \times 10^3$	$0.001 = 1 \times 10^{-3}$
$100 = 1 \times 10^2$	$0.01 = 1 \times 10^{-2}$
$10 = 1 \times 10^1$	$0.1 = 1 \times 10^{-1}$
$1 = 1 \times 10^0$	$1 = 1 \times 10^0$

Note: The exponent to which the power of 10 is raised is equal to the number of zeros to the right of 1.

Note: The exponent to which the power of 10 is raised is equal to the number of zeros to the left of 1 plus 1.

Multiplication

Rule: To multiply two numbers that are written in scientific notation (exponential form) you must **add** the exponents.

Using numbers larger than 1:

$$75 \times 1200 = 90{,}000$$

Scientific notation: $(7.5 \times 10^1) \times (1.2 \times 10^3) = 9 \times 10^4$

Addition of exponents: $1 + 3 = 4$

Using numbers less than 1:

$$0.75 \times 1200 = 900$$

Scientific notation: $(7.5 \times 10^{-1}) \times (1.200 \times 10^3) = 9 \times 10^2$

Addition of exponents: $(-1 + 3 = 2)$

$$0.75 \times 0.12 = 0.09$$

Scientific notation: $(7.5 \times 10^{-1}) \times (1.2 \times 10^{-1}) = 9 \times 10^{-2}$

Addition of exponents: $(-1) + (-1) = -2$

Division

Rule: To divide two numbers in scientific notation, you must **subtract** the exponents.

$$75{,}000 \div 1{,}200{,}000 = 0.0625$$

Scientific notation: $(7.5 \times 10^4) \div (1.2 \times 10^6) = 6.25 \times 10^{-2}$

Subtraction of exponents: $(4 - 6 = -2)$

$$7{,}500 \div .012 = 625{,}000$$

Scientific notation: $(7.5 \times 10^3) \div (1.2 \times 10^{-2}) = 6.25 \times 10^5$

Subtraction of exponents: $3 - (-2) = 5$

As you practice the use of scientific notation with large and small numbers, you will become more proficient and more comfortable with this system of scientific calculation.

Methods for the Preparation of Dilutions

In microbiology laboratories as in other science laboratories, solutions must be diluted to achieve a desired final concentration of the active material contained in that solution. A **solution** may be defined as a mixture of two or more substances (**solute**) in which the molecules of the solute are evenly distributed and will not separate on standing, or precipitate from the solution. Solutes are dissolved in a solvent or diluent, such as water, alcohol, or some other vehicle in which the solute is soluble. Solutions are usually referred to as stock solutions and may be diluted by a variety of methods, depending upon the experimental requirements. Some of these methods are listed as follows:

1. A **dilution factor** must be determined first in order to dilute a solution. This dilution factor tells us how many times a solution must be diluted and is calculated by dividing the **initial concentration (IC)** of the solution by the **final concentration (FC)** desired.

$$IC \div FC = \text{dilution factor}$$

   Example: You wish to dilute a 10% stock solution to a final concentration of 2%.

   $$10\% \div 2\% = 5 \,(\text{dilution factor})$$

   Take 1.0 ml of the 10% stock solution plus 4.0 ml of diluent (solvent), which equals a total of 5.0 ml. Thus each ml of the final solution will contain 2% solute.

2. Another method is used when a specific volume composed of a specific concentration is required.

   a. $\dfrac{IC}{FC} = 10 \,(\text{dilution factor})$

   b. $\dfrac{\text{volume needed}}{\text{concentration required}} = \dfrac{\text{amount of initial}}{\text{solution needed}}$

   c. volume needed − amount of initial solution = amount of diluent

Example: You have a 50% concentrated solution and you need 200 ml of a 5% solution.

   a. $\dfrac{IC}{FC} = \dfrac{50\%}{5\%} = 10 \,(\text{dilution factor})$

   b. $\dfrac{\text{volume needed}}{\text{concentration required}} = \dfrac{200 \,\text{ml}}{5\%} = 40 \,\text{ml}$

   c. 40 ml of 50% IC + 160 ml of diluent = 200 ml of a solution, such that each ml will contain 5% solute rather than the original 50% in the stock solution

3. The ability to prepare large dilutions is absolutely essential for work in the microbiology laboratory. This method requires that large dilutions be prepared in two steps.

   Example: A solution contains 1.0 g per ml of an active material and needs to be diluted to a final concentration of 1.0 µg per ml. A 1,000,000 (1×10^6)-fold dilution must be made. It is not practical to make such a dilution in one step since 999,999 ml of diluent would be required. This type of dilution is made as follows:

   a. Dilute 1 ml of the stock solution 1000 times:
      1.0 ml + 999 ml of diluent = 1000 µg/ml

   b. Dilute the solution containing 1000 µg/ml another 1000 times:
      1 ml of 1000 µg/ml + 999 ml diluent = 1.0 µg/ml

4. When working with large molecules, such as proteins, there will be times when you will be required to make large dilutions of the sample to be contained in a specific volume.

   Example: You need to make 50 ml of a 1/20,000 dilution of albumin.

   $$\dfrac{\text{final dilution}}{\text{volume needed}} = \dfrac{20,000}{50} = 400 \,(\text{dilution factor})$$

   a. $\dfrac{1.0 \,\text{ml of}}{\text{albumin}} + \dfrac{399 \,\text{ml of}}{\text{diluent}} = 1/400 \,\text{dilution}$

b.
$$\begin{array}{c} 1.0 \text{ ml of} \\ \text{a } 1/400 \\ \text{dilution} \end{array} + \begin{array}{c} 49 \text{ ml of} \\ \text{diluent} \end{array} = \begin{array}{c} 50 \text{ ml of a} \\ \text{soution; each ml} \\ \text{contains } 1/20,000 \\ \text{of albumin.} \end{array}$$

$$50\,(\text{volume}) \times 400\,(\text{dilution factor}) = 20,000$$

5. Perhaps the most useful type of dilution used in microbiology and immunology is the **serial dilution**. This is essential when small volumes of material are needed. This type of dilution procedure has many uses in the microbiology laboratory, especially for the determination of the total number of cells in culture (Experiments 20 and 21), the number of viral plaques found in suspensions of viruses (Experiment 39), the antibody titer (Experiment 72), and in other immunological studies. The procedure requires the use of dilution blanks containing a known volume of diluent (distilled water, saline, etc.) to which a specific volume of the sample is added. To facilitate the ease of calculations, dilutions are usually made in multiples of 10. For example: 1.0 ml of a sample is added to a 9.0-ml dilution blank (1.0 ml + 9.0 ml = 10) and is recorded as a 1:10 dilution.

It has been statistically determined that greater accuracy is achieved with very large dilutions made from a series of smaller dilutions. The procedure for the performance of a serial dilution has been explained and illustrated in Experiment 20. For the convenience of the student, it is illustrated in Appendix Figure 2.1:

1. All dilution blanks contain 9.0 ml of diluent.
2. A fresh pipette is used for each dilution, and the used pipettes are placed in a beaker of disinfectant.
3. After delivery of the sample, the tubes are mixed thoroughly before the next dilution is made.
4. Pippetting by mouth is not allowed. Only mechanical pipette aspirators may be used.

The stock solution in Appendix Figure 2.1 has been diluted 1 million times. In other words, 1.0 ml from Tube 6 will contain 1/1,000,000 of the sample contained in the stock solution.

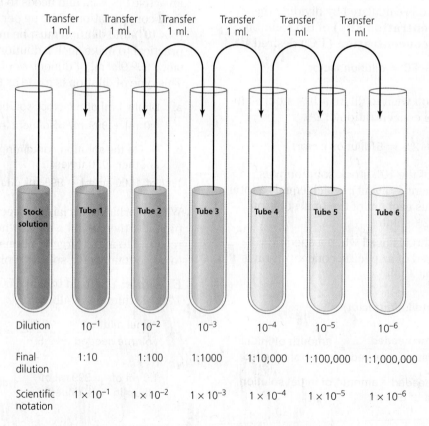

Appendix Figure 2.1 Serial dilution

Microbiological Media

The formulas of the media used in the exercises in this manual are listed alphabetically in grams per liter of distilled water unless otherwise specified. Sterilization of the media is accomplished by autoclaving at 15 lb pressure for 15 minutes unless otherwise specified. Most of the media are available commercially in powdered form, with specific instructions for their preparation and sterilization.

Ammonium sulfate broth (pH 7.3)

Ammonium sulfate	2.0
Magnesium sulfate · $7H_2O$	0.5
Ferric sulfate · $7H_2O$	0.03
Sodium chloride	0.3
Magnesium carbonate	10.0
Dipotassium hydrogen phosphate	1.0

Bacteriophage broth 10× (pH 7.6)

Peptone	100.0
Beef extract	30.0
Yeast extract	50.0
Sodium chloride	25.0
Potassium dihydrogen phosphate	80.0

Basal salts agar* and broth (pH 7.0)

0.5 M sodium diphosphate	100.0 ml
1.0 M potassium dihydrogen phosphate	100.0 ml
Distilled water	800.0 ml
0.1 M calcium chloride	1.0 ml
1.0 M magnesium sulfate	1.0 ml
Ammonium sulfate	2.0
*Agar	15.0

Note: Swirl until completely dissolved, autoclave, and cool. Aseptically add 10.0 ml of 1% sterile glucose.

Bile esculin (pH 6.6)

Beef extract	3.0
Peptone	5.0
Esculin	1.0
Oxgall	40.0
Ferric citrate	0.5
Agar	15.0

Blood agar (pH 7.3)

Infusion from beef heart	500.0
Tryptose	10.0
Sodium chloride	5.0
Agar	15.0

Note: Dissolve the above ingredients and autoclave. Cool the sterile blood agar base to 45°C to 50°C. Aseptically add 50 ml of sterile defibrinated blood. Mix thoroughly, avoiding accumulation of air bubbles. Dispense into sterile tubes or plates while liquid.

Brain heart infusion (pH 7.4)

Infusion from calf brain	200.0
Infusion from beef heart	250.0
Peptone	10.0
Dextrose	2.0
Sodium chloride	5.0
Disodium phosphate	2.5
Agar	1.0

Bromcresol purple dextrose fermentation broth (pH 7.2)

Bacto® casitone	10
Dextrose	5
Bromcresol purple (0.2%)	0.01

Bromcresol purple (0.2%) is made separately and filter sterilized. 5 ml is aseptically added to the medium.

Note: Autoclave at 12 lb pressure for 15 minutes.

Bromcresol purple lactose fermentation broth (pH 7.2)

Bacto casitone	10
Lactose	5
Bromcresol purple (0.2%)	0.01

Bromcresol purple (0.2%) is made separately and filter sterilized. 5 ml is aseptically added to the above medium.

Note: Autoclave at 12 lb pressure for 15 minutes.

Bromcresol purple maltose fermentation broth (pH 7.2)

Bacto casitone	10
Maltose	5
Bromcresol purple (0.2%)	0.01

Bromcresol purple (0.2%) is made separately and filter sterilized. 5 ml is aseptically added to the medium.

Note: Autoclave at 12 lb pressure for 15 minutes.

Bromcresol purple sucrose fermentation broth (pH 7.2)

Bacto casitone	10
Sucrose	5
Bromcresol purple (0.2%)	0.01

Bromcresol purple (0.2%) is made separately and filter sterilized. 5 ml is added to the medium aseptically.

Note: Autoclave at 12 lb pressure for 15 minutes.

Campy-BAP agar (pH 7.0)

Trypticase peptone	10.0
Thiotone™	10.0
Dextrose	1.0
Yeast extract	2.0
Sodium chloride	5.0
Sodium bisulfide	0.1
Agar	15.0
Vancomycin	10.0 mg
Trimethoprim lactate	5.0 mg
Polymyxin B sulfate	2500.0 IU
Amphotericin B	2.0 mg
Cephalothin	15.0 mg
Defibrinated sheep blood	10.0%

Note: Aseptically add the antibiotics and defibrinated sheep blood to the sterile, molten, and cooled agar.

Chocolate agar (pH 7.0)

Proteose peptone	20.0
Dextrose	0.5
Sodium chloride	5.0
Disodium phosphate	5.0
Agar	15.0

Note: Aseptically add 5.0% defibrinated sheep blood to the sterile and molten agar. Heat at 80°C until a chocolate color develops.

Crystal violet agar (pH 7.0)

Bacto beef extract	3
Bacto peptone	5
Bacto crystal violet	0.00014
Bacto agar	15

Note: 1.0 ml of a crystal violet stock solution may be added to the base medium. Stock solution: 14 mg of crystal violet dye dissolved in 100 ml of distilled water.

Decarboxylase broth (Moeller) (pH 6.0)

Peptone	5.0
Beef extract	5.0
Dextrose	0.5
Bromcresol purple	0.01
Cresol red	0.005
Pyridoxal	0.005
Distilled water	1000.0 ml

To make amino acid–specific medium, add one of the amino acids below; dispense in 3- to 4-ml amounts and autoclave at 121°C for 10 minutes.

L-lysine dihydrochloride or L-arginine monohydrochloride or L-ornithine dihydrochloride	10 g/l

Deoxyribonuclease (DNase) agar (pH 7.3)

Deoxyribonucleic acid	2.0
Phytane	5.0
Sodium chloride	5.0
Trypticase	15.0
Agar	15.0

Endo agar (pH 7.5)

Peptone	10.0
Lactose	10.0
Dipotassium phosphate	3.5
Sodium sulfite	2.5
Basic fuchsin	0.4
Agar	15.0

Eosin–methylene blue agar (Levine) (pH 7.2)

Peptone	10.0
Lactose	5.0
Dipotassium phosphate	2.0
Agar	13.5
Eosin Y	0.4
Methylene blue	0.065

Gel diffusion agar

Sodium barbital buffer	100.0 ml
Noble agar	0.8

Glucose acetate yeast sporulation agar (pH 5.5)

Glucose	1
Yeast extract	2
Sodium acetate (with 3H$_2$O)	5
Bacto agar	15

Glucose salts broth (pH 7.2)

Dextrose	5.0
Sodium chloride	5.0
Magnesium sulfate	0.2
Ammonium dihydrogen phosphate	1.0
Dipotassium hydrogen phosphate	1.0

Glycerol yeast extract agar supplemented with aureomycin (pH 7.0)

Glycerol	5.0 ml
Yeast extract	2.0
Dipotassium phosphate	1.0
Agar	15.0

Note: Aseptically add aureomycin, 10 μg per ml, to the sterile, molten, and cooled agar.

Grape juice broth

Commercial grape or apple juice	
Ammonium biphosphate	0.25%

Note: Sterilization not required when using a large yeast inoculum.

Hay infusion broth

Hay infusion broth preparations are prepared 1 week ahead of the laboratory session in which they will be used. Into a 2000-ml beaker place about 800 ml of water and two to three handfuls of dry grass or hay (obtained from a farm or storage barn). During the incubation period, the infusion should be aerated by passing air through a rubber tube attached to an air supply. This preparation is sufficient for a class and can be dispensed in 50-ml beakers.

Hektoen enteric agar (pH 7.1)

Peptic digest of animal tissue	12.0
Yeast extract	3.0
Bile salt	9.0
Lactose	12.0
Sucrose	12.0
Salicin	2.0
Sodium chloride	5.0
Sodium thiosulfate	5.0
Ferric ammonium citrate	1.5
Bromthymol blue	0.064
Acid fuchsin	0.5
Agar	13.5

Inorganic synthetic broth (pH 7.2)

Sodium chloride	5.0
Magnesium sulfate	0.2
Ammonium dihydrogen phosphate	1.0
Dipotassium hydrogen phosphate	1.0

KF broth (pH 7.2)

Polypeptone	10.0
Yeast extract	10.0
Sodium chloride	5.0
Sodium glycerophosphate	10.0
Sodium carbonate	0.636
Maltose	20.0
Lactose	1.0
Sodium azide	0.4
Phenol red	0.018

Lactobacilli MRS Broth Composition

Proteose peptone	10.0 g
Beef extract	10.0 g
Yeast extract	5.0 g
Dextrose	20.0 g
Sorbitan monooleate	1.0 g
Ammonium citrate	2.0 g
Sodium acetate	5.0 g
$MnSO_4 \times H_2O$	0.05 g
Na_2HPO_4	2.0 g
Deionized water	1000 ml

Note: Final pH of 6.5, autoclave 121°C to sterilize.

Lactose fermentation broth 1× and 2×* (pH 6.9)

Beef extract	3.0
Peptone	5.0
Lactose	5.0

*For 2× broth use twice the concentration of the ingredients.

Litmus milk (pH 6.8)

Skim milk powder	100.0
Litmus	0.075

Note: Autoclave at 12 lb pressure for 15 minutes.

Luria-Bertani (Miller) agar base (pH 7.0)

Pancreatic digest of casein	10.0
Yeast extract	5.0
Sodium chloride	0.5
Agar	15.0

Luria-Bertani (Miller) broth (pH 7.0)

Tryptone	10.0
Yeast extract	5.0
Sodium chloride	10.0

MacConkey agar (pH 7.1)

Bacto peptone	17.0
Proteose peptone	3.0
Lactose	10.0
Bile salts mixture	1.5
Sodium chloride	5.0
Agar	13.5
Neutral red	0.03
Crystal violet	0.001

Mannitol salt agar (pH 7.4)

Beef extract	1.0
Peptone	10.0
Sodium chloride	75.0
d-Mannitol	10.0
Agar	15.0
Phenol red	0.025

m-Endo broth (pH 7.5)

Yeast extract	6.0
Thiotone peptone	20.0
Lactose	25.0
Dipotassium phosphate	7.0
Sodium sulfite	2.5
Basic fuchsin	1.0

Note: Heat until boiling; do not autoclave.

m-FC broth (pH 7.4)

Biosate™ peptone	10.0
Polypeptone peptone	5.0
Yeast extract	3.0
Sodium chloride	5.0
Lactose	12.5
Bile salts	1.5
Aniline blue	0.1

Note: Add 10 ml of rosolic acid (1% in 0.2N sodium hydroxide). Heat to boiling with agitation; do not autoclave.

Milk agar (pH 7.2)

Skim-milk powder	100.0
Peptone	5.0
Agar	15.0

Note: Autoclave at 12 lb pressure for 15 minutes.

Minimal agar (pH 7.0) Minimal agar, supplemented with streptomycin and thiamine*

Solution A (pH 7.0)

Potassium dihydrogen phosphate	3.0
Disodium hydrogen phosphate	6.0
Ammonium chloride	2.0
Sodium chloride	5.0
Distilled water	800.0 ml

Solution B (pH 7.0)

Glucose	8.0
Magnesium sulfate • 7H$_2$O	0.1
Agar	15.0
Distilled water	200.0 ml

Note: Autoclave Solutions A and B separately and combine.

*To Solution B, add 0.001 g of thiamine prior to autoclaving. To the combined sterile and molten medium, add 50 mg (1 ml of 50 mg per ml) sterile streptomycin solution before pouring agar plates.

MR-VP broth (pH 6.9)

Peptone	7.0
Dextrose	5.0
Potassium phosphate	5.0

Mueller-Hinton agar (pH 7.4)

Beef, infusion	300.0
Casamino acids	17.5
Starch	1.5
Agar	17.0

Mueller-Hinton tellurite agar (pH 7.4)

Casamino acids	20.0
Casein	5.0
L-tryptophan	0.05
Potassium dihydrogen phosphate	0.3
Magnesium sulfate	0.1
Agar	20.0

Note: Aseptically add 12.5 ml of tellurite serum to the sterile, 50°C molten agar.

Nitrate broth (pH 7.2)

Peptone	5.0
Beef extract	3.0
Potassium nitrate	5.0

Nitrite broth (pH 7.3)

Sodium nitrite	2.0
Magnesium sulfate • 7H$_2$O	0.5
Ferric sulfate • 7H$_2$O	0.03
Sodium chloride	0.3
Sodium carbonate	1.0
Dipotassium hydrogen sulfate	1.0

Nitrogen-free mannitol agar* and broth (pH 7.3)

Mannitol	15.0
Dipotassium hydrogen phosphate	0.5
Magnesium sulfate	0.2
Calcium sulfate	0.1
Sodium chloride	0.2
Calcium carbonate	5.0
*Agar	15.0

Nutrient agar× and broth (pH 7.0)

Peptone	5.0
Beef extract	3.0
*Agar	15.0

Nutrient gelatin (pH 6.8)

Peptone	5.0
Beef extract	3.0
Gelatin	120.0

Peptone broth (pH 7.2)

Peptone	4.0

Phenol red dextrose broth (pH 7.3)

Trypticase	10.0
Dextrose	5.0
Sodium chloride	5.0
Phenol red	0.018

Note: Autoclave at 12 lb pressure for 15 minutes.

Phenol red inulin broth (pH 7.3)

Trypticase	10.0
Inulin	5.0
Sodium chloride	5.0
Phenol red	0.018

Note: Autoclave at 12 lb pressure for 15 minutes.

Phenol red lactose broth (pH 7.3)

Trypticase	10.0
Lactose	5.0
Sodium chloride	5.0
Phenol red	0.018

Note: Autoclave at 12 lb pressure for 15 minutes.

Phenol red sucrose broth (pH 7.3)

Trypticase	10.0
Sucrose	5.0
Sodium chloride	5.0
Phenol red	0.018

Note: Autoclave at 12 lb pressure for 15 minutes.

Phenylalanine agar (pH 7.3)

Yeast extract	3.0
Dipotassium phosphate	1.0
Sodium chloride	5.0
DL-phenylalanine	2.0
Bacto agar	12.0
Distilled water	1000.0 ml

Note: Completely dissolve ingredients in boiling water. Dispense in tubes, autoclave, and cool in slanted position.

Phenylethyl alcohol agar (pH 7.3)

Trypticase	15.0
Phytane	5.0
Sodium chloride	5.0
β-Phenylethyl alcohol	2.0
Agar	15.0

Potato dextrose agar (pH 5.6)

Infusion from potatoes	200.0
Bacto dextrose	20.0
Bacto agar	15.0

Sabouraud agar (pH 5.6) Sabouraud agar supplemented with chlortetracycline (Aureomycin)*

Peptone	10.0
Dextrose	40.0
Agar	15.0

*Aseptically add Aureomycin, 10 µg per ml, to the sterile, molten, and cooled medium.

Salt medium—*Halobacterium*

Sodium chloride	250.0
Magnesium sulfate • 7H$_2$O	10.0
Potassium chloride	5.0
Calcium chloride • 6H$_2$O	0.2
Yeast extract	10.0
Tryptone	2.5
Agar	20.0

Note: The quantities given are for preparation of 1-liter final volume of the medium. In preparation, make up two solutions, one involving the yeast extract and tryptone and the other the salts. Adjust the pH of the nutrient solution to 7. Sterilize separately. Mix and dispense aseptically.

SIM agar (pH 7.3)

Peptone	30.0
Beef extract	3.0
Ferrous ammonium sulfate	0.2
Sodium thiosulfate	0.025
Agar	3.0

Simmons citrate agar (pH 6.9)

Ammonium dihydrogen phosphate	1.0
Dipotassium phosphate	1.0
Sodium chloride	5.0
Sodium citrate	2.0
Magnesium sulfate	0.2
Agar	15.0
Bromthymol blue	0.08

Snyder test agar (pH 4.8)

Tryptone	20.0
Dextrose	20.0
Sodium chloride	5.0
Bromcresol green	0.02
Agar	20.0

Sodium chloride agar, 7.5% (pH 7.0)

Bacto beef extract	3.0
Bacto peptone	5.0
Sodium chloride	7.5
Bacto agar	15.0

Sodium chloride broth, 6.5% (pH 7.0)

Brain heart infusion broth	100.0 ml
Sodium chloride	6.5

Starch agar (pH 7.0)
Peptone	5.0
Beef extract	3.0
Starch (soluble)	2.0
Agar	15.0

Thioglycollate, fluid (pH 7.1)
Peptone	15.0
Yeast extract	5.0
Dextrose	5.0
L-cystine	0.75
Thioglycollic acid	0.3 ml
Agar	0.75
Sodium chloride	2.5
Resazurin	0.001

Tinsdale agar (pH 7.4)
Proteose peptone, No. 3	20.0
Sodium chloride	5.0
Agar	20.0

Note: Following boiling, distribute in 100-ml flasks. Autoclave, cool to 55°C, add 15 ml of rehydrated Tinsdale enrichment to each 100 ml, and mix thoroughly before dispensing.

Top agar (for Ames test)
Sodium chloride	5.0
Agar	6.0

Tributyrin agar (pH 7.2)
Peptone	5.0
Beef extract	3.0
Agar	15.0
Tributyrin	10.0

Note: Dissolve peptone, beef extract, and agar while heating. Cool to 90°C, add the tributyrin, and emulsify in a blender.

Triple sugar–iron agar (pH 7.4)
Beef extract	3.0
Yeast extract	3.0
Peptone	15.0
Proteose peptone	5.0
Lactose	10.0
Saccharose	10.0
Dextrose	1.0
Ferrous sulphate	0.2
Sodium chloride	5.0
Sodium thiosulfate	0.3
Phenol red	0.024
Agar	12.0

Trypticase nitrate broth (pH 7.2)
Trypticase	20.0
Disodium phosphate	2.0
Dextrose	1.0
Agar	1.0
Potassium nitrate	1.0

Trypticase soy agar (pH 7.3)
Trypticase	15.0
Phytane	5.0
Sodium chloride	5.0
Agar	15.0

Tryptone agar* and broth
Tryptone	10.0
Calcium chloride (reagent)	0.01–0.03 M
Sodium chloride	5.0
*Agar	11.0

Tryptone soft agar
Tryptone	10.0
Potassium chloride (reagent)	5.0 ml
Agar	9.0

Urea broth
Urea broth concentrate (filter-sterilized solution)	10.0 ml
Sterile distilled water	90.0 ml

Note: Aseptically add the urea broth concentrate to the sterilized and cooled distilled water. Under aseptic conditions, dispense 3-ml amounts into sterile tubes.

Yeast extract broth (pH 7.0)
Peptone	5.0
Beef extract	3.0
Sodium chloride	5.0
Yeast extract	5.0

Biochemical Test Reagents

Agarose gel, for Electrophoresis

Agarose	1 g
Tris-borate buffer (1×)	125 ml

Note: Melt agarose, being careful not to overboil. Cover loosely with foil and hold at room temperature, or place in 60°C waterbath until ready for use.

Barritt's reagent, for detection of acetylmethylcarbinol

Solution A

Alpha-naphthol	5.0 g
Ethanol, absolute	95.0 ml

Note: Dissolve the alpha-naphthol in the ethanol with constant stirring.

Solution B

Potassium hydroxide	40.0 g
Creatine	0.3 g
Distilled water	100.0 ml

Note: Dissolve the potassium hydroxide in 75 ml of distilled water. The solution will become warm. Allow to cool to room temperature. Add the creatine and stir to dissolve. Add the remaining water. Store in a refrigerator.

Biotin-histidine solution, for Ames test

L-Histidine HCl	0.5 mM
Biotin	0.5 mM
Distilled water	10.0 ml

Buffered glycerol (pH 7.2), for immunofluorescence

Glycerin	90.0 ml
Phosphate buffered saline	10.0 ml

Diphenylamine reagent, for detection of nitrates

Dissolve 0.7 g diphenylamine in a mixture of 60 ml concentrated sulfuric acid and 28.8 ml of distilled water. Cool and slowly add 11.3 ml of concentrated hydrochloric acid. Allow to stand for 12 hours. Sedimentation indicates that the reagent is saturated.

Endonuclease buffers

Buffer 1: EcoRI buffer

Tris-HCl (pH 7.5)	50 mM
$MgCl_2$	10 mM
NaCl	100 mM
Triton® X-100	0.02%
BSA	0.1 mg/ml

Buffer 2: HindIII buffer

Tris-HCl (pH 8.5)	10 mM
$MgCl_2$	10 mM
KCl	100 mM
BSA	0.1 mg/ml

Buffer 3: BamHI buffer

Tris-HCl (pH 8.0)	10 mM
$MgCl_2$	5 mM
KCl	100 mM
2-Mercaptoethanol	1 mM
Triton X-100	0.02%
BSA	0.1 mg/ml

Ferric chloride reagent

Ferric chloride	10.0 g
Distilled water	100.0 ml

Gram's iodine, for detection of starch
As in Gram's stain

Hydrogen peroxide, 3%, for detection of catalase activity
Note: Refrigerate when not in use.

Kovac's reagent, for detection of indole

p-Dimethylaminobenzaldehyde	5.0 g
Amyl alcohol	75.0 ml
Hydrochloric acid (concentrated)	25.0 ml

Note: Dissolve the p-dimethylaminobenzaldehyde in the amyl alcohol. Add the hydrochloric acid.

Loading Dye 6×, for gel electrophoresis

Glycerol (50%)	6 ml
Bromphenol blue (2%)	1 ml
Xylene cyanol (2%)	1 ml
Distilled water	1000 ml

Note: This can be stored in the refrigerator indefinitely.

McFarland Barium Sulfate Standards,
for API Staph-Ident procedure

Prepare 1% aqueous barium chloride and 1% aqueous sulfuric acid solutions. Using the following table, add the amounts of barium chloride and sulfuric acid to clean 15- × 150-mm screw-capped test tubes. Label the tubes 1 through 10.

Preparation of McFarland Standards

Tube	Barium Chloride 1% (ml)	Sulfuric Acid 1% (ml)	Corresponding Approximate Density of Bacteria (million/ml)
1	0.1	9.9	300
2	0.2	9.8	600
3	0.3	9.7	900
4	0.4	9.6	1200
5	0.5	9.5	1500
6	0.6	9.4	1800
7	0.7	9.3	2100
8	0.8	9.2	2400
9	0.9	9.1	2700
10	1.0	9.0	3000

Methyl cellulose, for microscopic observation of protozoa

Methyl cellulose	10.0 g
Distilled water	90.0 ml

Methylene blue stain (0.025%)

Methylene blue 1% stock solution (1 g + 99 ml distilled H_2O)	10 ml
Distilled water	390 ml

Methyl red solution, for detection of acid

Methyl red	0.1 g
Ethyl alcohol	300.0 ml
Distilled water	200.0 ml

Note: Dissolve the methyl red in the 95% ethyl alcohol. Dilute to 500 ml with distilled water.

Nessler's reagent, for detection of ammonia

Potassium iodide	50.0 g
Distilled water (ammonia-free)	35.0 ml

Add saturated aqueous solution of mercuric chloride until a slight precipitate persists.

Potassium hydroxide (50% aqueous)	400.0 ml

Note: Dilute to 1000 ml with ammonia-free distilled water. Let stand for 1 week, decant supernatant liquid, and store in a tightly capped amber bottle.

Nitrate test solution, for detection of nitrites

Solution A, Sulfanilic acid

Sulfanilic acid	8.0 g
Acetic acid, 5 N: 1 part glacial acetic acid to 2.5 parts distilled water	1000.0 ml

Solution B, Alpha-naphthylamine

Alpha-naphthylamine	5.0 g
Acetic acid, 5 N	1000.0 ml

Oxidase Reagent

Tetramethyl-*p*-phenylene diamine dihydrochloride	1.0 g
Distilled water	100 ml

Orthonitrophenyl-β-D-galactoside (ONPG),
for enzyme induction

0.1 M sodium phosphate buffer (pH 7.0)	50.0 ml
ONPG (8×10^{-4} M)	12.5 mg

***p*-Aminodimethylaniline oxalate,** for detection of oxidase activity

p-Aminodimethylaniline oxalate	0.5 g
Distilled water	50.0 ml

Note: To dissolve fully, gently warm the solution.

Phosphate-buffered saline, 1% (pH 7.2–7.4),
for immunofluorescence

Solution A

Disodium phosphate	1.4 g
Distilled water	100.0 ml

Solution B

Sodium dihydrogen phosphate	1.4 g
Distilled water	100.0 ml

Note: Add 84.1 ml of Solution A to 15.9 ml of Solution B. Add 8.5 g of sodium chloride and enough distilled water to make 1 liter.

Rabbit plasma, for detection of coagulase activity

Note: Store vials at 2°C to 8°C. Reconstitute by the addition of 7.5 ml of sterile water.

Sodium barbital buffer, for immunofluorescence

Sodium barbital	6.98 g
Sodium chloride	6.0 g
1 N hydrochloric acid	27.0 ml
Distilled water, to fill to 1000 ml	

Toluidine blue solution, 0.1%, for detection of DNase activity

1% toluidine blue solution	0.1 ml
Distilled water	99.9 ml

Tris-acetate buffer 10×

Tris base	48.4 g
Glacial acetic acid	11 g
EDTA (0.5 M)	20 ml
Distilled water	1000 ml

Note: Add ingredients to 1 liter volumetric flask and then add water to volume.

Tris-acetate buffer 1×

Tris-acetate buffer 10× (see previous entry)	100 ml
Distilled water	900 ml

Note: Buffer can be stored indefinitely at room temperature.

Tris-borate buffer 5×

Tris base	54 g
Boric acid	27.5 g
EDTA (0.5M, pH 8.0)	20 ml
Distilled water	1000 ml

Tris-borate buffer 1× working solution

Tris-borate buffer 5× (see previous entry)	200 ml
Distilled water	800 ml

Staining Reagents

Acid-Fast Stain

Carbol fuchsin (Ziehl's)

Solution A

Basic fuchsin (90% dye content)	0.3 g
Ethtyl alcohol (95%)	10.0 ml

Solution B

Phenol	5.0 g
Distilled water	95.0 ml

Note: Mix Solutions A and B. Add 2 drops of Triton X per 100 ml of stain for use in heatless method.

Acid Alcohol

Ethyl alcohol (95%)	97.0 ml
Hydrochloric acid	3.0 ml

Methylene blue

Methylene blue	0.3 g
Distilled water	100.0 ml

Capsule Stain

Crystal violet (1%)

Crystal violet (85% dye content)	1.0 g
Distilled water	100.0 ml

Copper sulfate solution (20%)

Copper sulfate ($CuSO_4 \cdot 5H_2O$)	20.0 g
Distilled water	80.0 ml

Fungal Stains

Lactophenol–cotton-blue solution

Lactic acid	20.0 ml
Phenol	20.0 g
Glycerol	40.0 ml
Distilled water	20.0 ml
Aniline blue	0.05 g

Note: Heat gently in hot water (double boiler) to dissolve; then add aniline blue dye.

Water-iodine solution

Gram's iodine (as in Gram's stain)	10.0 ml
Distilled water	30.0 ml

Gram Stain

Crystal violet (Hucker's)

Solution A

Crystal violet (90% dye content)	2.0 g
Ethyl alcohol (95%)	20.0 ml

Solution B

Ammonium oxalate	0.8 g
Distilled water	80.0 ml

Note: Mix Solutions A and B.

Gram's iodine

Iodine	1.0 g
Potassium iodide	2.0 g
Distilled water	300.0 ml

Ethyl alcohol (95%)

Ethyl alcohol (100%)	95.0 ml
Distilled water	5.0 ml

Safranin

Safranin O	0.25 ml
Ethyl alcohol (95%)	10.0 ml
Distilled water	100.0 ml

Negative Stain

Nigrosin

Nigrosin, water-soluble	10.0 g
Distilled water	100.0 ml

Note: Immerse in boiling waterbath for 30 minutes.

Formalin	0.5 ml

Note: Filter twice through double filter paper.

Spore Stain

Malachite green

Malachite green	5.0 g
Distilled water	100.0 ml

Safranin

Same as in Gram stain

Experimental Microorganisms

Cultures

BSL-1 Bacteria

Alcaligenes faecalis
Alcaligenes viscolactis
Bacillus cereus
Bacillus megaterium
Bacillus stearothermophilus
Branhamella catarrhalis (Moraxella catarrhalis)
Citrobacter freundii
Citrobacter intermedius
Clostridium butylicum (Clostridium beijerinckii)
Clostridium sporogenes
Corynebacterium kutscheri
Corynebacterium xerosis
Enterobacter aerogenes
Escherichia coli
Escherichia coli ATCC e 23724
Escherichia coli ATCC e 23725
Escherichia coli ATCC e 23735
Escherichia coli ATCC e 23740
Escherichia coli ATCC 39991
Escherichia coli ATCC 53100
Escherichia coli B
Halobacterium salinarum
Lactobacillus casei
Lactobacillus delbrueckii subsp. bulgaricus
Lactobacillus fermenti
Lactococcus lactis
Leuconostoc mesenteroides
Micrococcus luteus
Micrococcus varians
Moraxella catarrhalis (Branhamella catarrhalis)
Mycobacterium smegmatis
Neisseria sicca
Proteus inconstans
Proteus mirabilis
Proteus rettgeri
Proteus vulgaris
Pseudomonas fluorescens
Pseudomonas mallei
Pseudomonas savastanoi
Serratia marcescens
Staphylococcus epidermidis

Staphylococcus saprophyticus ATCC e 15305
Streptococcus bovis
Streptococcus salivarius subsp. thermophilus
Streptococcus var. Lancefield Group E
Streptomyces griseus
Treponema denticola

BSL-2 Bacteria

Based on publications from the American Type Culture Collection (ATCC) and Centers for Disease Control and Prevention (CDC).

Campylobacter jejuni
Enterococcus faecalis
Klebsiella ozaenae (Klebsiella pneumoniae subsp. ozaenae)
Klebsiella pneumoniae ATCC e 15574
Moraxella bovis
Neisseria mucosa
Pseudomonas aeruginosa
Salmonella typhimurium
Salmonella typhimurium ATCC e 29631 *(Salmonella enterica subsp. enterica* ATCC 29631)
Shigella dysenteriae
Staphylococcus aureus ATCC e 25923
Staphylococcus aureus ATCC e 27659
Staphylococcus aureus ATCC e 27660
Staphylococcus aureus ATCC e 27661
Staphylococcus aureus ATCC e 27691
Staphylococcus aureus ATCC e 27693
Staphylococcus aureus ATCC e 27697
Streptococcus agalactiae
Streptococcus mitis
Streptococcus pneumoniae
Streptococcus pyogenes ATCC 12385

Fungi

Alternaria sp.
Aspergillus niger
Candida albicans
Cephalosporium sp.
Cladosporium sp.
Fusarium sp.
Mucor mucedo
Penicillium notatum (penicillium chrysogenum)

529

Fungi (cultures) Ctd.

Rhizopus stolonifer
Rhodotorula rubra
Saccharomyces cerevisiae
Saccharomyces cerevisiae var. *ellipsoideus*
Schizosaccharomyces octosporus
Selenotila intestinalis

Viruses

T_2 coliphage

Prepared Slides

Bacteria

Aquaspirillum itersonii
Bacillus subtilis
Spirillum itersonii
Staphylococcus aureus

Fungi

Saccharomyces cerevisiae

Protozoa

Balantidium coli
Entamoeba histolytica
Giardia lamblia (Giardia intestinalis)
Plasmodium vivax
Trypanosoma gambiense

Archaea

Halobacterium salinarium

Other

Blood smear

Credits

Illustration Credits

All illustrations by Lachina unless otherwise noted.

33.1: From Harold S. Brown, *Basic Clinical Parasitology*, 4th ed. New York: Appleton-Century-Crofts, 1975.

56.2: Carolina Biological Supply Company/Precision Graphics.

66.2: Wampole Laboratories Division, Carter-Wallace, Inc., Cranbury, NJ.

Photo Credits

Cover: science photo/Shutterstock

2.2: James Cappuccino

4.1: Charles D. Winters/Science Source

6.1: L. Brent Selinger/Pearson Education

7.3a: L. Brent Selinger/Pearson Education

7.3b: Jennifer M. Warner/UNC Charlotte

7.3c: Michael Abbey/Science Source

8.1: L. Brent Selinger. Pearson Education

9.1a: James Cappuccino

9.1b: Centers for Disease Control and Prevention

9.2a-d: David B. Alexander/University of Portland

10.2: James Cappuccino

11.2: Steven R. Spilatro/Marietta College

11.4b: James Cappuccino

12.2a: L. Brent Selinger/Pearson Education

12.2b: James Cappuccino

13.1a-b: L. Brent Selinger/Pearson Education

13.2a: James Cappuccino

13.2b: L. Brent Selinger/Pearson Education

13.2c: James Cappuccino

13.3a: L. Brent Selinger/Pearson Education

13.3b: James Cappuccino

13.3c: James Cappuccino

14.2: David B. Alexander/University of Portland

17.3: L. Brent Selinger/Pearson Education

18.1a: Hausser Scientific

18.3: L. Brent Selinger/Pearson Education

18.4: L. Brent Selinger/Pearson Education

18.6: L. Brent Selinger/Pearson Education

19.3: L. Brent Selinger/Pearson Education

20.1: David B. Alexander/University of Portland

20.3: James Cappuccino

20.5a-b: L. Brent Selinger/Pearson Education

21.5: James Cappuccino

22.2: James Cappuccino

23.3: James Cappuccino

23.5: James Cappuccino

23.8: James Cappuccino

23.10: James Cappuccino

24.2: James Cappuccino

25.2: James Cappuccino

26.1: James Cappuccino

27.2: James Cappuccino

28.1a-b: David B. Alexander/University of Portland

28.1c: Brenda Grafton Wellmeyer/Department of Biology/Lone Star College - North Harris

29.1: David B. Alexander/University of Portland

30.2: L. Brent Selinger/Pearson Education

30.4: L. Brent Selinger/Pearson Education

32.1: Eric V. Grave/Science Source

32.2a: Biophoto Associates/Science Source

32.2b: M. I. Walker/Science Source

33.2a-b: Centers for Disease Control and Prevention (CDC)

33.3a: Centers for Disease Control and Prevention (CDC)

33.3b: Biophoto Associates/Science Source

33.4: Eric V. Grave/Science Source

33.5: Dr. Marilise B. Rott

33.6: Centers for Disease Control, Office on Smoking and Health

34.2: Jared Martin

34.3: Leonard Lessin/FBPA/Science Source

34.4: James Cappuccino

34.5: Perennou Nuridsany/Science Source

34.6: Biophoto Associates/Science Source

35.1a: James Cappuccino

35.1b: Biophoto Associates/Science Source

35.1c: John Durham/Science Source

35.2: Nguyen, Nhu

35.4: L. Brent Selinger/Pearson Education

37.1: Pearson Education

40.1: STERIS Corporation

42.2:   James Cappuccino
42.4:   James Cappuccino
43.2:   James Cappuccino
43.3:   Antimicrobial Test Laboratories LLC
44.1:   L. Brent Selinger/Pearson Education
47.2:   L. Brent Selinger/Pearson Education
47.3a-c: L. Brent Selinger/Pearson Education
48.2:   L. Brent Selinger/Pearson Education
57.4:   L. Brent Selinger/Pearson Education
57.6:   L. Brent Selinger/Pearson Education
59.2:   L. Brent Selinger/Pearson Education
60.2:   L. Brent Selinger/Pearson Education
60.3:   James Cappuccino
60.1a-b: L. Brent Selinger/Pearson Education
60.4:   James Cappuccino
60.5a-b: L. Brent Selinger/Pearson Education
61.1:   L. Brent Selinger/Pearson Education
61.2a-b: James Cappuccino
61.3:   James Cappuccino
61.4:   James Cappuccino
62.1a-c: L. Brent Selinger/Pearson Education
62.2:   James Cappuccino
62.3:   L. Brent Selinger/Pearson Education
62.4:   James Cappuccino
62.5:   James Cappuccino
63.1a-b: James Cappuccino
63.2:   James Cappuccino

64.2b: L. Brent Selinger/Pearson Education
64.3a-b: L. Brent Selinger/Pearson Education
65.1:   L. Brent Selinger/Pearson Education
67.2:   Becton Dickinson & Company (BD)
70.1:   LeBeau/Newscom

Text Credits

Table 42.2:   Clinical and Laboratory Standards Institute. *Performance Standards for Antimicrobial Disk Susceptibility Tests*, Tenth Edition, 2008.

Table 47.1:   pp 9–51, *Standard Methods for the Examination of Water and Wastewater*, 20th Edition (1998). M. J. Taras, A. E. Greenberg, R. D. Hoak, and M. C. Rand, eds. American Public Health Association, Washington, D.C. Copyright 1998, American Public Health Association, and Bacteriological Analytical Manual (BAM), 8th Edition, Food and Drug Administration, 1998.

Table 59.1:   *Assessment of Susceptibility to Dental Caries*, Courtesy of Difco Laboratories, Inc., Detroit, Michigan, Becton Dickinson & Company (BD

Table 61.2:   *API STAPH-IDENT Profile Register*, STAPH-IDENT, Analytab Products, Division of Sherwood Medical, Plainview, New York, Sherwood Medical Company

Table 66.1:   *Bacturcult: Interpretation of Colony Counts*, Wampole Laboratories Division, Carter-Wallace, Inc., Cranbury, NJ 08512, Carter-Wallace, Inc.

Index

A

Abbé condenser, 33
Abscesses, 17, 69, 435
Absidia spp.
   soil populations, 345
Abundance of growth (cultural
      characteristic), 25
Acetic acid bacteria, 156*f*
Acetobacter spp., 156*f*
Acetone, 156*f*, 344
Acetylmethylcarbinol, 171*f*
Acid-alcohol, 75, 528
Acid curd, 188*f*, 212*t*
Acid-fast stain, 75–80, 528
   clinical application, 80
   of mycobacteria, 77*f*
   principle, 75
   procedure, 76*f*, 77
Acidic stains, 47, 49*f*
   negative staining, 47
   picric acid in, 48*f*
Acidophiles, 115, 115*f*
Acids/alkylating agents, mechanism of action/
      uses, 306*t*, 307*t*
Acinetobacter spp., TSI reaction, 162*f*
Acne, 425
Acremonium spp., 351
Actinobacteria, 351
Actinomyces odontolyticus, dental caries
      and, 419
Actinomycetes, 234, 345
Adaptive enzymes, 365
Adaptive immunity, 491
Additive (indifferent) effect of drug
      combinations, 293, 294, 294*f*, 295
Adenosine triphosphate (ATP), 92, 150, 331
Adsorption
   in animal viruses, 259
   in bacteriophages, 257
Aerial mycelium, 235
Aerobes, 119
   vs Anaerobes, 120
Aerobic celllular respiration, 121*f*, 155*f*
Aerobic cellular respiration, 119
Aerobic microorganisms, 121*f*, 125*f*
   identifiying via growth distribution, 121*f*
   oxygen requirements, 119
   redox potentials, 125, 125*f*
Aerotolerant anaerobes, 119, 121*f*
   distribution of growth, 121*f*
   oxygen requirements, 119
African sleeping sickness, 225, 230
Ag glutinins, 491
Agar, 1, 2*f*
   blood, 100, 101, 424*f*, 444*f*, 451*f*
   chocolate, 424–425, 424*f*, 425*f*
   crystal violet, 99
   eosin-methylene blue, 100
   MacConkey, 100, 101*f*, 102
   mannitol salt, 99, 102*f*
   phenylethyl alcohol, 99, 100*f*
   sodium chloride (7.5%), 99
   as solid media, 2*f*
Agar deep tubes, 2, 2*f*
Agar plates, 2, 2*f*
   enrichment culture technique and, 359
   growth patterns on nutrient, 25
   Sabouraud, 236*f*, 425*f*
Agar plating method for cell counts, 133*f*

Agar slants, 2, 2*f*
   cultural characteristics, 26*f*
   cultural characteristics of selected bacteria,
      212*t*
   growth patterns on nutrient, 26*f*
Agarose gel, 407
   casting, 399*f*
   electrophoresis, 401
   set up of unit for DNA electrophoresis,
      401*f*
Agglutination, 491
Agglutination reactions, 497*f*
   febrile antibody test, 497, 497–502, 498
Agriculture, biotechnical applications, 387–388
AIDS (acquired immunodeficiency syndrome),
      244, 492, 507
Alcaligenes faecalis, 95, 188
   carbohydrate fermentation, 155–157
   cultural characteristics, 212*t*
   IMViC test series and, 171*f*, 173*f*
   litmus-milk reaction, 187–189, 212*t*
   nitrate reduction test, 193–194, 194*f*
   pH requirements, 116
   TSI agar test, 162*f*
   use of pyruvic acid by, 156f
Alcaligenes spp.
   as food-borne organism, 315
   TSI reaction, 162*f*
Alcaligenes viscolactis, 85
Alcohol production, biochemical pathway, 321*f*
Alcohols, mechanism of action/uses, 305*t*, 306*t*
Algae, pond water, 41*f*
Alkaline reaction in litmus milk reaction, 188,
      188*f*
Alkalophiles, 115, 115*f*
Alpha hemolysis, 443
   blood agar and, 101, 101*f*, 444*f*
   chocolate agar and, 424*f*
   identifying streptococcal pathogens
      by, 443
   identifying *Streptococcus pneumoniae*, 444*f*
Alternaria spp., 251*t*, 252*t*
Alternate streak-plate method, 16*f*
Amebic dysentery, 229*f*
American Society for Microbiology (ASM), 148,
      261, 278
American Type Culture Collection, 261
Ames, Bruce, 382
Ames test, 381, 382, 383*f*
   for identifying carcinogens, 381–386
Amine, 205
Amino acids, 151
   decarboxylase test, 205–207
   phenylalanine deaminase test, 205–210, 208*f*
Ammonia, phenylalanine deamination, 207
Ammonification by soil microbes, 343, 344*f*
Ammonium hydroxide, 157
Amoeba, 41*f*, 220*f*, 220*t*
Amoebic dysentery, 228*f*
Ampicillin, 291*t*
Amycolatopsis spp., 351
Amylase, 149
Anaerobes
   aerotolerant, 119
   cultivation of, 126*f*
   distribution of growth, 121*f*
   facultative, 119, 121*f*, 125*f*
   obligate, 119
   oxygen's toxicity to, 119
   vs aerobes, 120
Anaerobic, 121*f*

Anaerobic cells, 125*f*
Anaerobic cellular respiration, 155*f*
Analytical profile index (API) system, 459, 459*f*,
      461, 462, 464
Anapheles mosquito, 225, 228*f*
Animal feed/hides as source of food
      contamination, 315
Animal viruses, 259
Anionic agents, mechanism of action/uses,
      306*t*, 307*t*
Antibiotic serial dilution-plate setup, 300*t*
Antibiotics, 289
   as antimicrobial agents, 289*t*
   Bacillus spp. (bacitracin)., 344
   drug combination synergism, 292, 298
   effective, selection of, 291
   Kanamycins and, 344
   microbe-produced, 351–353
   Penicillium spp. (penicillin), 344
   prototypic, 289*t*
   resistance to, 372, 377, 377–380
   Streptomyces spp., 344
   testing new, 351
Antibodies, 491
   enzyme-linked immunosorbent assay
      (ELISA), 503, 503–506
   Tetracyclines and, 344
Antibody titer test, 497–498, 499*f*, 501
Antigen-antibody reactions
   agglutination and, 491
   precipitin formation, 492
Antigens, 491
   enzyme-linked immunosorbent assay
      (ELISA), 503, 503–506
   febrile, 497
Antimetabolites, 289
Antimicrobial agents
   antibiotics, 289, 289*t*, 291, 344, 351–353
   antiseptics, 277, 305–307*t*, 305–313
   disinfectants, 277, 305–307*t*, 305–313
Antimicrobial spectrum of isolates,
      determination of, 353–355
Antimocrobial-sensitivity disks, 292
Antiseptics, 277
   disc diffusion testing of, 311
   mechanism of actions/uses, 306*t*
   modified use dilution testing of,
      309–310, 312
   susceptibility test, 308
API (Analytical Profile Index) System, 459,
      459*f*, 461–462, 464
   strip, 460*f*
API STAPH-IDENT Profile Register, 439*t*–440*t*
Appendices, 515–530
Apples, fermentation of, 321–322, 321*f*
Aquaspirillum itersonii, 37, 38
Arborescent (cultural characteristic), 25, 26*f*
Archaea, 530
Arizona spp., TSI reaction, 162*f*
Ascomycetes, 233, 234*t*
Ascospores, 235*f*, 243
Ascus, 235*f*, 243
Aseptic inoculation and transfer, 9
Aseptic techniques, 3
Aspergillus fungi, 317
Aspergillus niger, 236*f*
   conidiophore and conidia of, 236*f*
   cultivation of
   cultivation of molds, 236*f*
   oxygen requirements, 120
   used to make citric acid, 344

Kirby-Bauer Antibiotic Sensitivity Test, 292–293
  on nutrient agar plates, 100*f*
  spontaneous mutation rate in, 363
  staining and, 57
  temperature and, 111
  TSI agar test, 162*f*
  in water, 331
Escherichia coli B, coliphage isolation, 267–272
Escherichia spp.
  as food-borne organism, 315
Essential metabolite, 289
Ester bonds, 149
Ethyl alcohol (95%), 68, 68*f*, 528
  mechanism of actions/uses, 305t, 306t
Ethylene oxide, mechanism of action/uses, 3*f*, 306t, 307t
Ethylhydrocupreine hydrochloride, 451
Eubacteriales, soil populations of, 345
Euglena, 41*f*, 220*f*, 220t
Excision repair system, 286
Exoenzymes. *see* Extracellular enzymes
Experimental microorganisms, 529, 529–530
  BSL-1 Bacteria, 529
  BSL-2 Bacteria, 529
  cultures, 529–530
  fungi, 529–530
  prepared slides, 530
  viruses, 529
Exponent, 515
Extended-spectrum b-lactamases (ESBLs), 300
Extracellular enzymes (exoenzymes), 147, 148*f*, 149–154
  pathogens and, 151
Eye conjunctiva, flora in, 423
Eye spot, 220t
Eyepiece lens of microscope, 33

F

Facultative anaerobes, 119, 125*f*
  carbohydrate fermentation by, 155–160
  distribution of growth, 121*f*
  oxygen requirements for, 119
  redox potentials, 125
Facultative thermophiles, temperature requirements, 110
Fasciolopsis buski, 327
Fastidious microorganisms, 93
Febrile antibody test, 497, 498, 501
Febrile antigens, 497
Febrile disease diagnosis, 497
Fecal contamination of water, 327, 337–340
  testing, 329–333
Fermentation, 321–322
  alcohol and, 321–322, 325
  as bioxidative pathway, 155*f*
  carbohydrate, 148*f*, 155–160
  cellular respiration vs., 155–160, 155*f*
  lactose, 187–189, 323–324
Fermentation studies, 245, 248
Fermenter, 101*f*
Ferrous sulfate, 162
Fertility factor (F factor), 371
Filamentous margin (cultural characteristic), 26*f*
Filiform (cultural characteristic), 25, 26*f*
Filtration
  asceptic technique and, 3*f*
  aseptic technique, 3*f*
  to control microbial growth, 278*f*
Final concentration (FC), 517
Fine-adjustment knobs, of microscope, 33, 34*f*
Fission, in yeasts, 243
Flagella, protozoan, 220t
Flat elevation (cultural characteristic), 26*f*, 27
Fleming, Alexander, 299
Flocculent growth, 26*f*, 27
Fluid thioglycollate medium, 127, 127*f*

Flukes, waterborne diseases and, 327
Fluorescent microscope, 32
Food handlers, as contamination source, 315
Food microbiology, 315–326
  bacterial count analysis, 317–320
  in cheese production, 344
  rold of Saccharomyces in, 243
  role of *Saccaromyces* in, 243
  in wine production, 243, 321–322, 321*f*
Food products, 317–320, 318*f*
Food utensil, as contamination source, 315
Food vacuoles, 220t
Forespore, 82*f*
Form (cultural characteristic), 25
  form, 26*f*
Formaldehyde, mechanism of action/uses, 306t, 307t
Formation of colored complex indicative of NO3- reduction, 194
Four-way streak-plate technique, 15*f*, 16*f*
Free-flowing steam, 279
Free-living protozoa, 219–225, 220t, 226t
Free radicals, 119
Free spores, 81, 82*f*
Fungal infection, diagnosis of, 251
Fungal Stains, 528
Fungi, 230–256, 234t, 251–255, 529–530, 530. *see also* Yeasts
  Blastomyces dermatitidis, 471*f*
  Candida albicans, 471*f*
  Coccidioides immitis, 471*f*
  cultivation of molds, 235–242
  identification of unknown, 251–255, 251t–253t. *see also* yeasts
  soil sample enumeration, 347*f*
  yeasts, 235–242, 243–250. *see also* Yeasts; yeasts
Fungi Imperfecti (Deuteromycetes), 233, 234t
Fungi, isolation of on solid media, 238
Fusarium spp., 252t
  as food-borne organism, 315
  morphology/microscopic appearance, 251t

G

Gametocytes, 226
Gamma hemolysis, 101, 101*f*
  identifying streptolococcal pathogens by, 443, 444*f*
Gamma radiation to control microbial growth, 278*f*, 285
Gas gangrene, 125
Gas production, detection of, 157
Gaseous requirement, 92
Gases
  in carbohydrate fermentation, 156*f*, 157
Gases, in lactose fermentation, 187
GasPak anaerobic system, 126*f*, 127
GasPak jar, 468*f*
GasPak system, 128*f*
Gastroenteritis, 457
Gel electrophoresis, 401, 407, 407–416
Gel loading scheme, 411lt
Gelatin hydrolysis, 148*f*, 150–151, 152, 212t
Gelatinase, 151
Generation time of bacterial growth, 139, 140*f*
Genetic engineering, 387–388. *see also* Biotechnology
  plasmids and, 398
Genetic variability mechanism, 363–364
Genetics (bacterial), 363–386
  chemical carcinogenicity screening, 381–386
  enzyme Induction, 365–370
  Isolation of a Streptomycin-resistant Mutant, 377–380
  isolation of mutants, 377–380
  recombination via conjugation, 371–376
Genital herpes, 507
  isolation of herpes simplex virus (HSV), 509–510, 513

Genital warts, 507
Gentamicin, 291t
Genus Identification of Unknown Bacterial Cultures, 211–215
Germination, 81, 82*f*
Giardia intestinalis:, 225, 228*f*
Giardia intestinalis:
  diarrhea and, 230
  waterborne disease and, 327
Glomerulonephritis, 443, 471
Glossina (tsetse fly), 225
Glucose, 93, 149
  in chemically defined media, 93
  degradation in Embden-Meyerhof (glycolytic) pathway, 156*f*
Glucose fermentation
  carbohydrate fermentation and, 155–160
  methyl red test, 170
  in saliva, dental caries and, 419, 420*f*
  triple sugar-iron agar test (TSI) and, 162*f*
Glucose fermentation by E. aerogenes, 171*f*
Glucose salts broth, 93
Glycerol, 149
Glycolic (Embden-Meyerhof) pathway, 155, 156*f*
  Embden-Meyerhof (glycolic) pathway, 162
Glycosidic bonds, 149
Golden Era of Microbiology, 417
Gonorrhea, 507
Governing Bodies for Laboratory Procedures, 278
Gradient-plate technique, 377
Grain stain, characteristics of selected bacteria, 212t
Gram-negative bacteria, 85, 471*f*
  Klebsiella pneumoniae, 471*f*
  Proteus vulgaris, 471*f*
  Pseudomonas aeruginosa, 471*f*
Gram-negative identification, 486*f*–487*f*
Gram-negative stain of E. coli, 67*f*
Gram positive bacteria, 471*f*
Gram-positive identification, 484*f*–485*f*
Gram-positive stain of streptococci, 67*f*
Gram stain, 67–74, 528
  microscopic observation of cells and, 68
Gram-stained cells, 67*f*
Gram staining, 71
  as diagnostic staining procedure, 69
  procedure, 70*f*
Gram's iodine, 68, 68*f*, 528
Grapes, fermentation of, 321–322, 321–322, 321*f*
Griffith, Fred, 389
Group A streptococci, 443, 445*f*
Group B streptococci, 443, 445*f*
Group D streptococci, 445*f*
Growth. *see* Microbial growth control
Growth curves, using to determine antimicrobial resistance, 140
Guinea worm *(Drancunculus medinenesis),* 327

H

Haemophilus, 291t
Haemophilus influenzae, 85, 389
Halogens, mechanism of action/uses, 306t
Hand washing, effectiveness of, 427–429
Hanging-drop preparation to view living microbes, 41–43, 43*f*
Heat, 41
  moist, to control microbial growth, 3*f*
Heavy metals, mechanism of action/uses, 306t
Helminths, 471
  diseases, 327
  Schistosoma haematobium, 471*f*
  Wuchereria bancrofti, 471*f*
Hemolydsis tests, to differentiate staphylococcal species, 436t
Hemolysins, 444
Hemolysis, 101*f*

M

MacConkey agar, 100, 101*f*, 102
Macrogametocytes, 226
Macronucleus, 220t
Maculopapular rash, 507
Magnification, linear, 35t
Magnification principles, 33
Magrogametocytes, 228*f*
Malachite green, 528
Malachite green stain, 81, 83
Malaria, 228, 229*f*
Maltose, 149
Mandelic acid, 357–362
Mannitol salt agar, 101*f*
   detecting skin flora on, 424–425, 425*f*
   to differentiate staphylococcal species, 436t
   inoculation procedure to cultivate
      bacteria, 102*f*
   plate showing a fermenter and nonfermenter
      organism, 425*f*
Manual of Antimocrobial Susceptibility Testing,
      American Society for Microbiology, 291
Margins (cultural characteristic), 25, 26*f*
Mass spectrometry, 483
Mastigophora, 217, 220t
   parasitic, 226t, 227t
Matrix-assisted laser desorption/ionization
      (MALDI), 483
Maximum growth temperature, 109
Mechanical barriers, immunity and, 491
Mechanical pipette aspirators, 5f
Media, 1–3, 2*f*, 9
   agar and, 1
   chemically defined, 93
   complex, 93
   differential/selective, 99–109, 101*f*
   enriched, 99–109
   measuring turbidity as indicator of growth,
      93–94
   for routine cultivation of bacteria, 99–201
   selective, 99, 100*f*
   solid, 2*f*
   specialized, 94
Medical microbiology, 417–490, 424*f*
   blood specimen analysis, 477–482
   Campylobacter isolation/presumptive
      identification, 467–470
   dental caries susceptibility test, 419–422,
      420*f*
   enrichment culture technique used in, 358
   genus identification of unknown culture,
      212*t*
   identifying enteric bacterial pathogens,
      457–465
   identifying normal skin flora, 424–425, 425*f*
   identifying normal throat flora, 423–424,
      424*f*
   identifying staphylococcal pathogens,
      435–442, 436t
   identifying streptococcal pathogens,
      433–450
   identifying *Streptococcus pneumoniae*, 444*f*,
      451–456
   species identification of unknown bacterial
      cultures, 483–490
   urine specimen analysis, 471–476
Meiosis, 371
Membrane filters, 337, 338, 340
   in quantitative water analysis, 337–342, 339*f*
Meningitis, 452
Merbak (acetomeroctol), mechanism of action/
      uses, 306t
Mercurial ointments, 306t
Mercurochrome (merbromin), mechanism of
      action/uses, 306t
Mercury bichloride, 306t
Mercury compounds
   inorganic, 306t
   mechanism of action/uses, 306t
Merozoites, 225, 226

Merozotes, 225
Merthiolate, mechanism of actions/uses, 306t
Merthiolate (thimerosal), mechanism of action/
      uses, 306t
Mesophiles, optimum growth temperature,
      110
Metabolic energy sources for microbes, 92
Metallic Elements, 92
Metaphen, mechanism of action/uses, 306t
Methecillin
   when testing staphylococci, 291*t*
Methods for the preparation of dilutions,
      517–518
Methyl red test, 148*f*, 162*f*, 169–170, 170*f*, 173*f*,
      212t
Methylene blue, 75
Methylene blue stain, 48*f*, 58, 528
Meuller-Hinton agar
   detecting throat flora on, 424*f*
Meuller-Hinton tellurite agar plate, 424*f*
MIC determination using a spectrophotometer,
      300–301
Microaerophiles, 119
   distribution of growth, 121*f*
   oxygen requirements, 119, 121*f*
Microaerophilic, 121*f*
   distribution of growth, 121*f*
Microbial flora, isolation of, 423–425
Microbial growth
   bacterial growth curves and, 139–146
   generation time and, 139, 140*f*
   nutrional requirements, 93–98
   pH needs, 115*f*
   pH requirements, 92
   temperature needs, 109–114, 109*f*
Microbial growth control, 277–313
   antiseptics, 277
   chemical methods, 277, 289–298, 305–313
      antiseptics, 277, 305–307t, 305–313
      chemotherapy agents, 289–298
      disinfectants, 277, 305–307t, 305–313
      penicillin activation/inhibition, 299–304
   moist heat, 279–284
   physical methods, 277–278, 277–288
Microbial types
   actinomycetes, 351
   molds, 351
   true bacteria, 351
Microbicidal effect, 277
Microbiologial media, 519–524
Microbiological analysis of urine specimens,
      471–476
Microbiology of food, microbial fermentation,
      321–326
Micrococcus luteus, 17, 19, 27, 63, 127
   biochemical activities and, 148
   IMViC test series and, 171*f*, 173*f*
Micrococcus spp., as food-borne organism,
      315
Microgametocytes, 226, 228*f*
Microincinerator, 2*f*
Microliters (ul) per digestion table, 407
Micronucleus, 220t
Microorganisms, 1, 25
   basic laboratory techniques for, 1–30
   biochemical activities of, 148–215, 148*f*
   cold-resistant, 110
   cultural characteristics, 1–30
   cultural characteristics of, 19
   fermentation, 321–326
   fermentation abilities, 155–160
   in food, 317
   motility of, 179–180
   on nutrient agar plates, 26*f*
   nutrional requirements, 93–99
   oxygen (atmospheric) requirements,
      119–124
Microscopes/microscopy, 31–46
   Abbé condenser, 33
   base of microscope, 34*f*
   body tube, 33

   brightfield, 31
   components of microscope, 33
   compound, 33, 34*f*
   darkfield, 31
   electron, 32
   essential features of various, 31–32
   examination of living organisms with a
      hanging-drop preparation or a wet
      mount, 41–43
   examination of stained cell preparations,
      33–38
   eyepiece lens, 33
   fluorescent, 32
   hanging-drop preparation, 41–43
   illumination, 33, 35
   light control, 34*f*
   linear magnification, 35t
   magnification, 33
   mechanical stage, 33
   nosepiece, 33, 34*f*
   numerical aperture, 34
   parfocal, 37
   phase-contrast, 32
   resolving power/resolution of lenses
      (microscope), 34
   stage, 33, 34*f*
   theoretical principles of, 33
   use and care of, 37
MicroTrak® Direct Specimen Test, 511
Milk
   breed smears and, 132
   litmus-milk reactions, 187–189
Minimal inhibitory concentration (MIC), 299
   determining using a plate reader, 301–302,
      303
   determining using a spectrophotometer,
      301–302, 303
   tube set-up, 300*f*
Minimum growth temperature, 109
Mixed culture, isolation of discrete colonies
      from, 21–22*f*
Moist heat
   as asceptic technique, 3*f*
   autoclaving, 280, 280*f*
   to control microbial growth, 3*f*, 278*f*,
      279–284
   pasteurization and, 278*f*, 280
   sterilization and, 279
   tyndalization and, 279
Molds
   cultivation, 235–242
   cultivation on solid surfaces, 237–238
   identification of, 251t–253t
   morphology, 235–242, 236*f*, 425*f*
   slide culture technique, 237–238, 239, 240
Molecular biology, 387
Moniliasis, 244
Mordant, 67, 68, 68*f*
   Gram's iodine as, 68, 68*f*
Morganella, 207
Morphological characteristics, 245
Mosquito *(Anopheles)*, 225, 228*f*
Most probable number (MPN) test, 329,
      331, 334*t*
Motility of microorganisms, detection of,
      179–180
Mouse virulence test, 452
Mouth, normal flora of, 220t, 419–420, 420*f*
MPN (most probable number) test, 334*t*
MPN presumptive test results for a water
      sample, 332*f*
Mucoid consistency (cultural characteristic), 25
Mucor food contaminant, 251t
Mucor mucedo, 236*f*
Mucor spp.
   morphology/microscopic appearance, 251t
Mueller-Hinton agar
   detecting normal skin flora on, 425
   detecting throat flora on, 424*f*
   Kirby-Bauer Antibiotic Sensitivity Test,
      292–293

material on which the microorganisms exist, 307

pH of environment, 92, 115–118, 307

temperature, 109–114, 307

Physical methods to control microbial growth, 277–288, 278*f*

electromagnetic radiation methods, 285–288

moist heat, 279–284

osmotic pressure, 277–278

Picric acid, 48*f*

chemical formation of, 48*f*

Pigmentation (cultural characteristic), 25

temperature and, 110*f*

Pipettes, 2*f*, 4

blow-out, 5f

mechanical aspirators, 5f

Plaque, 261

Plaque-forming unit (PFU), 261

Plasmids, 371, 391*f*, 397, 397*f*

electrophoration and, 401

genetic engineering and, 398

isolation of bacterial, 397–406

Plasmodium vivax, 225–226, 227, 229*f*

Plums, fermentation of, 321–322, 321*f*

Pneumococcus infections, 452

Pneumonia, lobar, 443, 451–452

Point mutations, 363

Polylinker, 397

Polymyxin, 289*t*

Population growth curve stages, 139*f*

Positive bile esculin test, 445*f*

positive confirmed test, 330*f*

Positive result, 330*f*

Possible MPN preesumptive test results, 332*f*

poultry, 468

Pour plate-loop dilution technique, 2*f*

Pour-plate technique, to count viable cells, 17, 132, 132*f*, 135

Povidone-iodine solution (Betadine®), 306t

Pre-erythrocytic stage, 225, 228*f*

Precipitin formation, 492

Precipitin reaction (ring test), 493–496, 493*f*

Precipitins, 492

Preparation of a streptomycin gradient plate, 377

Preparation of negative r result by use of a Pharyngeal Specimen, 504–505

Preparation of positive result by use of positive control, 504–505

Preparation of pure cultures, 485, 486, 489

Prepared slides, 530

archaea, 530

bacteria, 530

fungi, 530

protozoa, 530

presumptive test, 335

Presumptive test (water analysis), 329

preventing nosocomial infections (hospital acquired), 427

Primary stain

acid-fast stain, 75

defined, 67

differential staining, 81

gram stain, 68, 68*f*

spore stain, 81

Procedure for isolating bacterial plasmid DNA, 399*f*

Proctitis, NGU and, 510

Prophage, 259

Protein hydrolysis, 150*f*

Proteins

as microbial energy source, 158*f*

Proteolysis

as litmus milk reaction, 188, 188*f*

Proteolysis (Peptonization), 188

Proteus spp., 167, 180, 207, 457

amino acid utilization, 205

Proteus spp.

differentiating, 162

Proteus spp.

as food-borne organism, 315

TSI reactions, 162*f*

Proteus vulgaris, 42

biochemical activities, 148*f*

decarboxylase test, 205–206

gram negative bacteria, 471*f*

hydrogen sulfide test, 148*f*

IMViC test series and, 171*f*, 173*f*

Kirby-Bauer Antibiotic Sensitivity Test, 292–293

phenylalanine deaminase test, 205–206

TSI agar test, 162*f*

urease test, 183–184, 184*f*

Protozoa, 41*f*, 217–232, 530

Cilophora, 217

Entamoeba histolytica, 471*f*

free-living, 219–225

free living, 220t, 226t

Mastigophora, 217

parasitic, 225, 227

waterborne disease from, 327

pond water, 41*f*

Sporozoa, 217

taxonomy, 211

Trichomonas vaginalis, 471*f*

Providencia, 207

Pseudomanas spp., 201, 344

enrichment culture technique used in, 357–362

Pseudomonadales, soil populations of, 345

Pseudomonas aeruginosa, 27, 42, 188, 194

biochemical activities, 148*f*

endoenzymatic activities, 149–150

experiments

antibiotic-producing microbes and, 351–356

litmus-milk reaction, 187–189

nitrate reduction test, 193–194, 194*f*

gram negative bacteria, 471*f*

IMViC test series and, 171*f*, 173*f*

Kirby-Bauer Antibiotic Sensitivity Test, 292–293

Pseudomonas aeruginosa

oxidase test, 201–202, 202*f*

Pseudomonas aeruginosa

TSI reactions, 162*f*

Pseudomonas denitrificans, vitamins and, 344

Pseudomonas savastanoi, cultivation of, 111

Pseudomonas spp.

as food-borne organism, 315

oxidase test to differentiate, 201

TSI reactions, 162*f*

Pseudopods, 220t

Psychrophiles, optimum growth temperature, 110

Puerperal fever (childbirth fever), 443

Pure cultures, 2*f*, 15*f*, 18

apparatus used in, 2*f*

isolation of, from a spread-plate or streak-plate preparation, 18

isolation of, from spread-plate/streak-plate preparation, 22*f*

isolation techniques, 15–20

procedure for the preparation of, 20*f*

Pyelitis, 471

pylonephritis, 471

Pyrogallic acid technique, 126*f*

Pyruvic acid, variation in uses of, 156*f*

Q

Quaternary ammonium compounds, mechanism of action/uses, 306t, 307t

Quebec colony counter, 132, 133*f*

Quellung (Neufeld) reaction, 452, 453

Quinolone, 289*t*

R

R Plasmids, antibiotic resistance and, 398

Radiation

to control microbial growth, 3*f*, 278*f*

gamma, 278*f*, 285

ionizing forms of, 285

X-radiation to control microbial growth, 278*f*, 285

Radiation resistant organisms, 286

Raifampin, 289*t*, 291*t*

Raised elevation (cultural characteristic), 26*f*

Rapid plasma reagin (RPR) test, 507–508, 508*f*, 513

Rapid species identification, new molecular techniques for, 483

Rapid testing methods, for sexually transmitted diseases (STDs), 507–514

Rapid water analysis, 337

Reactions in triple sugar-iron agar, 163*f*

Reagents, 58, 83

Barritt's, 171*f*

Kovac's, 168*f*

Reagin, 507

Real image, 33

Recombinant DNA technology, 387–388. *see also* Biotechnology

Recovery period, 393

Redox (oxidation-reduction reactions), 125, 125*f*

Refractive index (microscope), 35

Refrigerators, 2*f*, 6

Rennet curd, 188*f*

rennin, 188*f*

Replication

in bacteriophages, 258

in viruses, 258

Reproduction, in yeasts, 235–242, 243

Resident flora, 427

Resistance (immunity), 491

Resistant mutations, searching for, 378

Resolving power/resolution of lenses (microscope), 34

Resorcinol, mechanism of action/uses, 305t

Respiration, 125

Respiratory tract

flora in, 423

infections of, 443

Respitory tract

infections of, 443

normal flora of, 423–424

Restriction analysis, 407–416

Restriction endonucleasis, 407

Reticuloendothelial system, 491

Rheumatic fever, 443

Rhizobium spp., as soil microbes, 344

soil populations, 345

Rhizoid (cultural characteristic), 25, 26*f*

Rhizopus, morphology/microscopic appearance, 251t

Rhizopus stolonifer, 236*f*

characteristics of, 234t

morphology/microscopic appearance, 251t

Rhodotorula rubra, 243*f*

Ring test: preceptin reactions, 493–496, 493*f*, 494*f*

Rockefeller Institute, 389

S

Sabouraud agar plate, 236*f*, 425*f*

detecting normal skin flora on, 425, 425*f*

mold colonies, 425*f*

yeast colonies, 425*f*

Sabouraud broth culture of *Saccharomyces cerevisiae*, 111

Sabouraud broth plate, 251t

Saccate (cultural characteristic), 26*f*, 27

Saccharomyces, 201

VDRL (Venereal Disease Research Laboratory,
 agglutination test, 507–508
Vegetative cells, 81, 82*f*
Vegetative hyphae, 235*f*
Vegetative mycelium, 235
Venereal Disease Research Lab (VDRL)
 agglutination test, 507–508
Vibrio choerae, 327
Vibrio spp., 57*f*
Virulent bacteriophage particles, 258
Viruses, 257–275, 529
 coliphage isolation, 267–272, 268*f*, 273–275
 cultivation and enumeration of bacterio-
  phages and, 261–266
 Herpes hominus (type II), 471*f*
 replication of, 258
 sexually transmitted chlamydial diseases
  caused by, 510–511, 513
Visualization of bacteria, and staining
 techniques, 79–89
Vitamins
 microbe-produced, 344
 required by microbes, 92
Voges-Proskauer Test, 148*f*, 170–171, 173*f*, 212t
Volvox, 41*f*
Vorticella, 41*f*, 220t

W

Water
 as decolorizing agent, 81
 nutritional requirements of microbes, 92
 as source of food contamination, 315
Water-iodine solution, 528
Water microbiology, 327–342
 membrane filter method analysis, 337–342,
  339*f*
 standard bacterial analysis, 329–336, 330*f*
Waterbaths, 2*f*
 for moist heat experiment, 281*f*
 shaking, 5–6
Wet mount, 42, 219–225
Wetting agents, mechanism of action/
  uses, 306t
Wine production, 243, 321, 321–322, 321*f*
Wine vs. water for health, 322
Wire loops, 2*f*, 4
Wuchereria bancrofti, 471*f*

X

X-radiation to control microbial growth,
 278*f*, 285

Y

Yeast extract, 93
Yeast extract broth, 93
Yeasts, 243*f*
 asexual reproduction, 243*f*
 cells, 243*f*, 321
 cultural characteristics, 235–242
 identification of, 253t
 importance to food industry, 243
 morphology, 235–242, 247, 425*f*
 opportunistic, 244
 pathogenic, 243
 reproduction, 235–243, 243*f*, 244*f*
 use of pyruvic acid by, 156*f*
 in wine production, 243, 321, 321*f*
 yeast infections, 244
Yogurt, 323

Z

Ziehl-Neelson method, 75
Zone of inhibition (Kirby-Bauer test),
 290, 291*t*
Zygomycetes, 233